ORTHODOX CHRISTIANITY AND THE STUDY OF NATURE

SCIENCE AND ORTHODOX CHRISTIANITY

VOLUME 4

Orthodox Christianity and the Study of Nature

Histories of Interaction

Edited by

KOSTAS TAMPAKIS AND RONALD NUMBERS

BREPOLS

This publication has been implemented within the framework of the project 'Science and Orthodoxy around the World', which was made possible through the support of a grant from the Templeton World Charity Foundation, Inc. The opinions expressed in this publication are those of the author(s) and do not necessarily reflect the views of Project SOW and the Templeton World Charity Foundation, Inc.

D/2024/0095/218
ISBN 978-2-503-61246-1
eISBN 978-2-503-61247-8
DOI 10.1484/M.SOC-EB.5.138050

Printed in the EU on acid-free paper.

This book is dedicated to Ron Numbers, a wonderful friend, groundbreaking scholar and valued mentor.

Contents

KOSTAS TAMPAKIS AND
RONALD NUMBERS

Introduction

Eastern and Western, Orthodoxies and Heterodoxies, Sciences and Religions[*]

In March 1891, Gregory V (1746–1821), Archbishop of Constantinople, New Rome and Ecumenical Patriarch of the Orthodox Christian Church, led the rest of the Holy Synod in publishing an Encyclical concerning which courses should be taught at Greek-speaking, Orthodox schools.[1] Among many other things, the Encyclical states:

> '[...] by comparison and by quotation, it is seen that it is more profitable and more necessary for the Nation (*τω Γένει*), to teach Grammar than to teach the mathematical and scientific (courses); [...] because what benefit will the young have if they focus on these studies, and learn numbers, and algebras, and cubes, and cubicocubes, and triangles, and triangularotriangles, and logarithms, and symbolic analysis, and the projected ellipses, and atoms, and voids, and vortices and forces, and attractions and gravities, and the characteristics of light, and the aurora borealis and optical things, and acoustics, and the drop of rain and to move the earth if they have a place to stand like Archimedes, and then be barbarians in their speech, and erroneous

* The editors would like to express their gratitude to Dr Sakorrafou, for her invaluable help in preparing this book. The introduction to the volume was written after our dear friend, colleague and teacher Ronald Numbers passed away in 2023. Ronald Numbers was a driving force in the creation of thise volume, and the introduction especially was a main topic of discussion among the editors and authors. In its current form, the introduction has not benefited from Ron's erudite and profound knowledge or from his witty discourse. However, it encompasses and encapsulates the ideas and themes that the editors had been discussing in the years before Ron's passing. Thus, Ron appears as one of the co-authors of this introduction. It is a shame that he did not get to see its final, printed form.

1 On the Encyclical and its importance, see Ef. Nikolaidis, *Science and Eastern Orthodoxy: From the Greek Fathers to the Age of Globalization* (Baltimore: John Hopkings Press, 2011), pp. 165–69; K. Lappas, 'Πατριαρχική Σύνοδος "Περί Καθαιρέσεως των Φιλοσοφικών Μαθήματων" τον Μάρτιο του 1821', *Mnemon*, 11 (1987), 123–35.

Orthodox Christianity and the Study of Nature, ed. by Kostas Tampakis and Ronald Numbers, Science and Orthodox Christianity, 4 (Turnhout: Brepols, 2024), pp. 9–19
BREPOLS PUBLISHERS
10.1484/M.SOC-EB.5.141379

> in their writings, and ignorant of their religion, and harmful to their Polities, and insignificant as patriots and unworthy of their ancestral talents?
> [...] (teaching these courses is an example) of a corrupt education that denies the required fasting and promotes playing dice, and cards, and other wicked diplomatic ways to cause divisions among the faithful [...]
> [...] and the novelty that has been introduced of giving classical (*παλαιά*) Greek names to the baptized babies of the faithful, that has come to our ears, we consider to be an insult to Christian name-giving, and indecent and inappropriate [...]'.[2]

The Patriarch was, at the time, in dire straits. As the religious leader of the Greek-speaking, Orthodox populations within the Ottoman Empire, the so-called *rum millet*, he was seen by the Ottoman authorities not only as a spiritual figure, but also as a political and intellectual leader.[3] Thus, the various irredentist and revolutionary movements that kept appearing in those same populations were of special concern to him. Many Greek-speaking scholars residing in Europe, such as Adamantios Korais (1748–1833) and Rigas Velestinlis (1757–1798), provided the intellectual basis for those movements, by translating political and philosophical Enlightenment texts into Greek. It was these ideas, from Newton to Locke and from D'Alembert to Descartes, that the Patriarch tried to combat with the Encyclical. Alas, it was not to be. In March 1821, the Greek Revolution sprung up, and by 1830, Greece was a sovereign state. The Sultan had Gregory V lynched after the Easter Liturgy of 1821, and his body was left to rot for two days in front of the main gate of the Patriarchate.[4] In one of the ironies in which public history so delights, Gregory's stern statue is to be found outside the main building of the University of Athens, very close to the statues of his intellectual rivals, Rigas and Korais.

If the reader expects this story to escalate into some kind of science-and-religion conflict narrative, s/he will be disappointed. Rather, we want to emphasize a different point: The good Patriarch had a definite idea of what the natural philosophy of his era was, while also having a very specific list of characteristics of Orthodox Christianity he wanted to preserve. Modern-day historians, of either science, religion, or both, cannot but envy his certainty. Historians of science have spent more than a quarter of a century arguing about what science is, when was it established in its current form, and whether it had a revolutionary

2 The Encyclical is found in K. Th. Dimaras, ed, *Ο Κοραής και η εποχή του* (Athens: Zacharopoulos, 1956), pp. 293–96. All translations are by the authors.

3 For a discussion of the millet system, see A. F. Ergul, 'The Ottoman Identity: Turkish, Muslim or Rum?', *Middle Eastern Studies*, 48. 4 (2012), 629–45.

4 M. Mazower, *The Greek Revolution: 1821 and the Making of Modern Europe* (New York City: Penguin Books, 2021), pp. 33–34.

beginning or not.[5] Thus, this volume uses the neutral, but imprecise, appellation 'Study of nature', and may some god have mercy on our epistemic souls. On the religious front, concepts are not that much clearer. Scholars of religion have long problematized the idea of religion itself, and historians have shown that the perceived realism of the term is in fact intimately tied with the disciplinary colonialism of the West.[6] Finally, recent historiography of science and religion has re-examined the historicity of the concepts, noting not only their shifting and fuzzy boundaries, but also their changing nature.[7] Even if, however, we take into consideration the more modest focus of this volume, our conceptual and analytical problems do not disappear, or even diminish. The collection of essays presented here aims to discuss histories of the interactions between the study of nature and Orthodox Christianity. What is, then, Orthodox Christianity?

The name itself often means little to people who would not identify themselves as Orthodox Christians or would not claim descent from a country where Orthodox Christianity is prevalent. In many spaces, to be a religious Orthodox at times invites 'wishes for Happy Passover' and questions on the absence of '*payot* sidelocks and a hat'.[8] The name that most Anglo-American scholars would recognize is the *Eastern Orthodox Church*, which is however different from the — Assyrian, Nestorian, Chaldean or Syrian — Church of the East in Iran and Iraq. It is also not be confused with the Non-Chalcedonian Churches of India, Syria, Egypt, Ethiopia and Armenia. These are the Oriental Churches, and together with the Eastern Orthodox Church, they are the Eastern Churches.[9]

'Eastern' and 'Oriental'. These are descriptors that historians have come to be deeply suspicious of. However, Orthodox and non-Orthodox theologians alike

---

5 On the Scientific Revolution, see St. Shapin, *The Scientific Revolution, second edition* (Chicago: University of Chicago Press, 2018), Fl. Cohen, *The Rise of Modern Science Explained: a Comparative History* (Cambridge: Cambridge University Press, 2015) and D. Wootton, *The Invention of Science: A new history of the scientific revolution* (London: Penguin, 2015). On the nature of science in History of Science and on the problem of its emergence, see P. Dear, 'What is the History of Science the History of? Early Modern Roots of the Ideology of Modern Science', *Isis*, 96.3 (2005), 390–406 and A. Cunningham, 'Getting the Game right: Some Plain Words on the Identity and Invention of Science', *Studies in History and Philosophy of Science Part A*, 19.3 (1988), 365–89.

6 See for example B. Nongbri, *Before Religion: A History of a Modern Concept* (New Haven: Yale University Press, 2013); T. Masuzawa, *The Invention of World Religions: Or, how European Universalism was Preserved in the Language of Pluralism* (Chicago: University of Chicago Press, 2005) and T. Fitzgerald, *The Ideology of Religious Studies* (Oxford: Oxford University Press, 2000).

7 See P. Harrison, *The Territories of Science and Religion* (Chicago: Chicago University Press, 2015) and B. Lightman, ed. *Rethinking History, Science, and Religion: an exploration of conflict and the complexity principle* (Pittsburgh: University of Pittsburgh Press, 2019).

8 J. A. McGuckin, *The Eastern Orthodox Church: A New History* (New Haven: Yale University Press, 2020), p. 2.

9 T. Ware, *The Orthodox Church: An Introduction to Eastern Christianity, Third Revised Edition* (London: Penguin Books, 2015), pp. 8–9.

often take pains to emphasize that we should banish the thought of pejorative associations.[10] Instead, we are assured that these were the churches that were situated in the *Provincia Orientalis* of the Roman Empire. They later grew apart from the Papacy in Rome as the entity known today as the 'Byzantine Empire', but whose people called themselves 'the Romans', went its own way. In any case, the end result is that a Church whose Creed and main Symbol of Faith describes it as 'One, Holy, Catholic, and Apostolic Church' is today known as the Eastern Orthodox Church in most of the Western World.

Demakopoulos and Papanikolaou offer a far more skeptical interpretation of the 'Eastern' descriptor, noting how it is used by modern Orthodox theologians as an anti-Western emblem, and by Western theologians as a more or less orientalistic term. In a passage that foreshadows many of the issues that we will shortly discuss in the context of this book, they ask:[11]

> 'Who is Western? Who is Eastern? Am I "Eastern" if I commune in an Eastern Orthodox parish in Toledo, Ohio? Am I "Western" if I commune at an Eastern-Rite Catholic parish in Kiev? What if I was baptized into the Eastern Orthodox faith as a child, but I've never learned an Eastern language or traveled outside of the United States — am I Eastern or Western? What if I am a convert to an Eastern or Western faith? [...]
> Perhaps even more problematic are the questions that surround the actual location of East and West. Is it a physical location? If so, it certainly is not a static one. [...] Is the location of East and West linked to a linguistic and/or cultural distinction?'

In both 'West' and 'East', historically produced geographical distinctions are seamlessly internalized as theological, epistemic ones. Such borders were only accentuated by fifty years of Cold War and the subsequent religious revivals in the formerly communist countries. Paschalis Kitromilides very poignantly, and justly, remarks about how 'iron curtains are imagined to be replaced by velvet curtains associated with the aesthetics of Orthodoxy.'[12] It is thus no wonder the histories of science and religion — again, individually and conjunctively — internalize those tacit distinctions. The magisterial nine-volume *Cambridge History of Christianity* tackles the history of all Christian denominations in a holistic way, bar one: the whole of Eastern Christianity, in its Eastern, Monophysite and Orthodox

---

10 McGuckin, *The Eastern Orthodox Church*, pp. 1–5.

11 G. Demacopoulos and A. Papanikolaou, 'Orthodox Naming of the Other: A postcolonial approach' in *Orthodox Constructions of the West*, ed. by G. Demacopoulos and A. Papanikolaou (New York: Fordham University Press, 2013), p. 1. See also Vasilios Makrides. 'Orthodox Christianity in the Context of Postcolonial Studies.' In *Politics, Society and Culture in Orthodox Theology in a Global Age*, ed. by Hans–Peter Grosshans and Pantelis Kalaitzidis (Leiden: Brill, 2022), pp. 338–67.

12 P. Kitromilides, *An Orthodox Commonwealth: Symbolic Legacies and Cultural Encounters in Southeastern Europe* (New York: Routledge, 2020), p. XIII.

dimensions, is neatly tacked into one volume.[13] Subsequent volumes, with themes such as 'World Christianities' or 'Enlightenment, Reawakening and Revolution' apparently did not need any Orthodox Christian chapters. The excellent *Blackwell Companion to Eastern Christianity* also unapologetically bites the epistemic bullet and is organized by Church: Bulgarian, Greek, Ethiopian, and so on.[14] The fact that many of these Churches are not, and have not been for centuries, in communion with each other, is obviously not a strong enough reason to differentiate them. Another example will, we believe, suffice: The only comprehensive book on the history of science and Orthodox Christianity, written by Efthymios Nicolaidis and published in 2011 in English and in 2018 in French and Greek, has different names depending on the language: It is *Science and Eastern Orthodoxy* in English, *Science et Orthodoxie* in French and *Επιστήμες και Ορθοδοξία* (Sciences and Orthodoxy) in Greek. Apparently, French and Greek audiences were not in danger of expecting Passover stories or mentions of *payot* sidelocks and hats.

Clearly, a discussion of the name of the Church does not allow us to clearly delineate it as a historical agent or as a historical axis of analysis. In fact, our cursory examination gave rise to a multiplicity of facets and constitutive dimensions, theological, spatiotemporal, political, and cultural. Instead of rejecting or ignoring this interpretive fruitfulness, we propose instead to bring it to the fore and consider its ramifications. Let us then pose the question once more: What is Orthodox Christianity?

From a theological perspective, Orthodox Christianity is first and foremost a family of self-governing Churches in communion — in its precise ecclesiological and liturgical sense — with each other and with the Ecumenical Patriarchate of Constantinople. These include, among others, the Ancient Patriarchates of Alexandria, Antioch and Jerusalem, the Patriarchates of Russia, Serbia, Georgia, Romania and Bulgaria, and the autocephalous Churches of Greece and Albania. Their common Creed recognizes the Seven Ecumenical Councils, the role of the Risen Christ, and the importance of what is called the Holy Tradition. Orthodoxy carries its two-thousand-year history as a banner, and it is beyond the scope of this short introduction to do full justice to its liturgical, ecclesiastical, and theological complexity.[15] All in all, Orthodox Christianity is the religion of 260 million people, of which 77% live in Europe, and whose communities can be found from Canada to Eritrea and from Australia to Moldova.[16] In their eyes,

---

13 M. Angold, ed, *The Cambridge History of Christianity: Eastern Christianity* (Cambridge: Cambridge University Press, 2006).

14 K. Parry, ed, *The Blackwell Companion to Eastern Christianity* (Chichester: John Wiley & Sons, 2010).

15 See for example Ware, *The Orthodox Church*, pp. 10–12 and the whole second part of the same book.

16 Pew Research Center, *Orthodox Christianity in the 21st Century* (Washington: Pew Research Center, 2017), pp. 6–10.

Orthodox Christianity is not *a* church, it is *the* Church, simply put the Church of Christ on Earth.

This leads us to consider one of the most important, and yet most often overlooked, facets of Orthodox Christianity: it is a historical network of people and institutions that has been active for millennia, but whose interconnectedness has remained mostly unnoticed in historical scholarship. And yet, one has but to look to find the significance of this network. Consider for example Meletius Metaxakis (1871–1935) who became in turn Archbishop of Athens, Ecumenical Patriarch of Constantinople and then Patriarch of Alexandria, in a span of the twenty-five years from 1910 to 1935.[17] What were the conditions and ramifications of a person occupying three of the most prominent positions within the Orthodox Church in three different continents? So far, we do not know. Alternatively, consider the importance of the Orthodox network during the seventeenth century. The brothers Ioannikios (1635–1717) and Sofronios Leichoudes (Likhud) (1652–1720) were born on the Ionian island of Cephalonia, then under Venetian rule. They were educated in the Greek College of Venice and then worked in the College of Katounios in Padua. In 1685, on the recommendation of the Patriarch of Jerusalem Dositheos (1641–1707), they were called to Moscow to manage the newly created Slavic Greek Latin Academy, under the aegis of the Patriarch Joachim (1620–1690).[18] Their bronze statue — of two Orthodox, Greek-speaking scholars, proteges of the Patriarch of Jerusalem, educated in Venice and employed in Russia — was unveiled in 2007 in front of the Epiphany Monastery in Moscow, and is still to be found there. A third and final case in point could be the life and works of the cleric, educator, philosopher, astronomer, and Fellow of the Royal Society, Eugenios Voulgaris (1716–1806). He was born in Corfu, studied in Ioannina and Padua, and then taught in Ioannina, in the Athonite Academy of the Vatopedi Monastery of Mount Athos, and in the Patriarchal Academy of Constantinople. In the spearhead of the intellectual movement that was later called the Neohellenic Enlightenment, Voulgaris was called by none other than Catherine II to Russia, to become a magnate of her court. To his illustrious accomplishments, Voulgaris added eventually the titles of court librarian, archbishop, member of the Imperial Academy of Science in St. Petersburgh and monk. Through him, many of the scholars who would later call for a liberation of the Greek-speaking, Orthodox population from the Ottoman Empire, came into contact with the works of Locke, Wolff and Leibniz.[19]

---

17 See for example, R. E. Güllü, 'Ottoman Rums and the Venizelos–Constantine Conflict after the Armistice of Mudros: the Election of Meletios Metaxakis as Patriarch', *Middle Eastern Studies*, 57.4 (2021), 499–515; Th. Tsironis, *Εκκλησία πολιτευομένη. Ο πολιτικός λόγος της Εκκλησίας της Ελλάδος (1913–1941)* (Athens: Epikentro, 2010).

18 Ef. Nicolaidis, *Science and Eastern Orthodoxy: From the Greek Fathers to the Age of Globalization* (Baltimore: Johns Hopkins University Press, 2011), pp. 142–45.

19 See for example S. Batalden, *Catherine II's Greek prelate: Eugenios Voulgaris in Russia 1771–1806* (New York, East European Monographs: 1982) and M. Patiniotis, 'The Burden of Translation:

Moscow and Venice. Padua and Jerusalem. Alexandria and Corfu. Constantinople (today Istanbul) and St. Petersburgh. Even by examining three cases over three centuries, we see the range of the mobility of scholars and their ideas, as they engaged with Natural Philosophy and Orthodox Christianity. It is of course a characteristic of any network to negotiate a diaspora of nodes. However, Orthodox Christianity is something more than that. Its spatiotemporal diffusion, over areas, peoples and, dare we say, cultures, creates a historiographical challenge for science and religion scholars, even more so since these same nodes also changed profoundly over time. It is a truism to point out, for example, the historical significance of the adoption (under the influence of the Byzantine Church) of Orthodox Christianity by the Kievan Rus' state, its importance for the emergence of the Tsardom of Russia and the role of the Church in the subsequent Russian Empire. One could, however, be entirely more circumspect in considering how Orthodox Christianity and the nascent scientific disciplines of the nineteenth century helped construct the Balkans. From a geographical designation used erroneously by the German geographer August Zeune in 1808, the Balkans became the crypto-colonial term that launched a thousand stereotypes, by natives and foreigners alike.[20] Or perhaps, we could consider how the cultural importance of the Greek Orthodox Church of America has impacted the more than a million Greek Americans in their engagement with science, modernity and secularism over the last two centuries.[21] In all those cases, and in the many others one could identify, we see another often ignored component of religion: Orthodox Christianity is not only, or even mainly, a dogmatic decision. For believers and unbelievers alike, Orthodoxy is first and foremost a lived experience, a cultural signifier, and a collective identity.[22] By focusing solely on its intellectual dimension, science and religion scholarship can easily lose track of the importance of these other facets. However, political and intellectual actors of each and every period and space knew otherwise.

And this brings us back to the embattled, and ultimately doomed, Patriarch Gregory V. Alongside the 'trangularotriangles' and 'cubicocubes', Gregory also condemned, with the same ferocity, the giving of classical names to baptized children. Gregory, alongside many of his intellectual antagonists within the Neohellenic Enlightenment, understood that Orthodox Christianity was a dominant

---

Eugenios Voulgaris and the Circulation of Knowledge in Eighteenth–century Europe', *Journal of History of Science and Technology*, 8 (2013), 48–68.

20 On the imaginary construction of the Balkans, see M. Todorova, *Imagining the Balkans*, Updated Edition (Oxford: Oxford University Press, 2019).

21 See for example A. Kitroeff, *The Greek Orthodox Church in America: a modern history* (Ithaca: Cornell University Press, 2020) and A. Mitsakos Bezzerides, El. Prodromou, eds. *Eastern Orthodox Christianity and American Higher Education: Theological, Historical, and Contemporary Reflections* (Notre Dame: University of Notre Dame Press, 2017).

22 J. Navarro, K. Tampakis 'Introduction. Science, Religion and Nationalism, or the Entanglement of Mythical Narratives,', in *Science, Religion and Nationalism: Local Perceptions and Global Historiographies*, ed. by J. Navarro and K. Tampakis (New York: Routledge, 2024), pp. 1–19.

cultural marker for most of his flock. To be an Orthodox, for most people, was not about having strong views about the nature of the Holy Trinity. It was rather a way to live, a way to take part in public life and a way to identify and be identified.[23] By giving classical Greek names to children instead of Christian ones, a rival cultural legacy was being created. No wonder that the most prominent Greek Enlightenment scholar, Adamantios Korais, was a great supporter of the practice.[24] Another man who understood the sociocultural role of religion was none other than Joseph Stalin. Despite the Marxist view of religion as something destined to wither on the vine, Stalin — not a man known for his mellowness — had no qualms about disbanding the prominent League of Militant Atheists and ending the church prosecution, so that the Russian Church could be used to mobilize national sentiments during the Second World War.[25] However, the sword of cultural influence cuts both ways. As some of the essays of this volume show, the 'national' importance of Orthodoxy can also be used against it. Stalin's successor, Nikita Khrushchev, only resumed the push for scientific atheism within Soviet Russia after the Church's usefulness for the war efforts diminished, as one more way to repudiate Stalin's legacy.[26] On the other side of the Iron Curtain, Archbishop Iakovos of North and South America (1911–2005) was one of the religious leaders to march alongside Martin Luther King, appearing next to him in the famous 1965 *Life* magazine cover. However, Iakovos did so, not with the approval of his Greek-American Orthodox flock, but rather with their tacit disagreement.[27] The same Archbishop would adopt the most rigid Cold War rhetoric while visiting the Greek military Junta in Athens in 1967, praising the rabidly anti-communist Archbishop of Athens Hieronymos (1901–1988), and describing the Junta-sponsored Holy Synod of the Church of Greece as the only other 'free Orthodox Church' in the world.[28] For the approximately 8000 leftist dissidents who were jailed by the Junta at the time, and for the more than 2000

---

23 It bears noting that, today, a fair number of atheists — one of the editors of this volume included — take part in cultural religious practices or even adopt spiritual tenets. See El. Ecklund, D. Johnson, *Varieties of Atheism* (Oxford: Oxford University Press, 2021), pp. 58–93.

24 Adamantios Korais, *Προλεγόμενα στους αρχαίους Έλληνες συγγραφείς*, v. B', (Athens: National Bbank of Greece Cultural Foundation, 1988) pp. 237–38.

25 See D. Peris, *Storming the heavens: the Soviet League of the Militant Godless* (Ithaca: Cornell University Press, 1998) and S. M. Miner, *Stalin's holy war: Religion, Nationalism, and Alliance Politics, 1941–1945* (Chapell Hill: University of North Carolina Press, 2003).

26 V. Smolkin, *A Sacred Space is Never Empty: A History of Soviet Atheism* (Princeton: Princeton University Press, 2018), pp. 57–83.

27 Kitroeff, *The Greek Orthodox Church*, pp. 125–127.

28 O. Koulisis, 'Democracy for Some: Greek-American Institutions and the Greek Junta, 1967–1974', *Commonwealth Review of Political Science*, 6.1 (2023), pp. 59–86; pp. 15–16.

who were actively being tortured, this was a national identity inverted, used to exclude and dehumanize, with Orthodox Christianity right at the middle of it.[29]

This volume aims to bring to the fore the richness, fruitfulness and depth of the scholarship that still awaits historians of science and religion to engage with. Thus, it is intended to act less as a map of an intellectual region and more as a call to action. We make no claim of having fully engaged with all the historical issues raised above. The essays are most certainly not meant to be seen as a comprehensive guide or complete descriptions of the Orthodox Christianity-study of nature entanglement. Instead, we have strived to engage with as many spaces and periods as possible within the short span of a single volume, and we are fully aware that the end result is one of 'points of light on a dark landscape'. We only hope that it will motivate others to change that.

This book is organized chronologically. The first paper, by Gerasimos Merianos, takes us back to the first centuries of Christianity itself, and up to the fifteenth century. Merianos focuses on how Byzantine alchemy, in itself an until recently neglected historiographical field, intersected and interacted with religious beliefs. The paper shows that not only Christian concepts were incorporated into alchemical discourses, but that 'saints, patriarchs, monks and theologians' engaged over the centuries with alchemy, to showcase its risks, discuss its merits or use it as an analogy for tricky theological subjects, such as deification. In the next paper, Nicholas Jacobson concentrates on the fifteenth century and the calendrical considerations of the Council of Florence-Ferrara. He shows how exponents of Greek astronomy such as Gemistos Pletho tackled considerations of time-keeping, and how these were addressed by the heads of the conciliar committee of the Council of Florence-Ferrara. In doing so, Jacobson's paper reexamines the importance of the regulative function of astronomical principles, especially for the ceremonies and rites of Christian communities, and argues that it is to the domains of law and jurisprudence that we should look for analogies, rather than internal virtues or doctrinal practices. Olga Čadajeva's paper transports us to the seventeenth century and to Russian, Ukrainian and Belarusian (Ruthenian) texts on cosmology. It tackles how novel cosmological ideas appeared in Muscovite scholarship, what controversies they created, and what attempts at a more Orthodox-conformant cosmology they spurred. As Čadajeva states, this was not a battle between 'undeducated primitives and modernists' but rather a process by which different scholars and clergymen navigated the cultural and intellectual changes of seventeenth-century Russian society.

Sandy Sakorrafou's paper takes us a century later and 3000 km southwards, by focusing on how Natural Theology was introduced, discussed, and negotiated by Greek-speaking, Orthodox scholars. Sakorrafou focuses on the works of Christodoulos Pamblekis (1733–1793), Veniamin Lesvios (1759–1824), and

29 J. Becket, *Barbarism in Greece: A Young American Lawyer's Inquiry into the Use of Torture in Contemporary Greece, with Case Histories and Documents* (New York: Walker and Company, 1970).

Neophytos Vamvas (1776–1855), as representative of the movement of the Neohellenic Enlightenment. Pamblekis, Lesvios and Vamvas did not necessarily have identical, or even similar, ideas on Orthodoxy and Natural Philosophy. Nevertheless, Sakorrafou details how these scholars navigated through a common intellectual inheritance, multiple Orthodox movements and Protestant expansionism, to come to terms with the new proto-scientific ideas that were becoming prevalent in Western Europe and elsewhere. Daniel Todes and Eleonora Filippova pick up the temporal thread in the early twentieth century and in the life, belief and work of the great Russian physiologist Alexei Ukhtomsky (1875–1942), remembered today for his theory of the dominant. Through a fine-grained analysis of Ukhtomsky's personal archive, Todes' and Filippova's paper shows how Ukhtomsky strived to integrate the 'two paths' of science and religion to a unitary Truth. In the end, Ukhtomsky's scientific findings originated in his theological views, and his religious development went hand-in-hand with his understanding of the mind-body dynamics.

The next two papers are set later in the twentieth century. Kostas Tampakis reverses the 'science versus religion' trope, and instead focuses on how science was used as a religious intellectual weapon in post-civil war Greece. His paper discusses how the Christian Union of Scientists, through its journal *Rays* (*Aktines*), enabled the creation of the largest anti-communist campaign within 1950s Greece. Thus, what started as an apologetic, anti-materialist, science-and-religion venture under the aegis of a powerful para-ecclesiastical organization later became one of the most powerful national Cold War projects on Greek soil. Finally, Christopher Howell's paper is set in the same period, but instead focuses on the translator, theologian, and scholar Philip Sherrard (1922–1995). A late convert to Orthodoxy, Sherrard is mostly known abroad for his translations of Greek poets, among them the Nobel laureates George Seferis and Odysseus Elytis. Howell shows how Sherrard saw his literary pursuits as a gateway to Orthodox Christianity and to Greece, in which he saw a way to save modern civilization from its own excesses. His views led him to become an influential, if contested, author of theological treatises, and one of the first to criticize modern science and technology from an Orthodox perspective. As Howell puts it succinctly, his work until the end of his life was meant to show 'the world to reject the desecrations of the Earth and humanity' and 'to remember that 'everything that lives is holy'.'

The final paper of this volume, by Vangelis Koutalis and Efthymios Nicolaidis, has a different theme and purpose. Its goal is to provide an overview of recent developments in the field of History of the relations between Orthodoxy and the Sciences. In the last fifteen years since the publication of the landmark book *Science and Eastern Orthodoxy*, and due to the efforts of the project 'Science & Orthodoxy around the World' hosted by the National Hellenic Research Foundation and funded by the Templeton World Charity Foundation, an increasingly large number of relevant publications have appeared. Koutalis and Nicolaidis present a guide through the relevant literature and discuss the evolution of the field and its relations with some of the most prominent works within the more general field of science and religion studies.

We have already stated that this volume is meant to provoke more historians to engage with the temporal, spatial, cultural, intellectual, spiritual and philosophical diversity that the interaction of Orthodox Christianity and the study of nature have created. The papers presented here discuss among other things, calendars and anticommunism, alchemy and poetry, cosmology, and natural theology. They range from Byzantium to twentieth-century Russia and from modern Greece to fifteenth-century Florence. Nevertheless, one cannot but notice the many missing spaces. Allow us the indulgence of using modern names for *topoi* who have millennia-old histories: Cyprus, Ukraine, Bulgaria, Ethiopia, Serbia, Georgia, Alexandria and Jerusalem, but also the Orthodox diaspora in Australia, Canada and the USA. With that in mind, we note that for Catholics and Orthodox alike, Wisdom, and her daughters Faith and Hope ('Σοφία', 'Πίστη' and 'Ελπίδα' in Greek) are considered Saints. We can be excused then for having faith that hopefully, this volume will prompt more of our colleagues to increase our historical wisdom in these matters.

GERASIMOS MERIANOS

Aspects of the Relationship between Alchemy and Christianity in Byzantium

▼ **ABSTRACT** Since Antiquity, alchemy has been closely intertwined with religious beliefs and practices. The designation 'sacred and divine art', employed by alchemical authorities in late antique and Byzantine times, is most revealing about this connection. The sacred art gradually adapted to a Christian cultural milieu, leading to the Christianization of alchemical ideas. This paper explores the interplay between alchemy and Christianity in Byzantium, providing a discussion of key themes that unveil complex dynamics. It argues that alchemical theorization converses, to an extent, with Christian theology, progressively integrating alchemy into a Christian philosophy of nature. Byzantine alchemical authors infuse their discipline with Christian values, virtues, and beliefs, presenting adherence to these principles as a prerequisite for accessing alchemical knowledge. Moreover, the paper reveals the 'infiltration' of alchemical ideas into Christian literature of various genres, as is evidenced by specific hagiographical and theological texts. Despite limited evidence, these works confirm that alchemy was not alien to the social and cultural milieus of certain Byzantine periods. Notably, this 'art' could be appealing to members of the clergy, even up to the highest ranks. The study also offers preliminary methodological insights, emphasizing the nuanced nature of the relationship between alchemy and Christianity. It advocates for a theologically informed analysis of Byzantine alchemical texts, cautioning against simplistic categorizations. This examination is intended to be a starting point for future explorations into the intricate connections between alchemy and Christianity in Byzantium.

Gerasimos Merianos • National Hellenic Research Foundation

Orthodox Christianity and the Study of Nature, ed. by Kostas Tampakis and Ronald Numbers, Science and Orthodox Christianity, 4 (Turnhout: Brepols, 2024), pp. 21–54
BREPOLS PUBLISHERS
10.1484/M.SOC-EB.5.141570

Introduction

Undoubtedly, the term 'alchemy' with reference to late antique or Byzantine chemical ideas and practices is used anachronistically. 'Alchemy' derives from the Latin term *alchimia* (in variants), which in turn originates from the Arabic word *al-kīmiyā'*,[1] an adaptation of the Greek term *chym(e)ia / chēm(e)ia*. Of course, the scope and breadth of the Greek (including the Byzantine), Islamic, and medieval Latin chemical traditions do not always coincide, and the term 'alchemy' carries undertones that seem to better apply to medieval Islamic or Renaissance European contexts. On the other hand, 'alchemy' has deep historiographical roots and, despite its apparent shortcomings, helps scholarly communities understand and communicate with each other. Acknowledging the issues associated with this term and recognizing that other alternatives have their own drawbacks, I have chosen to use 'alchemy' in this paper to denote the chemical art in Byzantium.

Alchemy in Byzantium follows, systematizes, and expands the Graeco-Egyptian chemical tradition, while from some point onward, it also interacts with the Islamic and medieval Latin ones. The available sources for Greek alchemy — both its Graeco-Egyptian and Byzantine branches — form two groups. The first consists of two papyri, the Leiden (P. Leid. X) and the Stockholm (P. Holm.) papyri, which date to around AD 300 and represent the primary, technical phase of Graeco-Egyptian alchemy.[2] The second group comprises manuscripts containing Greek alchemy's theoretical treatises and technical texts. They are eponymous, pseudepigraphal, or anonymous, dating from the AD first to the fifteenth century. This collection of texts, the so-called 'Greek Alchemical Corpus', is witnessed primarily in the following three codices: *Marcianus graecus* 299 (second half of the tenth century); *Parisinus graecus* 2325 (thirteenth century); and *Parisinus graecus* 2327 (1478).[3] The existence of partial collections in Antiquity, while possible, cannot be confirmed.[4] Except for the two papyri, the fact that our knowledge concerning Greek alchemy mainly derives from Byzantine manuscripts implies that we have inherited the Byzantine perspective on the Greek alchemical tradition.

---

1 S. Moureau, '*Min al-Kīmiyā' ad Alchimiam*. The Transmission of Alchemy from the Arab-Muslim World to the Latin West in the Middle Ages', *Micrologus*, 28 (2020), 87–141, on p. 89.

2 *Papyrus de Leyde. Papyrus de Stockholm. Fragments de recettes*, ed. by R. Halleux, Les alchimistes grecs, I (Les Belles Lettres: Paris, 1981).

3 R. Halleux, *Les textes alchimiques* (Brepols: Turnhout, 1979), pp. 60–62; M. Mertens, *Zosime de Panopolis, Mémoires authentiques*, Les alchimistes grecs, IV.1 (Les Belles Lettres: Paris 1995), pp. XX–XLIII; M. Mertens, 'Graeco-Egyptian Alchemy in Byzantium', in *The Occult Sciences in Byzantium*, ed. by P. Magdalino and M. Mavroudi (La Pomme d'or: Geneva, 2006), pp. 205–30, on pp. 207–9, 220–24. Particularly on *Marc. gr.* 299, see A. M. Roberts, 'Framing a Middle Byzantine Alchemical Codex', *Dumbarton Oaks Papers*, 73 (2019), 69–102; A. M. Roberts, 'A Greek Alchemical Epigram in Its Middle Byzantine Context', *Journal of the Warburg and Courtauld Institutes*, 83 (2020), 1–36; A. M. Roberts, 'Islamicate Alchemy in Greek Letters on the First Page of *Marcianus graecus* 299', *Byzantinische Zeitschrift*, 115.1 (2022), 341–49.

4 Mertens, 'Graeco-Egyptian Alchemy', pp. 220–21.

But what was 'alchemy' for the Byzantines? This is not an easy question since there is no single conception or term for it throughout the Byzantine era. For instance, *chym(e)ia / chēm(e)ia* (written with a variety of spellings and roughly corresponding to 'alchemy') is a major term used in both alchemical and non-alchemical literature. The same applies to the term *chrysopœia* (gold-making), while the labels 'sacred' and/or 'divine art' were mainly employed by alchemical authors.[5] It is indicative that in the *pinax* ('table of contents') of *Marc. gr.* 299, the earliest Greek surviving alchemical manuscript, the majority of titles describe the subject of the listed treatises either as 'sacred and divine art' or 'sacred art' or 'divine art'; however, the term *chrysopœia* appears in some titles too.[6] Concerning the complex terminology in alchemical and non-alchemical texts, it should be noted that there is a similar situation in Arabic literature (*al-kīmiyā'* being the most common term in non-alchemical literature).[7] To return to the question posed, the late tenth-century *Souda* lexicon provides a celebrated answer as to what alchemy was regarded as in the middle Byzantine era, defining *chēmeia* as 'the preparation of silver and gold'.[8] However, we know that, since Antiquity, alchemical objectives have been broader, including techniques for creating (or rather imitating) gems, pearls, and purple dye. Nevertheless, the narrow conception of alchemy as gold-making (*chrysopœia*) and silver-making (*argyropœia*) was relatively dominant in Byzantium. The diverse terminology implies the contradictions and multi-layered meanings characterizing alchemical practices in Byzantium.

In broad outline, alchemy pursues the comprehension and realization of the transformation of matter, the pinnacle of which is considered metallic transmutation, (ideally) a complete transformation of base metals into precious ones, especially gold. The alchemical methodology is based on the conceptual understanding of the natural principles of substances via the study and exegetical

5 G. Merianos, 'Alchemy', in *The Cambridge Intellectual History of Byzantium*, ed. by A. Kaldellis and N. Siniossoglou (Cambridge: Cambridge University Press, 2017), pp. 234–51, on pp. 238–40.

6 *Marc. gr.* 299, f. 2^{r-v}. For a transcription of the *pinax*, see *Catalogue des manuscrits alchimiques grecs* (hereafter: *CMAG*), ed. by J. Bidez et al., 8 vols (M. Lamertin: Brussels, 1924–1932), II, pp. 20–22. See also Roberts, 'Framing', pp. 88–90.

7 Moureau, '*Min al-Kīmiyā*'', pp. 89–90.

8 *Souda*, ed. by A. Adler, *Suidae lexicon*, 5 vols (Leipzig: B. G. Teubner, 1928–1938), *s.v.* Χημεία (X 280): ἡ τοῦ ἀργύρου καὶ χρυσοῦ κατασκευή, [...]; see also *s.v.* Διοκλητιανός (Δ 1156); *s.v.* Χειμεία (X 227). Cf. John of Antioch, ed. by U. Roberto, *Ioannis Antiocheni fragmenta ex Historia chronica*, Texte und Untersuchungen zur Geschichte der altchristlichen Literatur, 154 (Berlin and New York: W. de Gruyter, 2005), fr. 248, p. 428 = ed. by S. Mariev, *Ioannis Antiocheni fragmenta quae supersunt omnia*, Corpus Fontium Historiae Byzantinae (hereafter: CFHB), 47 (Berlin and New York: W. de Gruyter, 2008), fr. 191, p. 348; Constantine VII Porphyrogennetos, *Excerpta de virtutibus et vitiis*, pt. 1, ed. by Th. Büttner-Wobst and A. Gerard Roos (Berlin: Weidmann, 1906; repr. Hildesheim, 2003), John of Antioch, fr. 52 (165), p. 196; *Acta sanctorum, Julii*, II, July 8 (Antwerp: J. du Moulin, 1721), p. 557C.

analysis of the Greek alchemical and philosophical tradition and, at the same time, on the empirical understanding of matter, which is achieved in the laboratory.[9] Alchemy applies philosophical principles in practice, combining philosophy with laboratory operations, the 'know-why' with the 'know-how'.[10] The renowned Byzantine philosopher Michael Psellos (1018–late 1070s) shows that transmutation is a philosophical inquiry pertaining to the science of nature (τῆς περὶ τὴν φύσιν ἐπιστήμης).[11] John Kanaboutzes, in the fifteenth century, considers *chymia* to derive from 'the high and contemplative (θεωρητικῆς) philosophy on the nature of beings'.[12] Thus, both scholars connect alchemy to natural philosophy, a crucial observation concerning the topic of the present volume since it calls to mind that in the Byzantine period, as in Antiquity, philosophy and (what we call today) 'science' were one undivided discipline.[13] As for the Byzantine conception of natural philosophy, Börje Bydén notes that it is

> [...] influenced by that of late antique Neoplatonic Aristotelianism. [...] nature is considered to be an inner principle of change (*kinēsis*) and stability (*stasis*), and natural philosophy is the branch of theoretical philosophy that studies such entities as are subject to change in accordance with nature, in contradistinction to mathematics and theology, the objects of which are exempt from change. [...] The natural world is the world of sense-experience, extended in space and time; the higher reality constituting the domain of theology is accessible only to the intellect. On this division the Byzantine writers superimposed the Christian dichotomy of Creator and created world.[14]

It is hardly surprising then that Greek and Byzantine alchemical authors were called 'philosophers' and represented themselves as such rather than as

9 G. Merianos, 'The Christianity of the Philosopher Christianos. Ethics and Mathematics in Alchemical Methodology', *ARYS*, 20 (2022), 271–322, on p. 286.

10 Merianos, 'Alchemy', p. 243.

11 *CMAG*, VI, § 1, p. 26.7–8.

12 J. Kanaboutzes, *Commentary on Dionysios of Halikarnassos* 13, ed. by M. Lehnerdt, *Ioannis Canabutzae magistri ad principem Aeni et Samothraces in Dionysium Halicarnasensem commentarius* (Leipzig: B. G. Teubner, 1890), p. 10.26–29. For an English translation of the work, see A. Kaldellis, *Byzantine Readings of Ancient Historians: Texts in Translation, with Introductions and Notes* (London and New York: Routledge, 2015), pp. 113–70. On Kanaboutzes' discussion of alchemy, see S. Sakorrafou and G. Merianos, 'John Kanaboutzes' Commentary on Dionysios of Halikarnassos: A Perception of Alchemy in a Fifteenth-Century Greek Text', in *Scientific Cosmopolitanism and Local Cultures: Religions, Ideologies, Societies. Proceedings of the 5th International Conference of the European Society for the History of Science*, ed. by G. Katsiampoura (Athens: Institute of Historical Research, National Hellenic Research Foundation, 2014), pp. 86–94.

13 J. M. M. H. Thijssen, 'Natural Philosophy', in *Encyclopedia of Medieval Philosophy: Philosophy between 500 and 1500*, ed. by H. Lagerlund (Dordrecht: Springer, 2011), pp. 839–49, on p. 839. On the Byzantine concepts of 'science', see D. O'Meara, 'Conceptions of Science in Byzantium', in *Cambridge Intellectual History of Byzantium*, ed. by Kaldellis and Siniossoglou, pp. 169–82.

14 B. Bydén, 'Natural Philosophy, Byzantine', in *Encyclopedia of Medieval Philosophy*, ed. by Lagerlund, pp. 858–63, on p. 859.

'alchemists'. The most celebrated use of the rare term *cheimeutēs*, usually rendered as 'alchemist', is pejorative, even more so because it is attributed to the swindler John Isthmeos (see below).[15] On the other hand, the statesman and scholar Demetrios Kydones (*c.* 1324–*c.* 1397), in his accurate, partial translation of Thomas Aquinas' *Summa theologiae*,[16] chose to render the terms *alchimia* as *chēmeutikē* and *alchimicus* as *chēmeutos* / *chēmeutēs*. Interestingly, in the *Secunda Secundae* of the *Summa*, Aquinas examines whether alchemical silver or gold (ἄργυρον ἢ χρυσὸν χημευτόν) may be sold as natural.[17] He argues that selling gold or silver produced by alchemists (ὑπὸ τῶν χημευτῶν γεγενημένον) as genuine equals fraud since artificial metals do not possess the true species of gold and silver. Although Aquinas seems to doubt the possibility of creating real gold through alchemy, he emphasizes that if such an achievement were to occur, it would not be illicit to sell the resulting gold as genuine.[18]

---

15 V. Koutalis, M. Martelli, and G. Merianos, 'Graeco-Egyptian, Byzantine and Post-Byzantine Alchemy: Introductory Remarks', in *Greek Alchemy from Late Antiquity to Early Modernity*, ed. by E. Nicolaidis (Turnhout: Brepols, 2018), pp. 11–43, on pp. 31–37; O. Dufault, *Early Greek Alchemy, Patronage and Innovation in Late Antiquity* (Berkeley, CA: California Classical Studies, 2019), pp. 95–100. Particularly on the term *cheimeutēs*, see G. Merianos, 'Insights on Alchemy, Deception, and Artisanal Knowledge in Byzantium', in *Byzanz am Rhein. Festschrift für Günter Prinzing anlässlich seines 80. Geburtstags*, ed. by A. Bosselmann-Ruickbie et al. (Wiesbaden: Harrassowitz, 2024), pp. 145–64, on pp. 153–57. It should be noted that the conceptual world of the learned philosopher in Byzantium was not isolated from the broad contexts of 'occult sciences'. See R. Greenfield, 'Magic and the Occult Sciences', in *Cambridge Intellectual History of Byzantium*, ed. by Kaldellis and Siniossoglou, pp. 215–33, on p. 218; also P. Magdalino and M. Mavroudi, 'Introduction', in *Occult Sciences in Byzantium*, ed. by Magdalino and Mavroudi, pp. 11–37, on pp. 13–14.

16 On Kydones as translator of Aquinas, see, e.g., M. Plested, *Orthodox Readings of Aquinas* (Oxford: Oxford University Press, 2012), pp. 63–72; I. Polemis, 'Thomas Aquinas' Reception in Fourteenth-Century Byzantium', in *The Oxford Handbook of the Reception of Aquinas*, ed. by M. Levering and M. Plested (Oxford: Oxford University Press, 2021), pp. 38–52; as well as several studies in the following volumes: P. Ch. Athanasopoulos, ed., *Translation Activity in the Late Byzantine World: Contexts, Authors, and Texts* (Berlin and Boston, MA: W. de Gruyter, 2022); J. A. Demetracopoulos and Ch. Dendrinos, eds, *Thomas Latinus — Thomas Graecus: Thomas Aquinas and His Reception in Byzantium* (Athens: Artos Zois, 2022).

17 D. Kydones, *Thomas Aquinas'* Summa theologiae *in Greek*, 2a2ae 77 [a. 2 arg. 1], ed. by E. M. Kalokairinou, *Δημητρίου Κυδώνη Θωμᾶ Ἀκυϊνάτου: Σούμμα Θεολογική, ἐξελληνισθεῖσα* [2a2ae 57–79], Corpus Philosophorum Graecorum Recentiorum, II.18 (Athens: Foundation for Research and Editions of Modern Greek Philosophy, 2002), pp. 270.25–271.7. See Merianos, 'Insights on Alchemy', pp. 155–56.

18 D. Kydones, *Thomas Aquinas'* Summa theologiae *in Greek*, 2a2ae 77 [a. 2 ad 1], pp. 272.22–273.8. On alchemy in the thought of Aquinas, see N. Tarrant, 'Between Aquinas and Eymerich: The Roman Inquisition's Use of Dominican Thought in the Censorship of Alchemy', *Ambix*, 65.3 (2018), 210–31, on pp. 214–18; also J. Hallebeek, 'Artificial Gold, Sold as Natural. Francisco de Vitoria on a Latent Defect in the Merchandise', in *Ius Romanum — Ius Commune — Ius Hodiernum. Studies in Honour of Eltjo J. H. Schrage On the Occasion of His 65th Birthday*, ed. by H. Dondorp et al. (Aalen: Scientia, 2010), pp. 161–76, on pp. 162–63; F. Boldrini, 'All

Alchemy has been intertwined with religious beliefs since Antiquity. For instance, the Byzantine chronicler George Synkellos (d. after 810) quotes a passage from a lost work by the Graeco-Egyptian alchemical author Zosimos of Panopolis (late third–fourth century). Zosimos describes, drawing on Enochic literature (*1 Enoch*), how certain angels lusted after women and taught them every work of nature. As a result of their actions and their role in teaching humankind everything wicked, they were expelled from heaven. They also revealed the first alchemical book, entitled *Chēmeu*, whence also the art is called *chēmeia*.[19] Thus, Zosimos considers the origins and naming of alchemy as supernatural. The Zosimean passage is usually studied in parallel with a text known as *The Letter of Isis to Horus* (late second or early third century). There, alchemy appears as a secret knowledge of divine origin, revealed to the goddess Isis by the angel Amnaēl, who desires to make the revelation in exchange for sex; this is also reminiscent of the Enochic description of angelic desire for women and revelation to them of esoteric knowledge.[20] The syncretism of these texts, employing Jewish, Greek, and Egyptian religious and philosophical ideas, is apparent.

It is worth noting that both the Zosimean passage and the *Letter of Isis* emphasize the acquisition of alchemical knowledge through revelation. The association between gaining access to alchemical wisdom and divine revelation endures as alchemy evolves within a Christian framework. As will be shown later, certain

---

That Glitters Is Not Gold: False Jewellery and Its Juridical Regulation in Italy between the Late Middle Ages and the Early Modern Period', in *Faking It! The Performance of Forgery in Late Medieval and Early Modern Culture*, ed. by Ph. Lavender and M. Amundsen Bergström (Leiden and Boston, MA: Brill, 2023), pp. 52–74, on pp. 59–60.

19 G. Synkellos, *Selections from Chronographers*, ed. by A. A. Mosshammer, *Georgii Syncelli Ecloga chronographica* (Leipzig: B. G. Teubner, 1984), p. 14.2–14. See M. Martelli and M. Rumor, 'Near Eastern Origins of Graeco-Egyptian Alchemy', in *Esoteric Knowledge in Antiquity*, ed. by K. Geus and M. Geller (Berlin: Max-Planck-Institut für Wissenschaftsgeschichte, 2014), pp. 37–62; also M. Mertens, 'Sur la trace des anges rebelles dans les traditions ésotériques du début de notre ère jusqu'au XVIIIe siècle', in *Anges et démons*, ed. by J. Ries and H. Limet (Louvain-la-Neuve: Centre d'histoire des religions, 1989), pp. 383–98; K. A. Fraser, 'Zosimos of Panopolis and the Book of Enoch: Alchemy as Forbidden Knowledge', *Aries*, 4.2 (2004), 125–147; Ch. H. Bull, 'Wicked Angels and the Good Demon: The Origins of Alchemy According to the Physica of Hermes', *Gnosis: Journal of Gnostic Studies*, 3 (2018), 3–33. See also next footnote.

20 For the edition of the text, see M. Mertens, 'Un traité gréco-égyptien d'alchimie: la *Lettre d'Isis à Horus*. Texte établi et traduit avec introduction et notes' (unpublished doctoral dissertation, Université de Liège, 1983–1984) = M. Berthelot and Ch.-É. Ruelle, eds, *Collection des anciens alchimistes grecs* (hereafter: *CAAG*), 3 vols (Paris: G. Steinheil, 1887–1888), II, pp. 28–33. See M. Mertens, 'Une scène d'initiation alchimique: la "Lettre d'Isis à Horus"', *Revue de l'histoire des religions*, 205.1 (1988), 3–23; F. Lopes da Silveira, 'In the Melting Pot: Cultural Mixture and the Presentation of Alchemical Knowledge in the *Letter from Isis to Horus*', *Ambix*, 69.1 (2022), 49–64; M. Blanco Cesteros '(De)Constructing an Authoritative Narrative. The Case of *The Letter of Isis*', *ARYS*, 20 (2022), 227–69 (where additional bibliography is cited). See also previous footnote.

Byzantine alchemical authors regard participation in divine knowledge, of which alchemy is a part, as a gift of God. Comprehending matter requires illumination bestowed by divine grace to unravel the lessons the universe imparts about its divine design. This observation suggests that alchemy — as well as the sciences in general — and (Christian) religion in Byzantium were not seen as incompatible.

Moreover, it does not seem improbable to claim that the labels 'sacred art', 'divine art', or 'sacred and divine art' originated from a relevant notion linked to the Egyptian religious setting and entwined with religious beliefs and practices.[21] The 'sacred art' was gradually adapted to a Christian cultural milieu, and alchemical ideas were progressively Christianized. Concepts once associated with pagan gods were now connected with Christ and the Trinity.

Of course, the relationship between alchemy and religion in Byzantium is not limited to concepts. Alchemical experimentation, in particular, sometimes meets Christianity in everyday practices, such as measuring time. For instance, in specific recipes from the collection entitled *On the Most Noble and Illustrious Art of the Goldsmith*,[22] the time needed for an action to be completed is calculated by how long it takes to recite a *Pater noster*.[23] This practice was common in craft workshops in the West.[24] The reciting of the prayer defines a segment of time while also becoming a symbolic ritual for blessing the process.

More and more light is being shed on the relationship between alchemy and religion in medieval and early modern Western Europe.[25] However, similar efforts concerning other Christian cultures and periods are extremely limited. Specifically, the relationship between alchemy and religion in early Christian and Byzantine times has not received due attention.[26] Of course, aspects of the

---

21 M. Martelli, *The* Four Books *of Pseudo-Democritus* (Leeds: Maney Publishing, 2013), pp. 65–66.

22 This recipe collection is the focus of the interdisciplinary project 'Der griechische Traktat "Über die hochgeschätzte und berühmte Goldschmiedekunst". Edition und interdisziplinärer Kommentar', Leibniz-Zentrum für Archäologie, Mainz (https://www.leiza.de/forschung/projekt/der-griechische-traktat-ueber-die-hochgeschaetzte-und-beruehmte-goldschmiedekunst [21 September 2023]).

23 *CAAG*, II, pp. 322.13–14, 325.21–22, 329.18, 330.15, 331.1.

24 See P. H. Smith, 'The Codification of Vernacular Theories of Metallic Generation in Sixteenth-Century European Mining and Metalworking', in *The Structures of Practical Knowledge*, ed. by M. Valleriani (Cham: Springer, 2017), pp. 371–92, on p. 372.

25 See, e.g., S. Matton, 'Thématique alchimique et littérature religieuse dans la France du XVIIe siècle', *Chrysopœia*, 2 (1988), 129–208; S. Matton, 'Remarques sur l'alchimie transmutatoire chez les théologiens réformés de la Renaissance', *Chrysopœia*, 7 (2000–2003), 171–93; W. R. Newman, *Promethean Ambitions: Alchemy and the Quest to Perfect Nature* (Chicago, IL and London: The University of Chicago Press, 2004), pp. 82–97; Z. A. Matus, 'Alchemy and Christianity in the Middle Ages', *History Compass*, 10.12 (2012), 934–45; T. Nummedal, ed., *Alchemy and Religion in Christian Europe* = *Ambix*, 60.4 (2013); L. M. Principe, *The Secrets of Alchemy* (Chicago, IL and London: The University of Chicago Press, 2013), pp. 190–206.

26 To my knowledge, the only thematic volume on the subject is N. Borrelli and M. Martelli, eds, *Chemical Arts and Religion in Antiquity* = *ARYS*, 20 (2022), particularly the papers by O. Dufault, G. Merianos, and V. Carlotta.

Christianity of important figures, such as the seventh-century alchemical philosopher Stephanos of Alexandria (whose identity is still a topic of lively debate), have been discussed (see below).[27] However, the general characteristics of this multifaceted relationship and its evolution through time are far from sufficiently studied.

Concerning Byzantium, this situation can be attributed to several reasons, three of which I would like to emphasize. (i) The study of alchemy in Byzantium has been neglected or underappreciated until recently. Most synthesizing works on the history of Western alchemy discussed the phenomenon in Byzantium in passing, and usually until the seventh century at the latest, that is, until one of the centuries that are thought of as the chronological end of Late Antiquity. This disregard implies the long-lasting stereotypical view of Byzantine science (and culture, in general) as unoriginal and unimportant, whose most memorable contribution is preserving the body of scientific knowledge from Greek Antiquity.[28] (ii) On the other hand, alchemy as a research field in Byzantine Studies was nonexistent until some years ago. Indeed, for most Byzantinists, alchemy was (and often still is) identified as magic, not considered a 'serious' research interest. As a result, there was a lack of interest in and unfamiliarity with the subject. (iii) Partly due to the Byzantinists' limited interest, scholars of Late Antiquity and experts in Graeco-Egyptian alchemy were left alone in investigating aspects of Byzantine alchemy. The still blurry and conventional chronological limits of what is late antique and what is early Byzantine have contributed to this expansion.[29] But despite the valuable engagement of scholars focusing on Graeco-Egyptian alchemy with Byzantine texts, sometimes their approach has been too literary. Ideas and practices peculiar to Byzantium have often been decontextualized from their sociocultural milieu. Concerning our topic in particular, scholars focusing on Graeco-Egyptian alchemy occasionally encounter challenges when dealing with the complex issue of Christianity in the late Roman / early Byzantine Empire (depending on the perspective). For example, it is evident that from some point onward, almost any alchemical author in Byzantium might be generally characterized as 'Christian', but this label is used so broadly that it ends up meaning little more than 'post-pagan'.

The type of Christianity to which Byzantine alchemical authors adhered — for instance, from the era of the great Christological controversies to the triumph

27 See, e.g., M. K. Papathanassiou, 'Stéphanos d'Alexandrie. La tradition patristique dans son œuvre alchimique', in *Greek Alchemy*, ed. by Nicolaidis, pp. 71–97.

28 See M. Mavroudi, 'Science, Byzantine', in *The Encyclopedia of Ancient History. First Edition*, ed. by R. S. Bagnall et al. (Chichester: Wiley-Blackwell, 2013), pp. 6063–65; M. Mavroudi, 'Translations from Greek into Latin and Arabic during the Middle Ages: Searching for the Classical Tradition, *Speculum*, 90.1 (2015), 28–59, on pp. 33–36.

29 On these periodization issues, see A. Cameron, 'Late Antiquity and Byzantium: An Identity Problem', *Byzantine and Modern Greek Studies*, 40.1 (2016), 27–37.

of the iconophiles, that is, from the fourth to the mid-ninth century — is crucial in the attempt to discover the course and depth not only of the Christianization but also of the 'Byzantinization' of Greek alchemy.

What Does 'Christian' Mean? The Importance of Religious Terminology

Before turning to Byzantium proper, it should be stressed that in the syncretic environment of the late antique world, in regions such as Egypt, early Christianity was a far more complex phenomenon than we usually think it was. In these formative years, any reference to Christ within an alchemical text did not necessarily suggest that its author was close to what we label, by and large, as 'Christian orthodoxy' — or even 'Christianity' — especially if seen under the perspective of post-Nicene dogmatic developments.[30] The most striking example is that of Zosimos of Panopolis. Zosimos' work explicitly refers to Christ and salvation, but it has been argued that the Zosimean imagery, such as his baptismal metaphors, derives from a Gnostic milieu.[31] Olivier Dufault highlights that Zosimos' Christology, anthropology, and soteriology have elements in common with 'Classic Gnostic' theology and certain early Christian ideas.[32] In this context, a revealing example of Zosimos' Christology is the 'docetic' view in his *On the Letter Omega* that Jesus 'suffered nothing but instead showed death trampled under foot and thrust aside'.[33]

Dylan Burns has aptly noted that Zosimos' 'home of Panopolis [modern Akhmim in Egypt] was just across the Nile from present-day Nag Hammadi'.[34] This remark brings to mind that specific texts from the Coptic Nag Hammadi Library use alchemical imagery. For example, Regine Charron and Louis Painchaud argue that the extracanonical *Gospel of Philip*, which has often been labeled as

30 In the context of this paper, 'orthodoxy' refers to the official Byzantine Christianity that was primarily, but not exclusively, developed and formulated through the decrees of the seven Ecumenical Councils of the first Christian millennium.

31 K. A. Fraser, 'Baptised in *Gnōsis*: The Spiritual Alchemy of Zosimos of Panopolis', *Dionysius*, 25 (2007), 33–54.

32 O. Dufault, 'Was Zosimus of Panopolis Christian?', *ARYS*, 20 (2022), 135–170; also Dufault, *Early Greek Alchemy*, pp. 115–17. Not all modern researchers agree that Zosimos was 'Christian'; on the pro and contra arguments, see the secondary literature in Dufault, 'Was Zosimus of Panopolis Christian?', p. 145 n. 27.

33 Zosimos of Panopolis, *On the Letter Omega*, ed. by Mertens, *Zosime de Panopolis, Mémoires authentiques*, 1.13.125–26, p. 7 (see commentary on pp. 102–103 [n. 82]); Eng. transl. by Dufault, 'Was Zosimus of Panopolis Christian?', p. 159.

34 D. M. Burns, 'μίξεώς τινι τέχνῃ κρείττονι: Alchemical Metaphor in the *Paraphrase of Shem* (NHC VII,1)', *Aries*, 15 (2015), 81–108, on p. 89.

'Gnostic' or 'Valentinian',[35] includes a baptismal metaphor that seems to be closely connected with alchemical descriptions of dyeing or chemical baptism.[36] Likewise, Regine Charron points out that another apocryphal text, the *Apocryphon of John*, conveys a baptismal theology that evokes alchemical dyeing processes too. She concludes her relevant paper by underlining that 'the alchemical literature and the alchemical context may prove useful in elucidating certain obscure passages in the Nag Hammadi Library, thereby contributing to a better understanding of its esoteric aspects'.[37] Dylan Burns, in his turn, shows that the *Paraphrase of Shem*, also found in the Nag Hammadi Library, includes metaphors originating in the metallurgical practice of tincturing, which played a crucial role in the development of Graeco-Egyptian alchemy.[38] From the above discussion, it is evident that alchemical imagery and analogy were appealing to certain Christian circles that could hardly be characterized as 'orthodox' but fit well in the fluidity and diversity of early Christianity.

The religious vocabulary of an alchemical text — or a non-alchemical one that contains chemical references — might be crucial in approaching the type of its author's Christianity. For example, Hugo Lundhaug argues that the Nag Hammadi extracanonical texts, which are usually read in light of second- or third-century contexts, probably echo fourth-century, and sometimes fifth-century, doctrinal debates.[39] Consistent with this reasoning, Lundhaug stresses that the aforementioned apocryphal *Gospel of Philip* is characterized — among other theological features that include an anti-Arian polemic — by referencing the critical distinction between 'begetting' and 'creating'. This evokes the Nicene Creed, describing the Son as 'begotten, not made' (γεννηθέντα οὐ ποιηθέντα).[40] Thus, and according to Lundhaug's argumentation, the *Gospel of Philip* — which contains the baptismal metaphor that, according to Charron and Painchaud, is

---

35 See a synopsis of these views in H. Lundhaug, *Images of Rebirth: Cognitive Poetics and Transformational Soteriology in the* Gospel of Philip *and the* Exegesis on the Soul (Leiden: Brill, 2010), pp. 349–56. Lundhaug endorses the examination of this text as part of early Christianity rather than through the lens of a predetermined category of 'Gnosticism' (ibid., p. 414).

36 R. Charron and L. Painchaud, '"God is a Dyer". The Background and Significance of a Puzzling Motif in the Coptic *Gospel According to Philip* (CG II, 3)', *Le Muséon*, 114.1–2 (2001), 41–50. Cf. Lundhaug, *Images of Rebirth*, p. 255 n. 388.

37 R. Charron, 'The *Apocryphon of John* (NHC II, 1) and the Graeco-Egyptian Alchemical Literature', *Vigiliae Christianae*, 59 (2005), 438–56, on pp. 455–56.

38 Burns, 'Alchemical Metaphor'.

39 H. Lundhaug, 'Dating and Contextualising the Nag Hammadi Codices and Their Texts: A Multi-Methodological Approach Including New Radiocarbon Evidence', in *Texts in Context: Essays on Dating and Contextualising Christian Writings from the Second and Early Third Centuries*, ed. by J. Verheyden, J. Schröter, and T. Nicklas (Leuven: Peeters, 2021), pp. 117–42, on pp. 132–35.

40 H. Lundhaug, 'Begotten, Not Made, to Arise in This Flesh: The Post-Nicene Soteriology of the *Gospel of Philip*', in *Beyond the Gnostic Gospels: Studies Building on the Work of Elaine Pagels*, ed. by E. Iricinschi et al. (Tübingen: Mohr Siebeck, 2013), pp. 235–71, on pp. 244–48.

reminiscent of alchemical descriptions of dyeing — could be dated closer to what we call 'Byzantium'.

In a similar way, we can utilize the religious vocabulary in specific texts of the Greek Alchemical Corpus. For example, the obscure Byzantine alchemical philosopher and commentator called 'Christianos' (late sixth–eighth century) concludes his treatise *On Making Gold, Thirty Chapters* with an oath before the Holy Trinity.[41] Interestingly, the oath refers to '[...] the Son, who without change became man (ἀτρέπτως ἐνανθρωπήσας) for the glorying of the duality, which is kindred with His name [...]'.[42] The wording, and specifically the reference to 'duality', could allude to the formula ἐν δύο φύσεσιν, included in the Definition of Faith of the Council of Chalcedon (451). It means Christ is to be acknowledged 'in two natures' (divine and human), united without confusion (ἀσυγχύτως), change (ἀτρέπτως), division (ἀδιαιρέτως), or separation (ἀχωρίστως).[43] Given the absence of further details, it is difficult, however, to identify the exact brand of orthodoxy to which Christianos adhered. But let us assume that the above sentence in the oath indicates Christianos' type of Christianity (however, one should be cautious, as it might have been part of a later addition). At face value, the author appears to align with 'neo-Chalcedonian' Christianity, namely, the interpretation that prevailed around the Fifth Ecumenical Council.[44] We may further speculate that his 'neo-Chalcedonism' is perhaps indicated by elements of the sentence emphasizing the unity of Christ ('the Son'; 'kindred with His name'). Rather than presenting a rigid distinction of natures, duality here may be interpreted from the perspective of unity (hypostatic union). If this line of interpretation is valid, then it can be suggested that the phrase echoes Justinian's reconciliatory religious policy between Chalcedonians and non-Chalcedonians.[45] Another indication

41 On the author's name, see Merianos, 'Philosopher Christianos', pp. 273–75. For a detailed analysis of the oath, see ibid., pp. 287–93. On oaths and oath-taking in early Byzantine imperial politics, see the recent article by M. Wuk, 'The Uses of Oaths in Early Byzantine Imperial Politics: A Reconsideration', *Byzantinische Zeitschrift*, 116.3 (2023), 1035–72 (citing previous bibliography).

42 *CAAG*, II, p. 27.16: [...] ὁ υἱὸς ἀτρέπτως ἐνανθρωπήσας καυχήσει τῆς δυάδος οἰκοιωθὲν (οἰκειωθὲν *Marc. gr.* 299, f. 128ᵛ) ὀνόματι [...].

43 Concilium Oecumenicum Chalcedonense, Definitio fidei, ed. by E. Schwartz, *Concilium universale Chalcedonense*, v. 1.2, Acta Conciliorum Oecumenicorum, II.1.2 (Berlin and Leipzig: De Gruyter, 1933), p. 129.30–31: [...] ἐν δύο φύσεσιν ἀσυγχύτως ἀτρέπτως ἀδιαιρέτως ἀχωρίστως γνωριζόμενον, [...].

44 In a previous article (Merianos, 'Philosopher Christianos', p. 290), I referred to Christianos' Christian faith as 'Chalcedonian'. However, upon further reflection, the term 'neo-Chalcedonian' appears to be more appropriate and nuanced. I would like to express my gratitude to Doru Costache for the enlightening discussion we had on this topic.

45 On Christological debates over the doctrine proclaimed at the Council of Chalcedon, Justinian's ecclesiastical policy, and the Second Council of Constantinople (553), see, e.g., P. T. R. Gray, 'The Legacy of Chalcedon: Christological Problems and Their Significance', in *The Cambridge Companion to the Age of Justinian*, ed. by M. Maas (Cambridge: Cambridge University Press, 2005), pp. 215–238; R. Price, *The Acts of the Council of Constantinople of 553: With Related*

supporting this assessment is the use of ἀτρέπτως ἐνανθρωπήσας, which, in this exact form, is found in the *troparion* 'Ὁ Μονογενὴς Υἱός' (The Only-begotten Son), which bears a 'neo-Chalcedonian' character.[46]

Religious terminology and concepts can be used to approach the Christianity of an author and even provide hints for the dating of a text. Nevertheless, this is a slippery process that requires a firm grasp of Christian theological knowledge to recognize and disentangle the relevant terms and concepts. Since this is hardly possible for most historians of alchemy, the assistance of a theologian often proves to be the safest solution. In any case, it becomes evident that not every reference to Christ or the Trinity constitutes proof of an author's 'orthodoxy'.

Incorporating Christian Concepts into Alchemical Discourse: Some Examples

Around the seventh century, the Christianization of alchemy became systematic in Byzantium. The most celebrated case is the work entitled *On the Great and Sacred Art of Making Gold*, which is attributed to the elusive alchemical author Stephanos of Alexandria.[47] In the manuscript tradition, the work is divided into nine textual sections known as 'lectures' (*praxeis*), in which Stephanos comments on quotations from the great masters of the past, such as Pseudo-Demokritos (AD first century) and Zosimos of Panopolis. A distinct and easily recognizable Christian characteristic of the lectures is that most of them begin and/or end with prayers to God, referring mainly to the Logos and divine wisdom. Maria Papathanassiou has shown that Stephanos draws for the composition of the prayers mainly from the New Testament, the fourth- and fifth-century Church Fathers, and the influential Christian Neoplatonizing author Pseudo-Dionysios the Areopagite (late fifth or early sixth century).[48] It should be stressed that the Christian elements in Stephanos' work are not confined to these prayers.

Recently, Vincenzo Carlotta questioned the extent of the Christianization of Stephanos' alchemy. Carlotta concluded that Stephanos 'linked directly the pagan

---

Texts on the Three Chapters Controversy, 2 vols (Liverpool: Liverpool University Press, 2009), I, pp. 1–103.

46 S. Janeras, 'Le tropaire Ὁ Μονογενὴς dans les liturgies orientales et sa signification œcuménique', in *Liturgies in East and West: Ecumenical Relevance of Early Liturgical Development*, ed. by H.-J. Feulner (Zürich and Berlin: Lit, 2013), pp. 209–23, on p. 220. The *troparion* 'Ὁ Μονογενὴς Υἱὸς' combines words and phrases mainly from the Nicene-Constantinopolitan Creed (381) and the Chalcedonian Definition of Faith; see J. H. Barkhuizen, 'Justinian's Hymn Ὁ μονογενὴς υἱὸς τοῦ Θεοῦ', *Byzantinische Zeitschrift*, 77.1 (1984), 3–5.

47 For the *status quaestionis* on Stephanos, see Koutalis, Martelli, and Merianos, 'Alchemy', pp. 23–31 (citing previous bibliography). On Stephanos' Christianized alchemy, see V. Carlotta, 'Introducing Greek Alchemy to Christianity. Inclusion and Exclusion of Religious Elements in Stephanus's *Lessons*', *ARYS*, 20 (2022), 323–48.

48 Papathanassiou, 'Stéphanos d'Alexandrie'.

sources of the alchemical tradition with a form of Christian wisdom and moral virtue'. He 'provided his Christian audience with a justification for the study of the pagan authors of the alchemical tradition', but he did not embed 'new Christian notions into their works' or exclude 'any element bearing strong connections to the pagan origin of these sources.' Although alchemy was presented 'as a Christianised discipline whose contents were not yet truly harmonised with Christian beliefs and practices', it nonetheless 'remained established in a technical and philosophical tradition strongly grounded in the Greco-Roman religious landscape'.[49]

Carlotta's rationale is welcome, consistent with Christianization being a long and ongoing process. A Christian framework alone does not necessarily entail an in-depth restructuring of alchemical concepts to become compatible with Christian principles and teachings. On the other hand, in Stephanos' times, Christianity had been conversing with pagan philosophical and spiritual ideas for centuries. It is known, for instance, that Hellenic and Christian Neoplatonism interacted, as is evident in the Christian adoption and resignification of initially Platonic concepts (e.g., participation in the divine) and technical vocabularies (e.g., of theurgy).[50] Therefore, two approaches seem worth exploring in the future: (i) a theologically informed examination of Stephanos' work to clarify his mechanisms of Christianization, especially regarding his technical language, which may sometimes seem vaguely Christian or even pagan at first glance. (ii) A comparative study between Stephanos' Christianization of alchemy and similar attempts concerning pagan philosophical concepts and scientific disciplines in early Byzantium.

There is still more to be said about Stephanos' alchemical discourse regarding Christian ideas. For example, in his third lecture, known as the 'Letter to Theodoros', Stephanos makes a well-known distinction between the 'legendary alchemy' (μυθικὴ χημεία) and the 'secret and hidden' (μυστικὴ καὶ κρυπτή). He says the former is confounded in a multitude of words, while the latter is practiced through the word (λόγῳ; also meaning Logos) of the world's creation. Thus, the man who is godly-minded (θεόφρων) and born of God (θεογενής) may learn about the material (ἐνύλου) world through direct labor and teachings about God (θεολογιῶν) and secret words.[51] Pseudo-Demokritos has already introduced the distinction between 'legendary' and 'secret' alchemical discourse,[52] but Stephanos

---

49 Carlotta, 'Introducing Greek Alchemy', p. 345.

50 See, e.g., S. Klitenic Wear, 'The Philosophy of the Later Neoplatonists: An Interaction with Christian Thought', in *The Routledge Handbook of Early Christian Philosophy*, ed. by M. Edwards (London and New York: Routledge, 2021), pp. 313–27, on pp. 320–21.

51 Stephanos of Alexandria, *On the Great and Sacred Art of Making Gold* 3.21–25, ed. by M. K. Papathanassiou, *Stephanos von Alexandreia und sein alchemistisches Werk. Die kritische Edition des griechischen Textes eingeschlossen* (Athens: Cosmosware, 2017), p. 167. For English translations, see F. S. Taylor, 'The Alchemical Works of Stephanos of Alexandria', *Ambix*, 2.1 (1938), 39–49, on p. 39; Carlotta, 'Introducing Greek Alchemy', p. 332.

52 Pseudo-Demokritos, *On the Making of Purple and Gold: Natural and Secret Questions* 16.168–70, ed. and transl. by Martelli, *The* Four Books *of Pseudo-Democritus*, pp. 96–97: 'In fact, they believe that we are presenting a legendary rather than a secret discourse, so they do not carry

further develops the idea. In doing so, he draws on Christian theological thought and terminology.

Specifically the adjectives *theophrōn*, 'godly minded', and *theogenēs*, 'born of God', are quite revealing in this respect. The first one, *theophrōn*, is common in Christian literature of various genres. It is often used, for example, by Maximos the Confessor (580–662) — a contemporary of Stephanos — in his *Ambigua* to characterize great Christian authorities such as Gregory of Nazianzos (fourth century) and Pseudo-Dionysios the Areopagite.[53] The *Ambigua to John*, in particular, was written as a commentary, aiming principally at disentangling the meaning of perplexing passages in the orations of Gregory of Nazianzos. According to Maximos, 'Saint Gregory the Theologian was a man of profound thoughts but of comparatively few words' (βραχυλόγος ὑπάρχων καὶ πολύνους).[54] Commenting on an excerpt by Gregory, Maximos notes:

> What the great teacher *secretly* [emphasis mine] reveals through these words is known, I believe, only to those who have received his same gift of wisdom and knowledge. For having removed all the dark fluid of passions and every material attachment from their intellective eyes, they were able to perceive in all things the ray of true knowledge, and in the light of its simple, unitary principle, they withdrew their intellect from *multiplicity* [emphasis mine] (as much as was possible), and through *simplicity* [emphasis mine] of mind received the whole scientific knowledge of beings. But to men like me, the 'gloom of ignorance continues to eclipse the observation of the truth', [...].[55]

Maximos argues that the secretly revealed teachings of Gregory could be fully grasped only by those who have received the same gift of wisdom, whose intellect is detached from passions and material attachments. The ethical prerequisite for acquiring the divine gift of knowledge also appears in Byzantine alchemical writings and will be discussed shortly. For now, let us focus on the observation that Stephanos refers not only to the notion of secretly revealed knowledge, as shown above, but also that of deeper understanding achieved by reducing multiplicity to simplicity. In his ninth lecture, Stephanos addresses Emperor Herakleios (r. 610–641) and stresses that he will make his words clear, having expelled the multiplicity of enmattered things (τὴν πολυπληθείαν τῶν ἐνύλων ἀπωσάμενος).[56] In

---

out any close examination of the species [...]'. See Carlotta, 'Introducing Greek Alchemy', p. 332 and n. 15.

53 See, e.g., Maximos the Confessor, *Ambigua to John* 9.2, ed. and transl. by N. Constas, *Maximos the Confessor, On Difficulties in the Church Fathers, The* Ambigua, 2 vols (Cambridge, MA: Harvard University Press, 2014), I, p. 148; 12.2, I, p. 348; 14.4, I, p. 360; 21.3, I, p. 422; 23.4, II, p. 6; 26.3, II, p. 22; 32.7, II, p. 60; 37.10, II, p. 86; 42.2, II, p. 122; 42.20, II, p. 158; 71.7, II, p. 324; 71.11, II, p. 328; also *Ambigua to Thomas*, Prol. 5, I, p. 6.

54 Maximos the Confessor, *Ambigua to John*, Prol. 7, I, pp. 66–67.

55 Maximos the Confessor, *Ambigua to John* 45.2, II, pp. 192–93.

56 Stephanos of Alexandria, *On the Great and Sacred Art of Making Gold* 9.30–34, p. 214; see also 7.7–9, p. 199. Cf. Carlotta, 'Introducing Greek Alchemy', p. 328.

the fourth lecture, he notes his intention to unravel what has been said by ancient and virtuous men in an allegorical and intricately entangled manner.[57] Similarly, Christianos emphasizes the necessity to show that alchemy is not unlimited and incomprehensible in every way. His aim is twofold, to remove the obscurity of the various descriptions and writings, revealing that the method is only one, but also to avoid attracting the usual criticism of presenting an unlimited number of productions.[58]

To return to Maximos, Nicholas Constas explains that his 'task is to unravel the skein of that language [Gregory's] into an intelligible sequence of interlaced patterns'.[59] Thus, the concept of secret and hidden alchemy is analogous to the hidden and secret meanings of lofty theological thought, each requiring to be deciphered appropriately by a godly-minded (*theophrōn*) pursuer of knowledge.

Nevertheless, for Stephanos, the practitioners of 'secret alchemy' need not only be 'godly-minded' but also 'born of God'. This second adjective, *theogenēs*, is relatively rare, found, for example, in this exact form in Anastasios of Sinai (d. after 700).[60] However, it has already been pointed out that the term is encountered in Pseudo-Dionysios as a noun — that is, *theogenesia*, 'divine birth'. Actually, it is his favorite of the two words he uses to denote the sacrament of baptism (the other one is *phōtisma*, 'illumination').[61] In baptism, human beings enter into the deifying light of Jesus, but '[t]his "illumination" is both a birth into a new life and a death on the cross to sin'.[62] Terms such as *theophrōn* and *theogenēs* found in Stephanos reveal familiarity with and influence from Christian theological concepts. As for the idea of 'secret alchemy', while originating from alchemical literature (Pseudo-Demokritos), the discussion made implies that, at the same time, this and other alchemical ideas could be accommodated quite well by early Byzantine theological thought.

Stephanos introduces moral deliberations into alchemical practice,[63] yet he is not alone in doing so. Possibly (nearly) contemporary but hard-to-date authors, such as Christianos and the four alchemical poets, show that the conduct of the 'true philosopher' coincides with that of the 'true Christian', and moral life is a prerequisite for being granted access to divine knowledge.

Christianos, in his *On the Consistency of Gold*, comments on the vagueness and obscurity of the alchemical writings of the ancient philosophers and refers to the

---

57 Stephanos of Alexandria, *On the Great and Sacred Art of Making Gold* 4.17–21, p. 173; cf. 4.35–36, p. 174; 4.89–93, p. 176; 4.140, p. 177; 5.5–8, p. 181.

58 *CAAG*, II, pp. 417.14–418.5.

59 Constas, *Maximos the Confessor, On Difficulties*, I, p. xii.

60 Anastasios of Sinai, *Hexaemeron* 12.3.5.210, ed. and transl. by C. A. Kuehn and J. D. Baggarly, Orientalia Christiana Analecta, 278 (Rome: Pontificio Istituto Orientale, 2007), p. 470.

61 Koutalis, Martelli, and Merianos, 'Alchemy', p. 31. On these terms, see Ch. M. Stang, *Apophasis and Pseudonymity in Dionysius the Areopagite: 'No Longer I'* (Oxford: Oxford University Press, 2012), p. 99 and n. 64.

62 Stang, *Apophasis and Pseudonymity*, p. 102.

63 Koutalis, Martelli, and Merianos, 'Alchemy', p. 31.

way one must find to interpret them. He centers this pursuit around the idea that the human intellect has access to knowledge not by nature but by participation (κατὰ μετοχήν). Only if the pursuer of (alchemical) knowledge embraces a pure way of life and has a pure purpose will he receive the gift to participate in divine wisdom.[64] The notion of participation, but also the definition of likeness to God as the end of the spiritual and moral life, has a Platonic origin. In the Greek patristic tradition, the concept of 'participation' (*metochē*; also *methexis, metechein, metousia*) is essential and overlaps with concepts such as 'likeness' (*homoiōsis*) and 'deification' (*theōsis*).[65] Pseudo-Dionysios defines deification as 'assimilation (*aphomoiōsis*) to and union (*henōsis*) with God as far as possible' and bridges the concepts of likeness / virtue with those of participation / union.[66] Similarly, Stephanos repeats one of the six classical definitions of philosophy in his work, 'likeness (*homoiōsis*) to God as far as humanly possible'.[67]

Christianos incorporates in his treatise *On Making Gold, Thirty Chapters* a moral code containing the qualities required for a proper engagement with the sacred art. It is not a coincidence that this text precedes the oath before the Holy Trinity, discussed in the previous section. According to the code, one who pursues science (hence, the true philosopher-alchemist) must love God and Man, be prudent, not love money, despise lies and everything deceitful, wicked, and envious; he must be a true and faithful disciple of the Holy and Consubstantial and Coeternal Trinity. Whoever lacks such God-pleasing morals will deceive himself, rushing into unattainable goals, and will be rather harmed.[68] Therefore,

64 *CAAG*, II, pp. 397.15–398.18. See the analysis in Merianos, 'Philosopher Christianos', pp. 279–84.

65 N. Russell, *The Doctrine of Deification in the Greek Patristic Tradition* (Oxford: Oxford University Press, 2004), pp. 2, 77.

66 Pseudo-Dionysios, *On the Ecclesiastical Hierarchy* 1.3, ed. by G. Heil and A. M. Ritter, *Corpus Dionysiacum*, II: *Pseudo-Dionysius Areopagita, De coelesti hierarchia; De ecclesiastica hierarchia; De mystica theologia; Epistulae*, 2nd edn (Berlin and Boston, MA: De Gruyter, 2012), p. 66.12–19. On the passage, see A. Golitzin, *Mystagogy: A Monastic Reading of Dionysius Areopagita*, ed. by B. G. Bucur (Collegeville, MN: Cistercian Publications and Liturgical Press, 2013), pp. 250–52; D. Costache, 'Being, Well-Being, Being for Ever: Creation's Existential Trajectory in Patristic Tradition', in *Well-Being, Personal Wholeness and the Social Fabric*, ed. by D. Costache, D. Cronshaw, and J. R. Harrison (Newcastle upon Tyne: Cambridge Scholars Publishing, 2017), pp. 55–87, on pp. 69–70. On the Pseudo-Dionysian concept of deification, see also Russell, *The Doctrine of Deification*, pp. 248–62; M. Edwards, 'Dionysius the Areopagite', in *Routledge Handbook of Early Christian Philosophy*, ed. by Edwards, pp. 609–18, on pp. 614–15.

67 Stephanos of Alexandria, *On the Great and Sacred Art of Making Gold* 6.34–35, p. 188; see Merianos, 'Alchemy', p. 243 and n. 76. On this classical definition of philosophy and its Platonic origin, see, e.g., M. Trizio, 'Byzantine Philosophy as a Contemporary Historiographical Project', *Recherches de Théologie et Philosophie médiévales*, 74.1 (2007), 247–94, on pp. 274–75.

68 *CAAG*, II, p. 35.10–16. For a detailed analysis of the moral code, see Merianos, 'Philosopher Christianos', pp. 285–87, 293. In her article on the development of medical deontology codes in Antiquity, Giulia Ecca explores various medical and non-medical deontological texts, including Christianos' moral code and oath, and suggests a common thread between medical and alchem-

the moral code can also be interpreted as a warning to those who do not strive to live up to these ideals and an explanation of why an aspiring alchemist might fail in his endeavors. The alchemical work's success depends on the virtuous conduct of the practitioner, whom God's grace will enlighten. Along the same lines, Stephanos, in his fourth lecture, presents the pursuit of virtue as a prerequisite to acquiring alchemical knowledge grounded on faith, humility, and love of God.[69] Consequently, Christian ethics form a constituent part of alchemical methodology.

The fact that Christianos and Stephanos give such importance to personal purity and moral integrity (an ancient concept eventually identified as Christian) is not a banality. In Christianized alchemical thought, moral rectitude shapes the philosopher-alchemist's intellect and soul into an appropriate receptacle of divine grace, through which he could be illuminated and participate in divine knowledge. According to Stephanos, the visitation of grace requires, among other things, the renunciation of the world, the mortification of the body, and the praise of God.[70] Notably, the alchemical poem attributed to Archelaos goes so far as to describe the preparation of the alchemist's body and soul to receive the knowledge granted by grace in a manner reminiscent of the ascetic way of life.[71] These views coincide with the Christian understanding of the terms 'philosopher' and 'philosophy'. The Christian way of life, aiming at moral perfection, was considered 'true philosophy' and was often identified with the monastic ideal.[72]

It becomes evident that the quest for knowledge is tied to pursuing moral life, a classic philosophical aim. Stephanos' and Christianos' lists of virtues serve to identify an alchemist as 'worthy' or 'unworthy' based on his conduct, thus

---

ical writings of this nature ('La formazione dei codici di deontologia medica nell'antichità', *Galenos*, 16 [2022], 39–58, on pp. 51–52).

69 Stephanos of Alexandria, *On the Great and Sacred Art of Making Gold* 4.29–34, p. 174; also 1.42–43, p. 158. See Carlotta, 'Introducing Greek Alchemy', pp. 330–31.

70 Stephanos of Alexandria, *On the Great and Sacred Art of Making Gold* 8.125–145, pp. 211–12 (see commentary on p. 142). See also M. K. Papathanassiou, 'The Occult Sciences in Byzantium', in *A Companion to Byzantine Science*, ed. by S. Lazaris (Leiden and Boston, MA: Brill, 2020), pp. 464–95, on p. 492.

71 Pseudo-Archelaos, *On the Same Sacred Art, in Iambic Verse*, vv. 37–48, ed. by G. Goldschmidt, *Heliodori carmina quattuor ad fidem codicis Casselani* (Giessen: A. Töpelmann, 1923), p. 51. For an English translation of the poem, see Ch. A. Browne, 'Rhetorical and Religious Aspects of Greek Alchemy. Including a Commentary and Translation of the Poem of the Philosopher Archelaos Upon the Sacred Art, Part I', *Ambix*, 2.3–4 (1946), 129–37.

72 See A.-M. Malingrey, *'Philosophia'. Étude d'un groupe de mots dans la littérature grecque, des Présocratiques au* IV*e siècle après J. C.* (Paris: C. Klincksieck, 1961); Trizio, 'Byzantine Philosophy', pp. 274–75; O'Meara, 'Conceptions of Science', p. 171. According to Doru Costache, Gospel writers, early apologists, and early monastic authors portrayed Jesus as the perfect philosopher and, consequently, Christianity as a philosophical school ('The Teacher and His School: Philosophical Representations of Jesus and Christianity', in *The Impact of Jesus of Nazareth: Historical, Theological, and Pastoral Perspectives*, II, ed. by P. G. Bolt and J. R. Harrison [Macquarie Park: SCD Press, 2021], pp. 227–51).

delineating the field's moral boundaries. Outside of them, true knowledge cannot be obtained. In this sense, these and other authors contribute to forming the identity of the philosopher-alchemist in Byzantium.[73] Furthermore, the necessity for alchemical philosophers to establish a 'moral code' must not be irrelevant to the attempt made in alchemical texts to distinguish the true philosopher from the false one, the genuine pursuer of knowledge from the charlatan.

The significance of moral excellence in acquiring access to the divine mysteries of nature is inextricably associated with a vital concept mentioned above: alchemy as a gift of God, a giving that springs from God's generosity.[74] In this context, both Stephanos and Christianos cite James 1:17 verbatim, repeating that '[e]very good gift (δόσις) and every perfect gift (δώρημα) is from above, and comes down from the Father of lights'.[75] Stephanos further refers to the Father of lights ('I confess the grace of the illumination from above, which is given to us by the Father of lights'; and 'O rich gifts by the Father of lights!') and also labels the alchemical opus as 'God-given' (θεοδώρητος).[76] Likewise, the four alchemical poems, attributed to Heliodoros, Theophrastos, Hierotheos, and Archelaos,[77] include references to the concept of God-given alchemical knowledge. For example, in the poem ascribed to Theophrastos, the 'gift' that is 'divinely given' is mentioned.[78]

A historiographical comment should be made here. The concept of alchemy as a gift of God, expressed under the technical term *donum dei*, can be found in texts from different cultural contexts and is common in medieval and early modern alchemical works. The current consensus holds that the idea of alchemy as *donum dei* in medieval Latin authors is an intellectual loan from the Islamic alchemical tradition. According to Vladimír Karpenko, the concept originated in the Hellenistic world and was transmitted through Islamic alchemy to Latin Europe.[79]

---

73 Merianos, 'Philosopher Christianos', p. 293.

74 What follows on the concept of alchemy as a gift of God draws on Merianos, 'Philosopher Christianos', pp. 282–83.

75 Stephanos of Alexandria, *On the Great and Sacred Art of Making Gold* 4.1–2, p. 173; *CAAG*, II, p. 398.2–3. The English translation of James 1:17 is taken from the New King James Version.

76 See, respectively, Stephanos of Alexandria, *On the Great and Sacred Art of Making Gold* 1.47–48, p. 158; 7.188–89, p. 205; 2.66, p. 163.

77 These poems are considered to be the work of a single author and are dated to the seventh–eighth century; see J. Letrouit, 'Chronologie des alchimistes grecs', in *Alchimie: art, histoire et mythes*, ed. by D. Kahn and S. Matton (Paris and Milan: S.É.H.A. and Archè, 1995), pp. 11–93, on pp. 82–83, 88. M. D. Lauxtermann, *Byzantine Poetry from Pisides to Geometres: Texts and Contexts*, II (Vienna: Austrian Academy of Sciences Press, 2019), pp. 205–07, has redated them to the fifth or early sixth century, based on their metrical analysis.

78 Pseudo-Theophrastos, *On the Same Divine Art, in Iambic Verse*, vv. 210–11, ed. by Goldschmidt, *Heliodori carmina quattuor*, p. 41. For an English translation of the poem, see Ch. A. Browne, 'The Poem of the Philosopher Theophrastos Upon the Sacred Art: A Metrical Translation with Comments Upon the History of Alchemy', *The Scientific Monthly*, 11.3 (1920), 193–214.

79 V. Karpenko, 'Alchemy as *donum dei*', *Hyle — An International Journal for the Philosophy of Chemistry*, 4.1 (1998), 63–80. See also W. R. Newman, *Gehennical Fire: The Lives of George Starkey, An American Alchemist in the Scientific Revolution* (Cambridge, MA and London: Har-

Yet, as made evident above, the Byzantine alchemical literature also expressed the same idea, although significantly earlier than medieval Latin authors.[80] Thus, the historiography of the concept of *donum dei* in alchemy, which up until now ignored its presence in Byzantium, should take this evidence into consideration.

Saints, Patriarchs, Monks, Theologians, and Their Multifaceted Relationship with Alchemy

There is scattered evidence for the acquaintance — theoretical at least — of the priestly and monastic milieus with alchemy. A fifteenth-century (?) text has been handed down under a hieromonk's name: *Explanation of the Science of Gold-Making by the Hieromonk Kosmas*.[81] According to Andrée Colinet, who prepared the modern edition, the only part of the text that could be attributed to Kosmas can be found in the first three of its eleven paragraphs.[82] Paragraphs 4–8, in particular, are recipes excerpted from *On How to Make Gold* by Michael Psellos. The Psellian recipes in their new context suffer from accidental omissions and must have originated from a later manuscript tradition.[83] This work resembles a recipe collection ascribed to Kosmas and is not a single unified treatise written by one author. Still, this conclusion does not diminish the significance of the fact that a hieromonk is associated (as a partial author and/or copyist?) with an alchemical text.

Nevertheless, the interest in the sacred art of certain clergymen and monks exceeds the limits of manuscript production. It is indeed telling, for instance, that the very *Marc. gr.* 299 once belonged to the Greek cardinal Bessarion (d. 1472).[84] In what follows, a few known cases from texts of various literary genres are

vard University Press, 1994), pp. 3, 8–10, 12, 66, 98, 114, 181; Newman, *Promethean Ambitions*, pp. 84, 89; Ch. Crisciani, *Il papa e l'alchimia: Felice V, Guglielmo Fabri e l'elixir* (Rome: Viella, 2002), pp. 44–45; T. Nummedal, *Alchemy and Authority in the Holy Roman Empire* (Chicago, IL and London: The University of Chicago Press, 2007), pp. 27–30; Principe, *Secrets of Alchemy*, pp. 192–195, 199–200.

80 Karpenko, 'Alchemy as *donum dei*', pp. 65–66, 68, refers to the presence of this idea in Byzantine alchemy; however, he mentions only Stephanos of Alexandria and refers vaguely to this concept in his work.

81 Kosmas the Hieromonk, *Explanation of the Science of Gold-Making*, ed. by A. Colinet, *Recettes alchimiques (Par. gr. 2419; Holkhamicus 109) — Cosmas le Hiéromoine, Chrysopée*, Les alchimistes grecs, XI (Paris: Les Belles Lettres, 2010), pp. 66–76. For the dating of the author, see ibid., pp. CXXXVII–CXXXVIII.

82 Colinet, *Recettes alchimiques*, p. CXXXVII.

83 *CMAG*, VI, pp. 16–18; Colinet, *Recettes alchimiques*, p. CXXIX.

84 See, e.g., M. Mavroudi, 'Occult Science and Society in Byzantium: Considerations for Future Research', in *Occult Sciences in Byzantium*, ed. by Magdalino and Mavroudi, pp. 39–95, on pp. 82–83; B. Mondrain, 'Le cardinal Bessarion et la constitution de sa collection de manuscrits grecs — ou comment contribuer à l'intégration du patrimoine littéraire grec et byzantin en

examined to indicate that alchemy was not regarded as incompatible with the Christian faith, clergy, and literature, but also to showcase public perceptions of the 'sacred art'.

The Risks of chrysopœia

Patriarch Michael I Keroularios (1043–1058) is the most striking example of a Byzantine clergyman fervently engaged with alchemy while at the highest echelon of the ecclesiastical hierarchy. Michael Psellos was well acquainted with the patriarch's alchemical interests as he dedicated the letter-treatise *On How to Make Gold* to him.[85] Psellos says he originally planned to undertake a comprehensive systematic treatment and investigate every processing of matter, such as gems, pearls, etc. However, the patriarch's penchant for gold-making limited Psellos to this topic.[86] In compliance with the patriarch's request, Psellos listed recipes for making gold, gilding metals, brightening gold, doubling its mass, and making a gold dye.[87]

Years later, Psellos wrote another work also mentioning Keroularios' alchemical interests. This time, he did not intend to please the patriarch but to accuse him formally. The *Indictment of Keroularios* was composed in 1058 on behalf of Emperor Isaac I Komnenos (r. 1057–1059). However, it was never delivered since the patriarch died before the trial.[88] Apart from certain charges, such as heresy, tyranny, murder, and sacrilege, Psellos also refers to Keroularios' fascination with alchemy. He sought to transmute base metals to precious ones and studied alchemical literature. Nevertheless, the patriarch, Psellos says, never succeeded in producing gold but only made copper look like gold in color, implying that

---

Occident', in *'Inter graecos latinissimus, inter latinos graecissimus'. Bessarion zwischen den Kulturen*, ed. by C. Märtl, Ch. Kaiser, and Th. Ricklin (Berlin and Boston, MA: De Gruyter, 2013), pp. 187–202, on p. 194.

85 For what follows on Psellos, see Merianos, 'Alchemy', pp. 243–46, 250. See also G. Katsiampoura, 'Transmutation of Matter in Byzantium: The Case of Michael Psellos, the Alchemist', *Science and Education*, 17.6 (2008), 663–68; M. Bellini, 'Psellos' Petrified Root: Transmutations and Natural Wonders from Classical Antiquity to Byzantine Times', *Physis: Rivista internazionale di storia della scienza*, 55.1–2 (2020), 223–41; A. M. Roberts, 'Thinking about Chemistry in Byzantium and the Islamic World', *Journal of the History of Ideas*, 84.4 (2023), 595–619, on pp. 611–15.

86 *CMAG*, VI, § 5, pp. 30–32. For an Italian translation of this work, see F. Albini, *Michele Psello, La Crisopea, ovvero come fabbricare l'oro* (Genoa: ECIG, 1988).

87 *CMAG*, VI, §§ 7–13, pp. 34–40. Cf. Colinet, *Recettes alchimiques*, pp. 119–21 (nn. 24, 27, 29–34).

88 On the *Indictment*, see G. T. Dennis, 'A Rhetorician Practices Law: Michael Psellos', in *Law and Society in Byzantium, Ninth–Twelfth Centuries*, ed. by A. E. Laiou and D. Simon (Washington, D.C.: Dumbarton Oaks Research Library and Collection, 1994), pp. 187–97, on pp. 190–92; S. Papaioannou, *Michael Psellos: Rhetoric and Authorship in Byzantium* (Cambridge: Cambridge University Press, 2013), pp. 153–54, 218–19.

his alchemical endeavors were limited to imitation. Additionally, Keroularios did not know the proportions and did not weigh, a detail that emphasizes the lack of quantification in his operations.[89] Psellos implicitly distinguishes between greedy amateurs — such as Keroularios — and true philosophers with substantial knowledge and methodology — such as himself.

However, the patriarch's preoccupation with alchemy is not the actual accusation but a prelude to preparing the scene for leveling the actual charge. Psellos first admits his own past interest in alchemy and claims that he had already renounced this knowledge as foolery. Let us not forget that he had authored a work on making gold, which he dedicated to Keroularios. Then Psellos stresses that the patriarch's engagement with alchemy, though perhaps blameworthy, did not constitute a crime. As he reveals later, the actual accusation is that Keroularios processed gold clandestinely using state workers to decrease its fineness to his benefit. This act blatantly violated Byzantine legislation.[90] Psellos ingeniously builds on Keroularios' passion for gold to expose his greed; he thus depicts the patriarch resorting to the 'aurific art' to satisfy his desire for gold. Keroularios' case suggested that alchemy is an art providing techniques for counterfeiting and gilding. From what Psellos states, it seems that Byzantine authorities considered the theoretical study of alchemy relatively harmless. However, practices aimed at counterfeiting precious metals, which could be broadly regarded as 'alchemical', constituted a serious crime, especially when state craftsmen were involved. Keroularios' secret counterfeiting activities took place in an underground room of his. Although Psellos refers to craftsmen, not alchemists, the secret processing of gold in an underground room is evocative of the cliché of covert alchemical operations.

The underground laboratory recalls an earlier but also infamous patriarch, at least for the iconophiles, John VII Grammatikos (the Grammarian), the last iconoclast patriarch of Constantinople (837?–843).[91] It is worth making a digression to discuss this exciting figure. After the iconophiles' triumph and the images' restoration, iconophile sources depicted him as a sorcerer, necromancer, and diviner who made accurate predictions. In the canon written to celebrate

89 Michael Psellos, *Orationes forenses et acta*, ed. by G. T. Dennis (Stuttgart and Leipzig: B. G. Teubner, 1994), *Or. for.* 1.2670–2689 (*Indictment of Keroularios*). For a French translation of the passages of the *Indictment* relating to alchemy, with introduction and commentary, see *CMAG*, VI, pp. 71–89.

90 Psellos, *Or. for.* 1.2690–2732 (*Indictment*). On the relevant legislation, see M. F. Hendy, *Studies in the Byzantine Monetary Economy, c. 300–1450* (Cambridge: Cambridge University Press, 1985), pp. 320–28.

91 On John VII Grammatikos, see P. Magdalino, *L'Orthodoxie des astrologues. La science entre le dogme et la divination à Byzance (VIIe–XIVe siècle)* (Paris: Lethielleux, 2006), ch. III; P. Magdalino, 'Occult Science and Imperial Power in Byzantine History and Historiography (9th–12th Centuries)', in *Occult Sciences in Byzantium*, ed. by Magdalino and Mavroudi, pp. 119–62, on pp. 122–24, 128–29, 132–35; E. Nicolaidis, *Science and Eastern Orthodoxy: From the Greek Fathers to the Age of Globalization*, transl. by S. Emanuel (Baltimore, MD: The Johns Hopkins University Press, 2011), pp. 55–56.

the Triumph of Orthodoxy — probably by John's successor in the patriarchate, Methodios I (843–847) — it is stated that John should be called Pythagoras (regarded as the original arcane philosopher), Kronos, or Apollo and not by the name of the forerunner of Christ.[92] Three other ninth-century texts depict John Grammatikos in similar colors. Both George the Monk's universal chronicle and the *Life of St. Theodora the Empress* call John a 'master of diviners' and 'master of demons', among other labels, and explicitly accuse him of practicing dish-divining, while the *Synodicon Vetus* characterizes him as a 'dish-diviner'.[93]

In the tenth-century history known as *Theophanes Continuatus*, there are graphic illustrations of John's engagement with the occult, among which an operation concerning a particular kind of sorcery. When an infidel barbarian nation under three leaders invaded the empire, and Emperor Theophilos (r. 829–842) was disheartened, John proposed and executed a magical operation. It involved the decapitation of a three-headed statue in the Hippodrome at his command to avert the attack by the three barbarian leaders.[94]

The incident concerning the statues of the Hippodrome is followed in *Theophanes Continuatus* by a vivid account of John Grammatikos' secret activities in an underground laboratory placed in a suburban estate owned by his brother.

> Making frequent sojourns there, it is said that the patriarch fitted out for himself an underground dwelling place, [...]; and behind the door that opened to the outside, there was a postern which led thither those who desired by means of not a few stairs. That was his evil laboratory; and sometimes certain chamber-maids were kept there by him, monastics and women who did not otherwise fake beauty, with whom he would corrupt

92 J.-P. Migne, ed., *Patrologiae cursus completus. Series graeca*, v. 99 (Paris: J.-P. Migne, 1860), col. 1776C.

93 George the Monk, *Chronicle*, ed. by C. de Boor and P. Wirth, *Georgii Monachi Chronicon*, 2 vols (Stuttgart: B. G. Teubner, 1978), II, p. 798.19–22; A. Markopoulos, 'Βίος τῆς αὐτοκράτειρας Θεοδώρας (*BHG* 1731)', *Symmeikta*, 5 (1983), 249–85, on p. 261 (§ 5.9–13); *The Synodicon Vetus* 155.6, ed. and transl. by J. Duffy and J. Parker, CFHB, 15 (Washington, D.C.: Dumbarton Oaks Center for Byzantine Studies, 1979), pp. 130–31. See Magdalino, 'Occult Science', p. 133 and n. 59. On dish-divining in Byzantium, see M. Grünbart, 'Lekanomanteia in the Medieval Eastern Christian World', in *Prognostication in the Medieval World: A Handbook*, ed. by M. Heiduk, K. Herbers, and H.-Ch. Lehner, 2 vols (Berlin and Boston, MA: W. de Gruyter, 2021), II, pp. 846–48.

94 *Theophanes Continuatus* 4.7, ed. and transl. by M. Featherstone and J. Signes Codoñer, *Chronographiae quae Theophanis Continuati nomine fertur Libri I–IV*, CFHB, 53 (Boston, MA and Berlin: W. de Gruyter, 2015), pp. 220–22. See C. Mango, 'Antique Statuary and the Byzantine Beholder', *Dumbarton Oaks Papers*, 17 (1963), 53–75, on p. 61 (= *Byzantium and Its Image: History and Culture of the Byzantine Empire and Its Heritage* [London: Variorum Reprints, 1984], no. V); Magdalino, 'Occult Science', pp. 123–24. Magdalino argues that John Grammatikos' alleged operation, combined with evidence from other sources, indicates the emergence of 'the magic of statues', a form of occult science on the talismanic powers of statues that was peculiar to early ninth-century Byzantium ('Occult Science', pp. 134–35).

himself; and at other times divinations were practised through inspection of the liver and dish-divining and sorcery and necromancy, and he used these women as partners and assistants. Whereby it often happened that certain true things were foretold by him through the co-operation of the demons, not only to Theophilus [the emperor] but to others of like mind.[95]

With this reference to his 'evil laboratory', the digression about John Grammatikos concludes. The cases of Grammatikos and Keroularios present differences; the former is depicted to engage in sorcery and the occult (especially dish-divining); the latter, an alchemy enthusiast, to manipulate gold with the aid of state craftsmen. Paul Magdalino observes that the constancy of the invective directed against John Grammatikos may provide some credence to the iconophile portrayal of this patriarch as an occult scientist, which could be more than a stereotypical image.[96] The same can be said about the depiction of Keroularios as an enthusiast of the aurific art. Nonetheless, the accusation of practicing magic and other occult sciences, regardless of its veracity, was sometimes used as a political weapon in Byzantium against rivals or to defame those who had lost official favor.[97]

The element that links the activity of these two patriarchs from different periods is the secret underground laboratory used for illicit purposes and the imagery of clandestine operations. Such operations evoke the Roman and Byzantine legal discourse associating secrecy and magic; rituals performed privately or clandestinely (e.g., nocturnally), often involving sacrifices, were forbidden by law.[98] This observation implies a reason why alchemy could be regarded as 'magic' in the public imaginary. Psellos offers an indication of this, emphatically stressing in *On How to Make Gold* that the transmutation of matter happens by natural alteration, not spells, wonder-working, or other unspeakable deeds.[99] He accentuates alchemy's rational character and underlines that it operates within the range of nature, suggesting that those unfamiliar with its principles could associate it with suspicious operative arts. This is unsurprising, given that alchemy employs a 'mysterious' technical language abounding with code names. Additionally, the uninitiated could see the process of metallic transmutation as 'miraculous'.

But true miracles were only performed by holy miracle workers with the aid of God; otherwise, they were acts of magic achieved with the cooperation of demonic powers. Hagiographical literature is revealing in this respect. The *Life* of Saint Leo, bishop of Catania, narrates the deeds of the sorcerer Heliodoros,

95 *Theophanes Continuatus* 4.8, pp. 222–25. See Magdalino, 'Occult Science', p. 124.

96 Magdalino, 'Occult Science', pp. 133–34.

97 Greenfield, 'Magic', p. 220 and n. 17.

98 S. N. Trojanos, 'Magic and the Devil: From the Old to the New Rome', in *Greek Magic: Ancient, Medieval and Modern*, ed. by J. C. B. Petropoulos (London and New York: Routledge, 2008), pp. 44–52, on p. 46; K. A. Fraser, 'The Contested Boundaries of "Magic" and "Religion" in Late Pagan Monotheism', *Magic, Ritual, and Witchcraft*, 4.2 (2009), 131–51, on p. 133.

99 *CMAG*, VI, § 5, p. 30.16–18; cf. § 1, p. 26.8–11.

the anti-hero of the *vita* and the saint's archenemy. Notably, Alexander Alexakis has suggested and amply demonstrated that the longer version of the Greek *vita* (*BHG* 981b) is a ninth-century anti-Iconoclastic text, and Heliodoros is, in fact, a skillfully crafted allusion to John VII Grammatikos.[100] Once again, albeit implicitly, the notorious iconoclast patriarch is depicted as a powerful magician.

Let us return to the persona of Heliodoros. After making a pact with the Devil and receiving an assistant demon, he performs extraordinary feats. Most pertinent to our topic is that he transmutes stones into gold and wood into silver, a deed with dire economic consequences. His transformed products (in the form of coins and/or other objects?) circulate in the market of Catania; but after transactions take place, they are found to have returned to their original state and appearance. As a result, the citizens abstain from selling and buying, which leads to acute poverty and destitution.[101] Later in the narrative, as the sorcerer is taken to prison, he turns a stone into gold and tells the soldiers dragging him that if they let him go free, he will give them 'the sum of three pounds of gold'. The soldiers accept, but later, when they try to divide the alleged gold, they find that it is a stone weighing three pounds.[102]

The transformations performed by Heliodoros seem to allude to alchemical transmutation. However, there are two dissimilarities. First, alchemists are expected to transmute base metals, not wood and stones, into precious ones. Second, the achievement of Heliodoros pertains to illusion since its effect is short-term.[103] It is noteworthy that the temporary duration of the magical effect is a standard feature that distinguishes true miracles from magic.[104] On the other hand, we should not expect a non-specialized text to be accurate on materials and processes — especially when it aims at exaggerating the powers of Heliodoros — or to discern alchemy from magic. However, provided that Heliodoros has anything to do with alchemy, his gold- and silver-making probably fit in the

100 A. G. Alexakis, ed., *The Greek Life of St Leo Bishop of Catania (*BHG *981b)*, transl. by S. Wessel, Subsidia Hagiographica, 91 (Brussels: Société des Bollandistes, 2011), pp. 63–74, 81, 83, 85, 218 (9.4), 231 (14.1). See A. Acconcia Longo, 'La *Vita* di S. Leone vescovo di Catania e gli incantesimi del mago Eliodoro', *Rivista di studi bizantini e neoellenici*, 26 (1989), 3–98, for the shorter version of the Greek *Life* (*BHG* 981).

101 *Greek Life of St Leo* 15, pp. 158–59 (commentary on pp. 232–33). On Heliodoros, see ibid., pp. 87–110. The most recent study (including previous bibliography) is J. Van Pelt, 'From the *Acts of Peter* to the *Life of Leo of Catania*: Distinguishing Magic and Miracle', in *The Apostles Peter, Paul, John, Thomas and Philip with Their Companions in Late Antiquity*, ed. by T. Nicklas, J. E. Spittler, and J. N. Bremmer (Leuven: Peeters Publishers, 2021), pp. 55–80 (Heliodoros' gold-making is discussed on pp. 72–73); from an 'alchemical' perspective, see also G. Merianos and S. Sakorrafou, 'Μαρτυρίες περί αλχημείας στο Βυζάντιο σε μη αλχημικά κείμενα', in *Επιστήμη και Τεχνολογία: Ιστορικές και ιστοριογραφικές μελέτες*, ed. by Ei. Mergoupi-Savaidou et al. (Athens: Ekdotike Athenon, 2013), pp. 45–65, on pp. 47–48.

102 *Greek Life of St Leo* 31, pp. 178–179.

103 Alexakis, *Greek Life of St Leo*, p. 232.

104 Van Pelt, 'From the *Acts of Peter*', p. 73 and n. 60.

alchemical 'imitation' category rather than in that of true transmutation. The links between alchemy and counterfeiting are well-known in all alchemical traditions. For example, an Arabic text narrates that Byzantine traders paid a Muslim alchemist for counterfeit gold bars in the ninth century and then demanded their money back.[105] There is also the celebrated case of the swindler John Isthmeos, which will be discussed below.

In Byzantine hagiography, references to alchemy are not limited to the transmutational undertones of Heliodoros' sorcery. Modern literature has examined the analogies and differences between the *Life of Saint Leo of Catania* and the *Life of Saint Pankratios of Taormina*, including magical and alchemical metal-making.[106] In the latter *vita*, we encounter positive references to alchemy.[107] The text narrates the events of the mission and martyrdom of Saint Pankratios, a disciple of Saint Peter sent by the latter to evangelize Taormina in Sicily. The *Life* was written by a certain Evagrios, the supposed successor of Pankratios as bishop of Taormina; yet it was probably composed at the end of the seventh or the beginning of the eighth century, its provenance being Sicilian.[108] The *Life* also deals with the history of the city. That is why it incorporates the story of the mighty Tauros and the beautiful and wise Menia, an etiological myth on the origin of the name of Taormina, called *Tauromenion* or *Tauromenia* in Greek. As Alexander Kazdan notes, apart from forming a 'romance', this tale is also 'the first Byzantine chivalrous story'.[109]

---

105 L. Brubaker and J. Haldon, *Byzantium in the Iconoclast Era (ca 680–850): The Sources. An Annotated Survey* (Aldershot: Ashgate, 2001), p. 304.

106 Acconcia Longo, '*Vita* di S. Leone', pp. 55–61; A. Kazhdan, in collab. with L. F. Sherry and Ch. Angelidi, *A History of Byzantine Literature (650–850)* (Athens: Institute for Byzantine Research, National Hellenic Research Foundation, 1999), pp. 295–308; W. Brandes, 'Das Gold der Menia. Ein Beispiel transkulturellen Wissenstransfers', *Millennium*, 2 (2005), 175–227, particularly pp. 194–95.

107 The positive attitude toward 'alchemical' concepts is not confined to this hagiographical text. See, for example, the account of the miraculous transmutation of a shipload of tin and lead into silver through the agency of the icon of Christ Antiphonetes; P. Magdalino, 'Constantinopolitana', in *Aetos: Studies in Honour of Cyril Mango*, ed. by I. Ševčenko and I. Hutter (Stuttgart: B. G. Teubner, 1998), pp. 220–32, on pp. 220–24 (= *Studies on the History and Topography of Byzantine Constantinople* [Aldershot: Ashgate Variorum, 2007], no. VIII); Magdalino, *Orthodoxie des astrologues*, p. 39. The miracle allegedly occurred during the reign of Herakleios, which is not a trivial detail. There is evidence to suggest that the study of astrology and alchemy was actively cultivated during his rule. See Magdalino, *Orthodoxie des astrologues*, ch. II; Merianos, 'Philosopher Christianos', pp. 304–05.

108 C. J. Stallman-Pacitti, ed. and transl., *The Life of Saint Pankratios of Taormina*, ed. by J. B. Burke, Byzantina Australiensia, 22 (Leiden and Boston, MA: Brill, 2018), pp. 11–19. On the dating of the text, cf. the review of Stallman-Pacitti's edition by A. Alexakis, *Bryn Mawr Classical Review*, 2019.12.30 (https://bmcr.brynmawr.edu/2019/2019.12.30 [21 September 2023]).

109 Kazhdan, in collab. with Sherry and Angelidi, *History of Byzantine Literature*, pp. 305–07.

At a certain point in the myth, the pre-Christian Menia, a Macedonian descended from Alexander the Great and repeatedly called 'wise',[110] and Tauros the Canaanite, a descendant of Nimrod, are pursued by King Akylinos. The two heroes, along with their followers, find refuge in a place that would later be known as *Tauromenia*, where they found the city. Menia discovers, through one of her seers, a plot of land abounding with metal ores. She then calls upon the god of the Higher Providence and learns how to make significant quantities of pure gold, silver, bronze, and iron. She thus endows the newly-founded city with vast wealth.[111] Let us recall, at this point, the tradition of the angelic revelation of alchemy to a female recipient, as discussed in the Introduction.

In the *Life of Pankratios*, Menia is praised — and, through her, the beneficial aspect of alchemy — because she uses the alchemical knowledge to produce real metals (both base and precious ones) that make her city wealthy. Since Menia's alchemical activity serves the common good, there is no conflict in incorporating such a detail in a hagiographical text. Once again, a pleasing-to-God objective is crucial for the favorable conception of the alchemical opus. Menia is not accidentally labeled 'wise' in this respect. The feature of 'wisdom' is common in hagiographical contexts, but what makes the pre-Christian Menia wise is her divinely acquired wisdom put in service of the common good.

Menia produces real base and precious metals; on the contrary, the gold and silver created by Heliodoros are false, a diabolical deception aiming at disorder and corruption.[112] Heliodoros' case brings to mind that of John Isthmeos, labeled *cheimeutēs*. The chronicler John Malalas mentions that Isthmeos appeared in Antioch during the reign of Anastasios I (491–518) and tricked *argyropratai* (goldsmiths and/or bankers, lit. 'sellers of silver'), selling them counterfeit golden products, before he escaped to Constantinople, where he also conned many *argyropratai* until he was arrested and banished to Petra by the emperor.[113] In both

110 *Life of Saint Pankratios* 283, p. 422.9; 287, p. 426.18; 289, p. 428.27; 290, p. 430.11; 293, p. 432.28; 297, p. 436.10; 298, p. 436.22; 299, p. 438.12.

111 *Life of Saint Pankratios* 297–98, p. 436.10–23; cf. 126, p. 230.13–17. For an analysis, enriched by insights from the Greek alchemical literature, see G. Merianos, 'Alchemy in Byzantine Hagiography: The Cases of Menia and Heliodoros', in *Alchemy in Byzantium*, ed. by G. Merianos (forthcoming).

112 Acconcia Longo, '*Vita* di S. Leone', p. 57; Brandes, 'Gold der Menia', p. 195.

113 John Malalas, *Chronicle* 16.5, ed. by H. Thurn, *Ioannis Malalae Chronographia*, CFHB, 35 (Berlin and New York: W. de Gruyter, 2000), p. 323. See Merianos, 'Insights on Alchemy', pp. 153–57; also M. Martelli, 'L'alchimie en syriaque et l'œuvre de Zosime', in *Les sciences en syriaque*, ed. by É. Villey (Paris: Geuthner, 2014), pp. 191–214, on pp. 192–93; Merianos, 'Alchemy', pp. 248–49; Dufault, *Early Greek Alchemy*, 94–100 (passim). I render the term *argyropratai* as 'goldsmiths and/or bankers' following G. Dagron, 'The Urban Economy, Seventh–Twelfth Centuries', in *The Economic History of Byzantium: From the Seventh through the Fifteenth Century*, ed. by A. E. Laiou, 3 vols (Washington, D.C.: Dumbarton Oaks Research Library and Collection, 2002), II, pp. 393–461, on p. 435.

cases, everyday economic and market life in Byzantine cities is evidently at stake due to the circulation of counterfeit products.

The tendency to identify alchemy as a branch of magic has led to the modern lay misconception that the Byzantine Church persecuted alchemical study and practice. Perhaps, this is a projection of what is commonly thought of as the Church's stance toward the 'sacred art' in medieval Western Europe. However, things are not clear-cut even with regard to the Latin Church. For instance, churchmen in the West did not seem to have agreed on whether alchemy should have been considered sorcery or necromancy, meaning that there was no consensus on whether it was acceptable for Christians to practice alchemy or not.[114]

Concerning Byzantium, from the state's point of view, what was at stake with the alchemical practices pertained to counterfeiting or imitating coinage and precious metals. In other words, alchemy could pose a risk to the state when it involved counterfeiting techniques and products that could threaten its prerogatives.[115] There was a line that one should not cross. Psellos' *Indictment of Keroularios* shows that engagement with alchemy could not be used as a formal charge. However, obsession with *chrysopœia* might be employed to dispute a practitioner's moral integrity; the exclusive quest for metallic transmutation was potentially a sign of avarice. Psellos' two works do not disapprove of alchemy *per se*; using Keroularios' case, they showcase and criticize, respectively, the obsession with alchemy as an aurific art, as a source of personal enrichment. Nevertheless, as the discussion on the *Life of Saint Pankratios of Taormina* has shown, there was yet another perception of alchemy that focused on its beneficial aspect, shifting the view from personal enrichment to communal welfare and exalting the (Christian) ethics of the practitioner. Goals regarded as selfish, antisocial, and devoid of virtue made an alchemist no better than a sorcerer in terms of morality.[116] The above discussion strengthens the argument expressed earlier that Christian ethics and values contributed to shaping the (ideal) identity of the philosopher-alchemist in Byzantium.

Alchemy as an Analogy

In the late-eleventh century, the monk Philippos, known in scholarship as Monotropos (i.e., solitary, monk), composed the *Dioptra*, consisting of more than 7000 political verses organized in five books. The first, *Klauthmoi* (Laments), is a poem of repentance addressed to the soul. The other four books contain a dialogue between two fictitious personae, the *Psyche* (soul) and the *Sarx* (here: body) in the form of *erotapokriseis* (questions and answers); the *Psyche*, the mistress, poses the *Sarx*, her maid, questions about human nature and its place

114 Matus, 'Alchemy and Christianity', p. 939; Tarrant, 'Between Aquinas and Eymerich'.
115 Merianos, 'Alchemy', pp. 247–50.
116 Cf. Greenfield, 'Magic', pp. 218–19.

in the world created and perpetually maintained by God. *Dioptra* proved to be a 'best-seller', and its popularity continued beyond Byzantium. Notably, in the fourteenth century, it was translated into Middle Bulgarian Church Slavonic, a translation of immense popularity that was transmitted in approximately 200 manuscripts dating from the fourteenth to the nineteenth century. The topics covered in the Greek original must have been a significant factor in its appeal, as they were probably related to the concerns of educated, religious Byzantines.[117]

Unexpectedly, the fourth book of the dialogue refers to metallic transmutation. To explain how Christ deified the mortal human nature, the *Sarx* employs the analogy of the transmutation of lead into pure gold by a 'wise artificer'. Here the alchemical process is presented as the convergence between wise knowledge and skillful technical operation.[118] An abstract theological concept such as 'deification' required a suitable analogy to become concrete and relatable. Philippos' choice suggests that alchemical imagery was not unexpected in the Byzantine culture of the eleventh century. This speculation is enhanced by the fact that Philippos preferred a literary device that draws upon transmutation rather than traditional themes, such as metallurgy or medicine, as we will see shortly. Furthermore, the analogy itself shows that Philippos regarded transmutation as a credible possibility; otherwise, he would not have used it to portray deification as a truth.

Similar questions to the ones appearing in the *Dioptra* are discussed by the twelfth-century scholar Michael Glykas in ninety-five epistolary texts and essays of theological content. However, there does not appear to be any direct

---

117 Ei. Afentoulidou and J. Fuchsbauer, 'Philippos Monotropos in Byzantium and the Slavonic World', in *A Companion to Byzantine Poetry*, ed. by W. Hörandner, A. Rhoby, and N. Zagklas (Leiden and Boston, MA: Brill, 2019), pp. 331–52 (citing previous bibliography); Ei. Afentoulidou-Leitgeb, 'Philippos Monotropos' *Dioptra* and Its Social Milieu: Niketas Stethatos, Nikolaos III Grammatikos and the Persecution of Bogomilism', *Parekbolai*, 2 (2012), 85–107. The first volume of the Slavonic *Dioptra*'s critical edition, containing the prefaces and the first book, has been published; see H. Miklas and J. Fuchsbauer, eds, *Die kirchenslavische Übersetzung der Dioptra des Philippos Monotropos*, I: *Überlieferung Text der Programmata und des ersten Buches* (Vienna: Holzhausen, 2013).

118 Philippos Monotropos, *Dioptra*, ed. by Spyridon Lavriotes, *Ἡ Διόπτρα*, Ὁ Ἄθως 1.1 (Athens: P. L. Bergianites, 1920), pp. 133–34. This still remains (September 2023) the only, not critical, edition of the whole work. A modern critical edition is in preparation by Eirini Afentoulidou. It is important to note that a later hand incorporated an adapted version (in sixteen verses) of the relevant passage on transmutation from the *Dioptra* into one of the last folios of the third primary witness of the Greek Alchemical Corpus, *Par. gr.* 2327 (= **A**), f. 296^{r} (copied in 1478). This excerpt also appears in other later manuscripts, such as *Laur. Plut.* 86.16 (= **L**), ff. 291^{v}–292^{r} (copied in 1492). However, in these manuscripts, the text is found under the heading 'By John of Damascus, <excerpt> from the *Dioptra*' (Ἰωάννου τοῦ Δαμασκηνοῦ [Δαμασκυνοῦ **A L**], ἐκ τῆς *Διόπτρας*). See *CMAG*, I, p. 61; II, p. 58. It appears that the excerpt in question has developed its own manuscript tradition.

relationship between Monotropos' and Glykas' works.[119] Be that as it may, in text no. 32 of his collection, Glykas resorts to metallurgical imagery.[120] Answering the question as to why and for what reason Christ is called stone and pearl in the Scripture, Glykas starts his exposition by referring to Cyril of Alexandria (fourth–fifth century). In fact, Glykas' text is based on Cyril's *On Adoration and Worship in Spirit and Truth*, Book 3.[121] Of particular interest are the metaphors of Christ as 'stone made of tin' (λίθος [...] κασσιτέρινος) and 'talent of lead' (μολύβδου τάλαντον). Both phrases originate in Zechariah[122] and are directly associated by Cyril of Alexandria with Christ, an exegesis adopted by Glykas. Concerning the appellation 'stone made of tin', Glykas reproduces the idea that Christ is called thus because He united us through Himself with God the Father, and explains that tin has the ability to put together what is separated and divided. As for the metaphor of the 'talent of lead', Glykas, echoing Cyril, repeats that lead, when mixed with not approved silver (ἀδοκίμοις ἀργύροις), can draw impurities from silver and render it pure. In the same way, Christ takes away our sicknesses (νόσους) to purify us.

Although both Philippos and Glykas aim to make the concept of deification relatable, the comparison between their literary devices shows that the latter, following Cyril of Alexandria, draws upon 'metallurgical lore', while the former innovates by resorting to the concept of metallic transmutation. At the same time, Philippos' analogy retains an affinity to the customary metallurgical imagery. In this sense, it could be argued that metallic transmutation also fits within an all-encompassing patristic tradition employing analogies and metaphors from crafts and industry.

Alchemy and Theology: Indications of Convergence at a Deeper Level

But was the potentiality of convergence between Byzantine alchemy and theology exhausted in the transmutational analogy of Philippos Monotropos? A systematic study that has not yet been undertaken is required for this question to be answered. However, a simple glance at theological terminology suffices to shape the idea that there is more to be found. For instance, in mainly theological texts, we

119 Afentoulidou and Fuchsbauer, 'Philippos Monotropos', p. 334 and n. 12. On Glykas' collection of texts, see Ei.-S. Kiapidou, 'Chapters, Epistolary Essays and Epistles. The Case of Michael Glykas' Collection of Ninety-Five Texts in the 12th Century', *Parekbolai*, 3 (2013), 45–64.

120 Michael Glykas, *Chapters on the Difficulties of the Holy cripture*, ed. by S. Eustratiades, *Εἰς τὰς ἀπορίας τῆς θείας Γραφῆς κεφάλαια*, 2 vols (Athens: P. D. Sakellariou, 1906; Alexandria: Patriarchal Press, 1912), I, no. 32, pp. 367–68.

121 J.-P. Migne, ed., *Patrologiae cursus completus. Series graeca*, v. 68 (Paris: J.-P. Migne, 1864), coll. 296C–300B.

122 Cf. Zech. (LXX) 4:10 and 5:7, respectively.

encounter the verb *metastoicheioō* and the noun *metastoicheiōsis*. As Andrew Louth comments, *metastoicheioō* literally means 'transelement' (but it is also rendered as 'change the elementary nature of' or 'transform'). It is used among the Fathers to characterize the resurrection body (Gregory of Nyssa), the transformed state of Christ's humanity (particularly Cyril of Alexandria), and the transformation in the eucharistic elements (Gregory of Nyssa).[123]

To offer some examples of the patristic use of the term, Cyril of Alexandria, in one of the several instances that he uses *metastoicheioō* in *Against Nestorius*, states with regard to the soteriological impact of the Incarnation:

> For the Word of God the Father was born according to the flesh in the same way as ourselves, so that we too might be enriched with a birth which is from God through the Spirit, no longer being called children of flesh but rather, having been transformed (μεταστοιχειούμενοι) into something that transcends nature, being called sons of God by grace. For the Word, by nature and in reality the only-begotten and true Son, became like one of us.[124]

Gregory of Nyssa (fourth century) uses the term in his *Catechetical Oration* within the context of Eucharistic theology. Utilizing medical concepts, Gregory refers to the Eucharistic bread as a drug in which the healing remedy of the Incarnation is materialized.[125] More precisely, the Eucharistic bread is presented as the antidote for humankind suffering from the poison taken in the Garden of Eden: '[t]hose who have been tricked into taking poison offset its harmful effect by another drug'.[126] Gregory concludes his instruction on the Eucharist by saying that God united Himself with our mortal nature to deify humanity. He implants Himself in all believers through His flesh, constituted by bread and wine. He confers incorruptibility on humankind 'by the power of the blessing, through

---

123 A. Louth, *St John of Damascus: Three Treatises on the Divine Images* (Crestwood, NY: St Vladimir's Seminary Press, 2003), p. 67 n. 27. See H. G. Liddell, R. Scott, and H. S. Jones, *A Greek-English Lexicon. Ninth Edition with a Revised Supplement* (Oxford: Oxford University Press, 1996), *s.v.* μεταστοιχειόω; G. W. H. Lampe, *A Patristic Greek Lexicon* (Oxford: Clarendon Press, 1961), *s.vv.* μεταστοιχειόω, μεταστοιχείωσις.

124 Cyril of Alexandria, *Against Nestorius* 3.2, ed. by E. Schwartz, *Concilium universale Ephesenum*, v. 1.6, Acta Conciliorum Oecumenicorum, I.1.6 (Berlin and Leipzig: W. de Gruyter, 1928), p. 59.10–14; Eng. transl. by N. Russell, *Cyril of Alexandria* (London and New York: Routledge, 2000), pp. 161–62.

125 J. D. Penniman, 'Blended with the Savior: Gregory of Nyssa's Eucharistic Pharmacology in the *Catechetical Oration*', *Studies in Late Antiquity*, 2.4 (2018), 512–41, on p. 532.

126 Gregory of Nyssa, *Catechetical Oration* 37, ed. by E. Mühlenberg, *Gregorii Nysseni, Oratio Catechetica*, Gregorii Nysseni Opera, III.IV (Leiden: E. J. Brill, 1996), p. 93.8–9; Eng. transl. by E. R. Hardy and C. C. Richardson, *Christology of the Later Fathers* (Philadelphia, PA: The Westminster Press, 1954), p. 318.

which He transforms (μεταστοιχειώσας) the nature of the visible elements [bread and wine] into that immortal body'.[127]

It is fascinating then that John of Damascus (seventh–eighth century), in his *Three Treatises on the Divine Images*, uses the verb *metastoicheioō* after a simile of medico-pharmaceutical character:

> [...] but now, since the divinity has been united to our nature, as a kind of lifegiving and saving medicine (φάρμακον), our nature has been glorified and its very elements transformed (μετεστοιχειώθη) into incorruption.[128]

John likens the divinity to a drug, *pharmakon*, through which human nature is transformed. However, those acquainted with the Greek alchemical tradition would probably identify *pharmakon* as a chemical term that sometimes denotes a liquid substance for dyeing metals. More specifically, this dyeing 'drug', a coloring agent, was of two kinds, yellow and white, and was used to make metals gold and silver, respectively.[129] The transformation of human nature from corruptible to incorruptible would be reminiscent of a base metal's 'transformation' into pure gold, whose 'nature' is also 'incorruptible' as it is resistant to corrosion. The term *pharmakon* — another piece of evidence for the affinity of alchemy to medicine —, close to the verb *metastoicheioō*, creates connotations that hardly go unnoticed. It should be stressed that the common thread here is medical imagery and terminology, which serve as the medium enabling theology and alchemy to connect through a shared source of analogy.

The above 'alchemical' reading of John of Damascus' excerpt aims to illustrate that a modern, aware reader would recognize analogies between theological concepts and the notion of metallic transmutation. Nevertheless, it remains to be determined whether there is concrete evidence demonstrating that Byzantine chemical discourse and theory also reflected concepts or employed vocabulary derived from Christian theological works beyond what has been discussed in the section on the incorporation of Christian concepts into alchemical discourse; for

---

127 Gregory of Nyssa, *Catechetical Oration* 37, p. 98.6–7; Eng. transl. (adapted from) Hardy and Richardson, *Christology*, p. 321.

128 John of Damascus, *Three Treatises on the Divine Images* 2.10.109–15, ed. by B. Kotter et al., *Die Schriften des Johannes von Damaskos*, 13 vols (Berlin and New York: W. de Gruyter, 1969–2019), III, p. 101 (cf. 3.9a.109–15, p. 101); Eng. transl. (adapted from) Louth, *John of Damascus*, p. 67. John of Damascus uses the verb μεταστοιχειόω also in *Against the Nestorians* 33.5–6, in *Schriften des Johannes von Damaskos*, IV, p. 275; *Oration on the Nativity of the Holy Theotokos Mary* 1.15, in *Schriften des Johannes von Damaskos*, V, p. 169.

129 For the term *pharmakon* in Pseudo-Demokritos, see M. Martelli, '"Divine Water" in the Alchemical Writings of Pseudo-Democritus', *Ambix*, 56.1 (2009), 5–22, on p. 13; Martelli, *The* Four Books *of Pseudo-Democritus*, pp. 71, 221 (n. 35), 225 (n. 49), 227 (n. 54), 238–39 (n. 3). For the same term in Stephanos of Alexandria, see M. Papathanassiou, 'Stephanus of Alexandria: Pharmaceutical Notions and Cosmology in His Alchemical Work', *Ambix*, 37.3 (1990), 121–33, on pp. 121–22, 124; Papathanassiou, *Stephanos von Alexandreia*, pp. 110–11, 132, 134–35.

instance, ideas and technical language related to mixture and transformation. By this, we should not necessarily expect Byzantines to have made explicit references, for example, to the Eucharistic transformation as a metaphor for alchemical transmutation, as Melchior Cibinensis did in the early sixteenth century, or to have viewed the Eucharist as a metaphysical type of chemical process, as, for example, Michael Maier, Robert Fludd, and Jacob Böhme did in the early seventeenth century.[130] Fludd, in particular, also regarded the death and resurrection of Christ not just as an alchemical process but as '*the* alchemical process, the process that all others must imitate'.[131]

In Byzantine alchemical texts, an intriguing example emerges, wherein technical phraseology subtly alludes to the work of a renowned theologian and Church Father. In the sixth-century alchemical work attributed to Olympiodoros, we encounter the phrase Ὢ παραδόξου κράσεως! ('O unexpected blending!').[132] This exclamation serves as an expression of admiration concerning the unity of opposites achieved through the secret art of *chymeia*.[133] What makes it even more fascinating is its origin in Gregory of Nazianzos' *On the Theophany* (Or. 38). In this oration, the Cappadocian Father emphasizes the union of Christ's two complete and distinct natures, divine and human, within a single person:

> […] coming forth as God, along with what he had taken on; one from two opposites, flesh and Spirit — the one of which shared divinity, the other of which was divinized. O new mixture! O unexpected blending (ὢ τῆς παραδόξου κράσεως)![134]

---

130 U. Szulakowska, *The Sacrificial Body and the Day of Doom: Alchemy and Apocalyptic Discourse in the Protestant Reformation* (Leiden and Boston, MA: Brill, 2006), ch. 2.

131 B. Janacek, *Alchemical Belief: Occultism in the Religious Culture of Early Modern England* (University Park, PA: The Pennsylvania State University Press, 2011), p. 58.

132 I am grateful to Vangelis Koutalis for bringing this phrase to my attention.

133 *CAAG*, II, p. 94.16–21: Εἰ καὶ τὴν κεκρυμμένην τέχνην τῆς χυμείας ἐπίσταται, φησὶν πρὸς αὐτοὺς ὅτι «πῶς μεταβολὴν νῦν ὁρῶ; πῶς τὸ ὕδωρ καὶ τὸ πῦρ, ἐχθρὰ καὶ ἐναντία ἀλλήλοις καὶ <πρὸς τὴν> ἀντιπαράθεσιν πεφυκότα εἰς τὸ αὐτὸ συνῆλθον ὁμονοίας καὶ φιλίας χάριν» καὶ τὰ ἑξῆς. Ὢ παραδόξου κράσεως! πόθεν ἥτις ἡ τῶν ἐχθρῶν ἀπροσδόκητος φιλία; See the textual emendations proposed by Letrouit, 'Chronologie', p. 50. Berthelot and Ruelle (*CAAG*, III, p. 102 n. 4) state that '[t]out ce passage montre combien les phénomènes chimiques avaient excité l'admiration des premiers observateurs et revêtu dans leur esprit et dans leurs écrits une forme poétique. C'est le premier germe des poèmes alchimiques'. On Olympiodoros, see C. Viano, 'Olympiodorus and Greco-Alexandrian Alchemy', in *Olympiodorus of Alexandria. Exegete, Teacher, Platonic Philosopher*, ed. by A. Joosse (Leiden and Boston, MA: Brill, 2021), pp. 14–30 (citing previous bibliography).

134 Gregory of Nazianzos, *On the Theophany* (Or. 38) 13.25–28, ed. by C. Moreschini, French transl. by P. Gallay, *Grégoire de Nazianze, Discours 38–41*, Sources Chrétiennes, 358 (Paris: Les Éditions du Cerf, 1990), pp. 132–34: […] προελθὼν δὲ Θεὸς μετὰ τῆς προσλήψεως, ἓν ἐκ δύο τῶν ἐναντίων, σαρκὸς καὶ Πνεύματος, ὧν τὸ μὲν ἐθέωσε, τὸ δὲ ἐθεώθη. Ὢ τῆς καινῆς μίξεως, ὢ τῆς παραδόξου κράσεως. Eng. transl. by B. E. Daley, *Gregory of Nazianzus* (London and New York: Routledge, 2006), p. 123.

Gregory's conception of the Christological union as 'mixture' or 'blending' has sparked various interpretations concerning the philosophical foundation of the mixture theory that informs his perspective — scholars have debated whether his understanding aligns more closely with Aristotelian, Stoic, or Neoplatonic philosophy or if it adheres to no ancient philosophical model at all.[135] However, this analysis exceeds the scope of this paper.

It is worth noting that Justinian's *Letter on the Three Chapters*, dating to the mid-sixth century, refers to Gregory the Theologian and quotes the relevant passage from his work *On the Theophany* where the exclamations 'O new mixture! O unexpected blending!' are included.[136] This quotation raises the possibility that the Justinianic text could have been an indirect source of the phrase ὢ τῆς παραδόξου κράσεως for Olympiodoros' contemporary or nearly contemporary alchemical text. Be that as it may, the significant fact remains that an alchemical text incorporates phraseology related to 'blending' that can be traced back to Gregory of Nazianzos.

The presence of references to transformation, transmutation, and change within late antique or Byzantine theological texts, particularly in the context of deification, suggests that these writings could effectively accommodate chemical concepts while bestowing them with the necessary theological authority. The texts and ideas that could have provided the theological justification for alchemical transmutation already existed in Byzantium. Significantly, allusions such as the one found in Olympiodoros' alchemical work indicate that this potential was indeed utilized in Byzantine chemical literature, in specific texts, and to some degree at least. However, systematic exploration is required to examine the actual extent, reasons, and implications of such a confluence of ideas.

Conclusion

The treatment and discussion of topics in this paper can be likened to navigating through beacons rather than following a map. Furthermore, the investigation of alchemical and non-alchemical literature was far from exhaustive. However, the discussion is intended as a starting point for framing the topic, focusing on sub-themes such as issues of historiography and terminology; the convergence of certain alchemical and theological concepts; the relation of the priestly and the monastic milieu with the sacred art; the use of alchemy as a literary device.

135 See, e.g., B. A. Harris, 'Gregory Nazianzen's Christology of Mixture', *The Journal of Theological Studies*, 73.2 (2022), 731–77. Harris offers a helpful overview of the rival interpretations and argues that Gregory conceives of the Christological union in terms of Neoplatonic mixture theory.

136 Justinian, *Letter on the Three Chapters* 43, ed. by E. Schwartz, *Drei dogmatische Schriften Iustinians*, 2nd edn, ed. by M. Amelotti, R. Albertella, and L. Migliardi (Milan: A. Giuffrè, 1973), p. 104.27.

We can discern various levels in the relationship between alchemy and Christianity in Byzantium. Alchemical theorization, aiming to clarify ancient ideas, conversed with Christian theology to gradually transform alchemy into an integral part of a Christian philosophy of nature. Byzantine alchemical authors, such as Stephanos and Christianos, clad a discipline conceived as a gift of God with Christian values, virtues, and beliefs. Adhering to them was regarded as a requirement to receive access to alchemical knowledge. In this way, Christian ethics were closely intertwined with the proper alchemical methodology.

On the other hand, it seems that, from some point onward, alchemical ideas and imagery found their way into Christian literature of various genres. So far, we only have indications of this development due to the scarcity of non-alchemical sources explicitly referring to the sacred art. It is no coincidence that we have focused only on two texts, namely the *Life of Saint Pankratios of Taormina* and Philippos Monotropos' *Dioptra*. Although the evidence is meager, both works utilize alchemical notions and imagery in a manner that suggests alchemy was not alien to the sociocultural milieus of their time. Notably, Monotropos' reference to metallic transmutation in the late eleventh century is nearly contemporary to the two Psellian works related to alchemy. This remark contextualizes Monotropos' reference and shows that alchemical ideas circulated within lay and ecclesiastical circles in eleventh-century Byzantium. Regarding the examined *vitae* — and given that hagiography was a popular genre of Byzantine literature — perceptions of (*Life of Saint Pankratios*) or allusions to alchemy (*Life of Saint Leo*) in saints' *Lives* reflect its presence in the broader Christian culture. At the same time, these texts demonstrate that the consideration of alchemy as a beneficial art or not in the public imagination was shaped by the presence or absence of Christian ethics.

The Christian texts of different genres discussed in this study indicate that alchemy was not deemed dissonant with the Christian religion, its ideas, and its writings. Nevertheless, the examples of Patriarch Michael Keroularios and the sorcerer Heliodoros — a real figure and an imaginary one (the latter probably alluding to John Grammatikos) — imply that alchemical practice could instigate tensions that affected the Byzantine Church and society.

Finally, this examination of alchemy and Christianity in Byzantium leads to some preliminary methodological observations regarding the study of their relationship. Categories such as 'Christianity' and especially 'orthodoxy' may appear vague or misleading when discussing alchemy in specific periods and contexts. Furthermore, the degree to which particular authors Christianized alchemy does not depend only on the presence of Christian ideas. However, to accurately evaluate such attempts and uncover unforeseen connections, it necessitates a theologically informed analysis of Byzantine chemical texts.

NICHOLAS JACOBSON

Religion and Lawgiving in Astronomical Tables

Tables of Time and the Council of Florence-Ferrara (1431–1449)

▾ **ABSTRACT** This paper considers the links between the domain of knowledge and practice known as civil science (*politike episteme*) and astronomical table-making activities of such exponents of Greek astronomy as Gemistos Pletho (*c.* 1355–*c.* 1452) and his students. It focuses on the ways in which timekeeping practices and the record of historical memory embedded within Gemistos Pletho's tables were part of his radical religious reform program, the witness to which remains partially extant in his *Book of Laws* (*Nomoi*). It then goes on to examine a Latin calendar reform proposal submitted before the pope at the Council of Florence-Ferrara, and explores the efforts of the heads of the conciliar committee, Hermannus Zoest (d. 1445) and Nicholas of Cusa (1401–1464), to accommodate Latin timekeeping to the methods of their Greek Orthodox counterparts. Finally, it concludes by considering some alternative categories within the realm of law and jurisprudence that are salient when considering the boundaries between science and Orthodoxy in the late Byzantine era.

Introduction

The period from 1437 to 1440, during which time the Emperor John VIII Palaiologos (1392–1448) led a diplomatic mission to Florence and Ferrara in an effort to reconcile the doctrines of the Byzantine Church with the Latin rite, has

Nicholas Jacobson • University of Wisconsin-Madison

Orthodox Christianity and the Study of Nature, ed. by Kostas Tampakis and Ronald Numbers, Science and Orthodox Christianity, 4 (Turnhout: Brepols, 2024), pp. 55–75
BREPOLS PUBLISHERS 10.1484/M.SOC-EB.5.141380

come to signify the last gasp of a dying order.[1] The Byzantine state had become encircled by rival polities that threatened its sovereignty. Chief among these was the Ottoman Sultanate, which, by the 1430s, controlled the Anatolian peninsula to Constantinople's east, as well as the provinces of Thessaloniki, Thrace, and Kosovo to its west.[2] John VIII's bid to end a nearly four-century long doctrinal schism between the Greek and Latin Churches in exchange for Western military aid initially resulted in the promulgation of an official papal bull of union, *Laetantur caeli*.[3] This victory was to be short-lived, however. When John VIII's delegation returned home to Constantinople, the Greek people, stoked by the Orthodox fervor of the local clergy, rejected the agreement.[4] It would be revived again only very briefly in 1452 just before the Byzantine capital fell to the conquering forces of Sultan Mehmet II (1432–1481).[5]

John's inability to put the provisions of *Laetantur caeli* into practice in any meaningful sense revealed the general impotence of the imperial court and the clerical elite to impose sweeping religious reforms in the face of a reactionary public. The few delegates to Florence and Ferrara who had vociferously opposed doctrinal union during the 1440s, such as Mark Eugenikos (1394–1445), metropolitan of Ephesus, and Antonios of Herakleia (fl. 1430–1450), were championed by the people.[6] In turn, these clerics regarded themselves as preservers of a Hellenic Christian identity, which served as a bulwark against foreign intrusions. To this end, they had also supported the monastic Hesychast movement, which emerged specifically to combat theological innovations emanating from the Latin West.[7]

Their commitment to doctrinal uniformity notwithstanding, however, there is another sense in which the efforts of conservative clerical leaders like Mark of Ephesus reflect (albeit negatively) the profound multiplicity and cosmopolitanism of the late Palaeologan era. The military weakness of the Palaeologi and the political dismemberment of their empire over the course of the fourteenth and fifteenth centuries had palpable effects on economic patterns throughout the eastern Mediterranean. Destabilizing though it may have been for Byzantine society, the breaking of imperial monopolies of silk, grain, and salt in the Aegean and Black

---

1 S. Lazaris, 'L'empereur Jean VIII Paléologue vu par Pisanello lors du concile de Ferrare–Florence', *Byzantinische Forschungen*, 29 (2007), 293–324.

2 S. Runciman, *The Fall of Constantinople, 1453* (Cambridge: Cambridge University Press, 1965), pp. 42–43.

3 'The Decree of Union *Laetentur caeli*, 6 July 1439', in J. Gill, *The Council of Florence* (Cambridge: Cambridge University Press, 1959), pp. 412–15.

4 Basileus Theodore Doukas' *Chronicle* as quoted in A. E. Siecienski, *The Filioque: History of a Doctrinal Controversy* (Oxford: Oxford University Press, 2010), p. 171.

5 Siecienski, *The Filioque: History of a Doctrinal Controversy*, pp. 171–72.

6 Siecienski, *The Filioque: History of a Doctrinal Controversy*, pp. 159–61.

7 E. Nicolaidis, *Science and Eastern Orthodoxy: From the Greek Fathers to the Age of Globalization* (Baltimore: Johns Hopkins University Press, 2011), pp. 93–100.

Seas resulted in a brisk trade, particularly with Venetian merchants.[8] Furthermore, although the aftermath of the Fourth Crusade and the Latin occupation of Byzantium led to a privileged commercial relationship with the Republic of Venice, exponents of Jewish, Genoese, Arab, and Turkish cultures also became more integrated into this nexus of eastern Mediterranean exchange.[9] The consequences of such an open market redounded, in turn, to the intellectual realm as countless manuscripts preserving both ancient and contemporary sources were circulated through these same trade networks.[10]

This unique moment in Byzantine economic and political history created the conditions for what Edmund Fryde, echoing Steven Runciman, has called the Palaiologan Renaissance.[11] In the 1410s and 1420s, this cultural openness led to the forging of bonds among future Byzantine intellectual leaders who would go on to hold radically divergent positions on matters of religion and state. For instance, in such a milieu, the aforementioned Mark of Ephesus found himself to be a fellow pupil with those who would later become his future adversaries at Ferrara, Bessarion (1403–1472) and Isidore of Kiev (1385–1463).[12] All of them studied astronomy and philosophy at Mistra under the tutelage of the renowned philosopher, Gemistos Plethon (*c.* 1355–*c.* 1452), who would also attend the Council of Florence in the 1430s at the invitation of the emperor.[13]

A look at Gemistos' own intellectual pedigree reveals the eclectic range of interests available to an astronomer and philosopher sitting at the cross-roads of the Mediterranean. Committed to the re-establishment of a Neoplatonic philosophical tradition at Mistra, he was also a devotee of Persian illuminationist thought. Perhaps as a result of this, he further considered himself an heir to the spiritual traditions of Zoroastrianism, and believed philosophy had been communicated

---

8 A. E. Laiou, 'Monopoly and Privileged Free Trade in the Eastern Mediterranean (8th–14th century)', in *Chemin d'outre-mer: études d'histoire offertes à Michel Balard*, ed. by D. Coulon, C. Otten-Froux, P. Pages, and others (Paris: Publications de la Sorbonne, 2004), pp. 511–26.

9 For the importance of Jewish merchants in Byzantine trade despite the outsize influence of Venice, see J. Holo, *Byzantine Jewry in the Mediterranean Economy* (Cambridge: Cambridge University Press, 2009), esp. pp. 14–22; for the freedom of trade between the French, Genoese, Venetians, Byzantine Greeks, and Jewish intermediaries within the bounds of the Ottoman Empire, see G. T. Dennis, 'The Byzantine-Turkish treaty of 1403', *Orientalia Christiana Periodica*, 33 (1967), 72–88.

10 As an example of the increased permeability of linguistic and cultural boundaries during the fifteenth century, and its consequence for the transmission of astronomical manuscripts in the eastern Mediterranean, see Robert Morrison's case study of Moses Galeano in 'A Scholarly Intermediary between the Ottoman Empire and Renaissance Europe', *Isis* 105.1 (March 2014), 32–57.

11 Runciman, *The Last Byzantine Renaissance* (Cambridge: Cambridge University Press, 2008); E. B. Fryde, *The Early Palaeologan Renaissance (1261–c. 1360)* (Leiden: Brill, 2000).

12 A. L. Caudano, 'Le calcul de l'éclipse de soleil du 15 avril 1409', *Byzantion*, 73.1 (2003), 215–18.

13 C. M. Woodhouse, *George Gemistos Plethon : The Last of the Hellenes* (Oxford : Oxford University Press, 1986), pp. 171–79 ; F. Masai, *Plethon et le platonisme de Mistra* (Paris : Société d'édition Les Belles Lettres, 1956), pp. 315–26.

to the Greeks by *magi* from Persia. He appears to have been exposed to these doctrines as a disciple to the nebulous intellectual figure Elissaios, who is purported to have served as a scholar at the chief palace of the Ottoman Empire, either in Adrianople (Thrace) or Bursa (Bithynia).[14]

The precise identity of Elissaios remains unknown, but the future Patriarch Gennadios II Scholarios (1400–1473), from whom the sole account of Elissaios' identity is preserved, claimed he was a Jewish physician and Averroist philosopher who had apostatized from his faith in favor of a renovated paganism.[15] Niketas Siniossoglou has suggested that Elissaios may have been a part of an esoteric and utopian political network that connected the universalist teachings of Hurufi and Bektashi Sufism with certain mystical branches of Kabbalism.[16] Whatever the case may be, it is clear that such political and intellectual strains drawn from this sort of philosophical religion had strong resonances with the program that Gemistos espoused at his school in Mistra.

In time, Mark of Ephesus and his protégé, Gennadios, deemed Gemistos Plethon's writings to be heretical.[17] Gennadios argued his religious positions were dangerously subversive to the Orthodox order, and he did everything in his power to silence the pagan philosopher.[18] Nevertheless, Gemistos' prominence at the Council of Florence and Ferrara and his school's presence at the center of Palaeologan humanism point to the remarkable blending of ideologies in the Late Byzantine Empire, whether Latin, Orthodox, Jewish, Muslim, or pagan.

What united the many students of Gemistos Plethon was their deep interest in the study of mathematics and astronomy. What is more, their religious diversity lends credence to an observation that Bernard Goldstein has made about centers of learning where the astral sciences were studied in the Mediterranean world. Goldstein explained that such schools and translation hubs constituted a 'neutral zone' where the moral and political commitments of various participants could be set aside for the sake of a common intellectual enterprise.[19] With this said, however, part of the allure of the science for those who studied and compiled

14 Woodhouse, *George Gemistos Plethon*, 23–28; Efraim Wust, 'Elisha the Greek: A Physician and Philosopher at the Beginning of the Ottoman Period', *Pe'amim*, 41 (1989), 49–57.

15 N. Siniossoglou, 'Sect and Utopia in Shifting Empires: Plethon, Elissaios, Bedreddin', *Byzantine and Modern Greek Studies* 36.1 (2012), pp. 38–39.

16 Siniossoglou, 'Sect and Utopia in Shifting Empires', p. 49.

17 M. C. Briel, *A Greek Thomist: Providence in Gennadios Scholarios* (Notre Dame, IN: University of Notre Dame Press, 2020), pp. 67–84.

18 J. W. Barker, 'Gennadios II Scholarios, *c.* 1405–1472: Scholar and Patriarch of Constantinople', in *Encyclopedia of Greece and the Hellenic Tradition*, ed. by G. Speake (Abingdon: Taylor and Francis, 2021), pp. 650–51.

19 B. Goldstein, 'Science as a "Neutral Zone" for Interreligious Cooperation', *Early Science and Medicine*, 7 (2002), 290–91; for the relevance of Goldstein's concept of neutral zones of study in the late medieval Eastern Mediterranean context, see A. Bardi, 'Scientific interactions in colonial, multilinguistic and interreligious contexts: Venetian Crete and the manuscript *Marcianus latinus* VIII.31 (2614). A preliminary study', *Centaurus*, 63.2 (May 2021), 339–52.

astronomical tables in the 1430s lies precisely in what the numbers in these tables could reveal about the significant political events that were unfolding at that time. Regardless of their religious persuasion, scholars increasingly looked to astronomy to grant them insights into the cyclical rhythms of history, which, in turn, seemed to govern the establishment, mutation, and destruction of the rites and religious laws guiding society.

As we shall see, in this regard astronomical tables were far from politically or doctrinally neutral. The powerful possibility of using astronomical knowledge to mold and change society was not lost on even the most conservative exponents of Orthodoxy. The instability of the Byzantine polity and the uncertainty of the empire's future led to an integration of works of science and political theory, which were then turned toward questions of reorganizing the laws, rituals, and ceremonies guiding Byzantine life. Students of Gemistos' school would impress similar attitudes upon Latin reformers in Florence and Ferrara during their communication there in the 1430s. And even the most bitter opponents of Gemistos' Neoplatonism, including Gennadios Scholarios, would fall back on astronomical chronology as a way to make sense of Orthodoxy's disorienting transition from the dominant faith to a subordinate religion after the Ottoman conquest.[20]

Lawgivers and Sages in Gemistos Plethon's Tables of Time

Gemistos Plethon has been interpreted as an aberration on the margins of Byzantine thought by his contemporaries and modern historians alike (although it must be added that there is a debate as to whether or not he was an actual practicing pagan or merely a Hellenophile).[21] When Gemistos was a young man, he enjoyed a prestigious career as a senator in Constantinople, but even at this early stage in his life there were those who suspected him of heresy.[22] Some have interpreted the emperor's decision, around the year 1410, to send Gemistos away from the metropole to serve his young son, Theodore (*c.* 1396–1448), at Mistra as being a kind of *de facto* exile for his controversial beliefs. If that is the case, however, the emperor did not make this explicit, and Gemistos continued to enjoy the

20 M. Jugie, L. Petit, et X. Siderides, *Oeuvres complètes de Georges Scholarios*, Tome 3 (Paris: Maison de la Bonne Presse, 1930), p. XXI.

21 For arguments of his paganism see, Nicolaidis, *Science and Eastern Orthodoxy*, p. 122; N. Siniossoglou, *Radical Platonism in Byzantium: Illumination and Utopia in Gemistos Plethon* (Cambridge: Cambridge University Press, 2011), pp. 418–26; Woodhouse, *George Gemistos Plethon*, p. 378; Ch. Alexandre, *Traité des lois ou recueil des fragments, en partie inédits, de cet ouvrage* (Paris: J. Vrin, 1982), pp. II–III, LI–LIII; for arguments of his Orthodoxy, see V. Hladky, *The Philosophy of Gemistos Plethon: Platonism in Late Byzantium, Between Hellenism and Orthodoxy* (Abingdon: Taylor and Francis: 2017), pp. 269–85; E. Wind, *Pagan Mysteries in the Renaissance* (New Haven: Yale University Press, 1958), p. 44n2.

22 W. J. Hanegraaff, *Esotericism and the Academy: Rejected Knowledge in Western Culture* (Cambridge: Cambridge University Press, 2012), p. 31.

honors of a senator while also being appointed a civil judge.[23] Indeed it would seem that, from the 1410s to the 1430s, Gemistos was better known for his new astronomical tables that he had designed than for any radical opinion that he may have espoused regarding law and religion.[24] His most controversial treatise, a literary work of political theory entitled *The Laws* (*Nomoi*), was not publicly disseminated until after his death.[25]

Gemistos' tables were innovative in their own right. What would have first struck a user as novel was the manner in which the temporal periods of planetary motion were arranged. Rather than tabulating planetary positions in years marked off from the beginning of the Byzantine calendar, Gemistos determined the values in the argument of the tables based upon the position of the first mean syzygy of every year. Such values were recorded annually for 19 years in the table, thus giving mean values over the total duration of time known as the Metonic cycle.[26] The tables further included mean syzygy positions for the first month of every nineteenth year counted out to 497 years.[27] All the other mean planetary motions were found at the same times as were the syzygies.[28]

The format of Gemistos' tables would have appeared promising to those who first received them, as recording the dates and positions of the luminaries over the course of Metonic cycles was a useful tool in determining the date of Easter by synchronizing the annual tropical motion of the Sun with the synodic periods of the Moon (19 tropical years = 235 synodic months). Since the resurgence and proliferation of Byzantine astronomical material began at the start of the fourteenth century, organizing astronomical tables to help the user fix the date of Easter was often at the forefront of the mind of the calculators who designed them.[29] A popular predecessor of Gemistos' work was a collection of syzygy tables, called the *New Tables*, which were compiled by a certain monk named Isaac Argyros in 1368.[30] Argyros took the parameters for his syzygy tables from Ptolemy's *Almagest* and the *Handy Tables*, whereas Gemistos' values were drawn from a Hebrew version of the *zīj* of al-Battānī's.[31] But both sets provided canon

23 Gregory the Monk, *Monodia to sopho didaskalo to Gemisto* in *Patrologiae Graecae*, vol. 160, ed. by J. P. Migne (Turnholt: Brepolis, 1965), p. 817; Hieronymos Charitonymos, *Encomium Plethonis*, *PG*, vol. 160, ed. by Migne, p. 807c. See Siniossoglou, *Radical Platonism in Byzantium*, p. 8.

24 George Gemistos Plethon, *Manuel d'Astronomie*, ed. and trans. by A. Tihon and R. Mercier (Louvain-la-Neuve: Academia Bruylant, 1998).

25 George Gemistos Plethon, *Traité des lois ou recueil des fragments, en partie inédits, de cet ouvrage*, ed. and trans. Ch. Alexandre (Paris: J. Vrin, 1982).

26 R. Mercier, 'The Astronomical Tables of George Gemistus Plethon', *Journal for the History of Astronomy*, 29 (1998), 117–18.

27 Mercier, 'The Astronomical Tables', p. 119.

28 Mercier, 'The Astronomical Tables', p. 123.

29 Nicolaidis, *Science and Eastern Orthodoxy*, p. 117.

30 Nicolaidis, *Science and Eastern Orthodoxy*, p. 112.

31 A. Tihon, 'The Astronomy of George Gemistus Plethon', *Journal for the History of Astronomy*, 29 (1998), 115; Mercier, 'The Astronomical Tables', p. 119.

directions and a format that privileged the syzygies of the Sun and the Moon as the primary phenomena of investigation. In this way, Gemistos' work sat well within the norm of the general research interests of his contemporaries.

In fact, the extant manuscript record of Gemistos' tables, though it is thin, reveals the level to which astronomically inclined members of the Orthodox elite had taken an interest in his work. One of the two manuscripts that contains the most mature form of Gemistos' tables was copied in between 1433 and 1446 by Bessarion who studied the tables carefully, and attempted with difficulty to use them (more on this shortly).[32] An even earlier version of the tables, which contains different parameters and appears to have been compiled some 20 years before the final work, was copied by Bessarion's teacher, a high-ranking official at the Byzantine court named John Chortasmenos.[33] These tables have been dated to a period some time near the year 1410, and were set for the meridian of Constantinople.[34] Thus they were likely prepared by Gemistos before his departure for Mistra. Chortasmenos had a proclivity for astronomy as well, as has been made clear by certain syzygy calculations of his that have been preserved, including a calculation for a solar eclipse dated to 15 April 1409.[35] It was perhaps due to this period in which Chortasmenos had developed an enthusiasm for Gemistos' work that he would be induced later to send his own students, Bessarion and Mark of Ephesus, to go to Mistras to learn astronomy.[36]

There was one area in which Gemistos' tables very clearly broke with the conventions of his contemporaries, however, and that was in his decision to abandon the Orthodox calendar and to alter the epochs from which his tabulated mean syzygies were to be calculated. Gemistos did not make these changes wholly explicit in his canon to the tables, which caused confusion for at least one of his users from whom we have a letter requesting clarification. The letter's author was none other than Gemistos' former student Bessarion who wrote him from Italy where he had been permanently residing after the Council of Florence and Ferrara.[37] In the letter, Bessarion expressed his concerns that, even though the canons told the reader the date for beginning the tables of syzygy and planetary

---

32 Sinaïticus Sanctae Catharinae Monasterii, MS 534 (2124), ff. 2–15v.

33 Vatican, Biblioteca Apostolica Vaticana, MS codex Urbinas gr. 80, ff. 25–28v.

34 Tihon, 'The Astronomy of George Gemistus Plethon', p. 110.

35 Caudano, 'Le calcul de l'éclipse de soleil du 15 avril 1409', pp. 215–18.

36 Caudano, 'Astronomy and Astrology', in *A Companion to Byzantine Science*, ed. by S. Lazaris (Leiden: Brill, 2020), p. 226; Bardi, 'Bessarione a lezione di astronomia da Cortasmeno', *Byzantinische Zeitschrift*, 111.1 (2018): 1–37.

37 The most complete witness to the letter is found in Vat. gr. 1002, ff. 97–98v. For a modern transcription and translation of the text, see 'Correspondance entre Bessarion et Pléthon à propos des tables astronomiques', in *Manuel d'Astronomie*, ed. and trans. by A. Tihon and R. Mercier (Louvain-la-Neuve: Academia Bruylant, 1998), pp. 118–27.

motion, they did not give precise directions on how to use the Orthodox calendar to arrive at the values found in the tables.[38]

In order to flesh out Bessarion's confusion, it is necessary to explain the chronological element of the astronomical tables more fully. Gemistos explained in his canon that a year of 12 months, alternating between 30 and 29 days each, should be calculated out for each year of his calendar.[39] Then a period of 11 days was to be added at the end of each year. After three years an embolism or intercalary month would accrue as a thirteenth month to ensure that the lunar calendar kept apace with the seasons of the tropical year.[40] All of this conformed with the efforts of Byzantine computists who applied the same method to the Julian year in order to determine the date of Easter. The difference, however, could be found in the beginning point or radix for the year. Gemistos explained that the values for planetary motion in the tables all possessed radices set to midnight on the night of the first full moon after the winter solstice in the year 1433 for the meridian of Mistra.[41] By tying his radices to the lunar cycle in this way, Gemistos' calendar could vary drastically from the date of the solstice or equinox point that the Orthodox epoch was set to. Thus, Gemistos was operating with a different calendar system than the Orthodox Christian one. His radices did parallel, however, the ancient Athenian custom of beginning their sacred calendar at the beginning of a lunar phase following the solstice.[42]

It is clear then that what Bessarion had discovered and discussed in his letter was not an oversight in the canon instructions, but a political and religious statement on the part of their author. Gemistos' political aspirations are made more clear when the chronological portions of his tables are compared to preserved fragments of his *Laws*, which were completed sometime after his return from the ecumenical council in Italy. In Chapter 21 of the first book of the *Laws*, which is entitled 'Περὶ θεῶν θεραπείας' (On the worship of the gods), we find the description of the same calendric system as the one described in the canon, this time characterized as the traditional method of timekeeping for the Athenians, as well as the proper way to keep the annual ceremonial rites befitting the ancient pantheon of gods.[43]

Gemistos' proposal to dispense with the Orthodox calendar — subtly implied by the numbers in his astronomical tables, dramatically explicit in the *Laws* —

38 'Correspondance entre Bessarion et Pléthon à propos des tables astronomiques', in Tihon and Mercier, pp. 118–19.

39 *Manuel d'Astronomie*, in Tihon and Mercier, pp. 40–41.

40 *Manuel d'Astronomie*, in Tihon and Mercier, pp. 42–43.

41 *Manuel d'Astronomie*, in Tihon and Mercier, pp. 40–41.

42 A. Jones, 'Greco-Roman Astronomy and Astrology', in *The Cambridge History of Science*, vol. 1, ed. by A. Jones and L. Taub (Cambridge: Cambridge University Press, 2018), p. 381.

43 Gemistos Plethon, *Traité des lois*, ed. and trans. by Alexandre, pp. 58–63. Note that, although François Masai has shown the chapters in Alexandre's edition are not presented in their original sequence (Book 1, Chapter 21 should, in fact, be Book 3, Chapter 36), I have kept Alexandre's ordering for ease of reference.

speaks not only to the subversive quality of his scientific work, but also to the powerful rhetorical weight that the epochs and eras in astronomical tables could really hold. The choices a calculator made as to when they would mark the beginning of the revolutions of planetary cycles could have substantial doctrinal significance. Most Byzantine tables took 1 September 5509 BCE as the starting point for their radices.[44] This date was believed to mark the beginning of the world. Furthermore, it was determined through serious considerations of the judgments of the Church fathers and the authority of the chronological information found in the Bible.[45] The link between this chronologically significant date and the Byzantine calendar thus shored up imperial legitimacy by associating it with God's creative power.

As was the case with many tablemakers, Gemistos included a range of dates that were religiously and politically significant all of which a practitioner could theoretically use as starting points for planetary motion. In addition to the date for the creation of the world, we find other eras that were commonly presented in tables of time. There appears the beginning of the reign of the Babylonian ruler Nabonassar (747 BCE), which was the epoch by which Ptolemy measured planetary motion in the *Almagest*.[46] He also includes the era of Alexander the Great, measured from 331 BCE, the era of the Roman emperor Antoninus starting in 137 CE, and the era of the Arabic civilization calculated from the *hijra* in 622 CE.[47] More unique to his own project, Gemistos also furnished the traditional epoch of the ancient Athenian calendar, starting with the first Olympiad in 775 BCE, and the foundation of Greek civilization associated with the Dorian invasions in 1103 BCE.[48]

To Gemistos, and to tablemakers generally, the starting points for these many eras were not simply included in the tables to ease the efforts of users in converting between calendar systems (indeed Gemistos had rendered such a feature impossible in his own work). Instead, they represented the foundations of empires, the cultivation of the common good within societies, and the promulgation of new religious laws.[49] In the *Laws*, Gemistos argued that lawgivers and sages had contributed more than any other class to the improvement of humankind, and that they had done so by providing for the common good and offering answers to spiritual questions. He wrote:

---

44 C. Ph. E. Nothaft, *Dating the Passion: The Life of Jesus and the Emergence of Scientific Chronology (200–1600)* (Leiden: Brill, 2011), pp. 65–68.

45 R. Dean Ware, 'Medieval Chronology: Theory and Practice', in *Medieval Studies: An Introduction*, 2nd ed., ed. by J. M. Powell (Syracuse, NY: Syracuse University Press, 1992), p. 262.

46 G. J. Toomer, *Ptolemy's Almagest* (New York: Springer–Verlag, 1984), pp. 9–14.

47 Mercier, 'The Astronomical Tables', p. 121.

48 Mercier, 'The Astronomical Tables', p. 121.

49 As a point of comparison in the Latin West, John of Saxony, *Les Tables Alphonsines avec les Canons de Jean de Saxe*, ed. and trans. by Emmanuel Poulle (Paris: Éditions du CNRS, 1984), pp. 107–21.

Ἡμεῖς μὲν δὴ ἡγεμόνας ποιούμεθα τῶν λόγων, ἕνα μὲν νομοθετῶν καὶ σοφῶν τὸν αὐτὸν παλαιότατον, ὧν ἀκοῇ ἴσμεν, Ζωροάστρην, Μήδοις τε καὶ Πέρσαις καὶ τῶν ἄλλων τοῖς πλείστοις τῶν πάλαι κατὰ τὴν Ἀσίαν, τῶν τε θείων καὶ ἄλλων καλῶν τῶν πλείστων ἐπιφανέστατον γενόμενον ἐξηγητήν. Μετὰ δὲ τοῦτον, ἄλλους τε καὶ Εὔμολπον, ὃς τὰς Ἐλευσινίας τελετὰς Ἀθηναίοις ἐπὶ τῇ τῆς ἡμέτερας ψυχῆς ἀθανασίᾳ κατέστησε, καὶ Μίνω Κρητῶν γενόμενον νομοθέτην, καὶ Λυκοῦργον Λακεδαιμονίων, ἔτι δ' Ἴφιτόν τε καὶ Νουμᾶν, οἷν ὁ μὲν τὰς Ὀλυμπιᾶσιν ἐς Δία τὸν μέγιστον θεὸν σὺν Λυκούργῳ τούτῳ κατέστησεν ἁγιστείας, ὁ δὲ Ῥωμαίοις ἄλλων τε συχνῶν καὶ τῶν περὶ θεοὺς καὶ οὗτος μάλιστα ἁγιστειῶν νομοθέτης ἐγεγόνει.'

> (In our opinion, here are the guides that we select among the lawgivers and sages: First, the most ancient whose name comes to us is Zoroaster who, with the greatest insight revealed to the Medes, Persians, and most of the ancient peoples of Asia, the truth about divine things and that of the most pressing questions. After him comes, among others, Eumolpos, who established the Eleusinian mysteries among the Athenians in order to teach them about the immortality of the soul; Minos, the lawgiver of Crete; and Lykourgos, the lawgiver of the Spartans. Let us add Iphitos and Numa, the first of whom along with Lykourgos, established the Olympic games in honor of Zeus, the greatest among the gods. The second gave to the Romans a great number of the laws, many of which relate to the gods, and notably to religious ceremonies.)[50]

Just as the *Laws* explored the political and spiritual significance of the calendar drawn up in his canon and syzygy tables, so too did they offer a political and spiritual appraisal of the new epochs that Gemistos had introduced in the chronological portions of the tables. The Dorian invasions of 1103 BCE were supposed to have laid the groundwork for the major constitutions in Crete, Athens and Sparta described in the *Laws*. And the epoch of the Olympiads is referenced both in terms of its political and its ceremonial import.

Above all, it is the Persian prophet Zoroaster who is given pride of place in Gemistos' list of lawgivers and sages. From the perspective of Gemistos' own spiritual commitments, Zoroaster stood behind much of the theology that he had come to embrace. To him, Gemistos had ascribed authorship of the *Chaldean Oracles*, a work that he knew had been studied closely and revered by such Neoplatonist philosophers as Porphyry, Iamblichus, and Proclus.[51] It was from Zoroastrian priests that Gemistos believed the traditions of the Orphic and Eleusinian mystery schools had been imported to Greece.

Interestingly, it was the last Zoroastrian king, Yazdegerd III, whose era (632 CE) was memorialized in many of the most prominent astronomical tables

50 Gemistos Plethon, *Traité des lois*, ed. and trans. Alexandre, pp. 30–31.

51 R. Majercik, *The Chaldean Oracles: Text, Translation, and Commentary* (Leiden: Brill, 2015).

that circulated throughout the Byzantine Empire in the fourteenth and fifteenth centuries.[52] This was not because of any general reverence for the Persian religious founder among Byzantine astronomers, but rather it was an accident of the history of the Greek renaissance of the astronomical sciences shortly after the restoration of the Byzantine Empire at the end of the thirteenth century. It was Gregory Chioniades' (*c.* 1240–*c.* 1320) journey to the Ilkhan capital of Tabriz and his subsequent return to Byzantium with al-Tūsī's *Zīj-i Ilkhānī* and the astronomical tables of al-Khāzinī that marked the beginning of this revival.[53] Chioniades' efforts were memorialized by George Chrysococces and expanded upon in his influential *Persian syntaxis* (1347), which continued to use Persian tables and the Persian calendar to measure planetary motions.[54] The pervasiveness of the Persian calendar in many of the most influential works of Byzantine astronomy has led the scholar Alberto Bardi to suggest that it may have even stifled astronomy's cultivation among the more nationalistic and conservatively orthodox elements of the Byzantine intellectual classes.[55]

For Gemistos Plethon, however, the links drawing together various non-Orthodox rulers and lawgivers found in the tables of time in no way posed a challenge to his theological predilections. In fact, they intriguingly paralleled his ambitious utopian visions as they were laid out in the letters that he drafted to his patron, Despot Theodore of Morea, in 1416 and, two years later, to Theodore's father Emperor Manuel II (1350–1425).[56] These two letters, sometimes called the *Memorandum to the Despot Theodore* and the *Memorandum to the Emperor Manual*, offer up sweeping proposals for land and tax reform, the shoring up of the military through land grants to soldiers in exchange for their service, and an abolition of monasticism through the dispossession of their lands.[57] All of these aims were framed within the larger purpose of returning Byzantine society back to its original Hellenic form, which he claimed could be found in the ancient constitutions of the Dorians from the Peloponnese.[58] The city of Constantinople

52 A. Bardi, 'The Reception and Rejection of 'Foreign' Astronomical Knowledge', in *Finding, Inheriting or Borrowing? The Construction and Transfer of Knowledge in Antiquity and the Middle Ages*, ed. by D. Berrens, J. Althoff, and T. Pommerening (Bielefeld, Germany: Transcript, 2019), p. 173.

53 D. Pingree, *The Astronomical Works of Gregory Chioniades*, vols 1–2 (Leiden: Brill, 1986).

54 U. Lampsides, 'Georges Chrysococces, le médecin, et son oeuvre', *Byzantinische Zeitschrift*, 38 (1938), 312–22.

55 A. Bardi, 'The Reception and Rejection of 'Foreign' Astronomical Knowledge', p. 178; see also, Bardi, *Persische Astronomie in Byzanz: Ein Beitrag zur Byzantinistik und zur Wissenschaftsgeschichte* (Munich: Utzverlag, 2020), pp. 23–29, 33, 62, 92.

56 For modern editions of the two memoranda, see *Patrologiae cursus completus, series graecolatina*, ed. by J. P. Migne (Paris: J. P. Migne, 1857–66), p. 160, 821 A–B–840 A–C; Sp. Lambros, *Παλαιολόγεια καὶ Πελοποννησιακά*, vol. 3 (Athens: B. N. Grigoriadis, 1912–30), pp. 246–65.

57 For a discussion of this, see Woodhouse, *George Gemistos Plethon*, pp. 103–05.

58 Woodhouse, *George Gemistos Plethon*, p. 107.

itself, he claimed, was founded by Dorians and was, therefore, a part of this heritage.[59]

Thus the political and religious connotations of astronomical tables of time mattered, and certainly were used to this effect in the work of Gemistos Plethon. In a very real way, Gemistos' tables knit together the science of the stars with his science of the state. They were designed specifically to facilitate the transition from Orthodox Christianity into a new era guided by the Eleusinian rites of Eumolpos, which he believed had hitherto guided the spiritual life of ancient Athens, and the Dorian constitutions, which he claimed undergirded the law and custom of ancient Crete and Sparta.[60]

Astronomical Tables and Ecumenism at the Council of Florence

In our consideration of tables of time in the foregoing section, we have presented two ways in which such astronomical material influenced religious belief and practice. In one way, the tables were quite straightforward in providing the methods for organizing sacred time according to specific liturgical calendars. By directing the hours, days, and years that regulated the celebration of the sacraments, these tables effectively functioned as an extension of the Orthodox rite they described. In another sense, however, these same tables of time, containing the eras and calendars of myriad creeds across time and space, virtually insisted that their user be aware of the inherent variety of religious expression throughout the world. Insofar as there was any unity at all, it would have to be in the mind of the astronomical adept who had developed the expertise to have mastered the methods of timekeeping for all sacred communities with the capacity to navigate between them.

In this second sense, an astronomer occupied a dual role as one who could preserve the liturgical practice of the religious community while also having the power to transcend it. In practice, this duality was embraced by a person like Gemistos Plethon who was able to skillfully handle the tension, as a brief perusal of the accounts of his comportment at the ecumenical council makes clear. Reports from his Florentine and Ferraran hosts quite frequently amounted to expressions of awe at his erudition in science and philosophy.[61] He clearly left an indelible impression. After an interview with Cosimo de' Medici (1389–1464), the duke became so curious about Gemistos' doctrines, that he intimated to him his wishes to acquire important manuscripts from Morea. He would famously

59 Woodhouse, *George Gemistos Plethon*, p. 108.

60 Tihon, 'The Astronomy of George Gemistus Plethon', pp. 112–113; George Gemistos Plethon, *Manuel d'Astronomie*, ed. and trans. by A. Tihon and R. Mercier, pp. 239–40.

61 See the *Proemium* in Marsilius Ficino, *Plotini Divini Illius E Platonica Familia Philosophi* (Solingen: Soter, 1540), ff. AAiir–AAiiv.

go on to found his own Platonic academy in Florence devoted to studying and translating the teachings of Plato and his disciples into Latin.[62]

At the same time, Gemistos remained a close confidante of John VIII, and endeared himself to the patriarch by taking the Orthodox side in inter-religious debates over matters of Christian doctrine. It appears that Gemistos was willing to argue for strict Orthodox positions if only for the higher purpose of upholding the imperial interest, and both the emperor and the patriarch were grateful for this.[63] This has led some historians, such as Vojtech Hladky to conclude that Gemistos was a devout Orthodox Christian throughout his life.[64] Another delegate in attendance at the council, George of Trebizond (1395–1486), viewed Gemistos' approach much more skeptically, and summarized what he characterized as Gemistos' hypocrisy in an account written a few years after the latter's death:

> Audivi ego ipsum Florentiae, venit enim ad concilium cum Graecis, asserentem unam eandemque religionem, uno animo, una mente, una praedicatione, universum orbem, paucis post annis esse suscepturum. Cumque rogassem, Christine an Machumeti? Neutram inquit, sed non a gentilitate differentem. Quibus verbis commotus, semper odi, et ut venenosam viperam pertimui, nec videre aut audire amplius potui. Percepi etiam a nonnullis Graecis, qui ex Peloponenso huc profugerunt, palam dixisse ipsum antea quam mortem obisset, iam fere triennio, non multis annis post mortem suam, et Machumetum et Christum lapsum iri, et veram in omnes orbis oras veritatem perfuluram.
>
> (I myself heard him at Florence (for he had come with the Greeks to the Council) asserting that in a few more years the whole world would accept one and the same religion with one heart, one mind, one proclamation. And when I asked him 'Christ's or Muhammad's?' he said: 'Neither; but it will not differ much from paganism.' I was so shocked by these words that I hated him ever after and feared him like a poisonous viper, and I could no longer bear to see or hear him. I heard, too, from a number of Greeks who escaped here from the Peloponnese that he openly said before he died (now almost three years ago) that not many years after his death Muhammad and Christ would collapse and the true truth would shine through every region of the globe.)[65]

---

62 A. M. Field, *The Origins of the Platonic Academy of Florence* (Princeton: Princeton University Press, 1988), p. 12.

63 Hladky, *The Philosophy of Gemistos Plethon*, pp. 239–50.

64 See note 21 above.

65 George of Trebizond, *Comparationes phylosophorum Aristotelis et Platonis* (Venetiis: per Iacobum Pentium de Leuco, 1523), f. V6v, as trans. by J. Monfasani in 'Platonic Paganism in the Fifteenth Century', in *Reconsidering the Renaissance*, ed. by M. A. de Cesare (Albany, NY: State University of New York Press, 1992), pp. 59–60, with small modifications and expansions.

Gemistos may have gone further than most by frankly predicting that the end of the era would result in the extinction of the two dominant contemporaneous religions of the world. Nevertheless, the reality was that the only reason the ecumenical council was being convened was because so many were anticipating that such a scenario was possible. The many Latin-speaking clerics in attendance were attempting to prevent a schism in the West as well, with many of the delegates having just arrived in Italy absolved from their previously anti-papal sentiments. Furthermore, Armenian Christians, Jacobites, Copts, Nestorians, and Ethiopians had also come.[66] It truly did appear as though a world-wide reckoning was at hand.

Having seemingly unified the many sects of Christianity within an ecumenical council the likes of which had not been witnessed since the early Patristic period, the head of the Latin Church, Pope Eugenius IV (1383–1447), approached his unifying task pragmatically. He strove to wrest from his non-Latin co-religionists as many concessions and compromises relevant to rites and doctrines as he could. Liturgical practice and the organization of the calendar were chief areas of discussion in this regard. In fact, four months after the council had decreed the union between Latin and Orthodox churches, it was also propounding an agreement with the Armenian church to reform their calendar.[67]

This conciliar bull, called the *Decretum pro Armenis*, though it is uncompromising in its expectations that the Armenian church change their liturgical calendar, is of some interest because it assumes general agreement between the Latin and Orthodox position.[68] Furthermore, several of the Latin delegates at the council had been involved in ongoing efforts to ensure that Western methods for determining the feast of Easter were in agreement with the calendar of the Greeks. In doing so, they enlisted the help, not of Gemistos himself, but one of his former students Isidore of Kiev. The committee of delegates that had been tasked with addressing the issue of calendar reform first convened in Basel three years before John VIII and his delegation would join the council.[69] Like their Greek counterparts, Latin clerics were eager to use astronomical tables to reckon the proper date for celebrating Easter. There was a general agreement, moreover, that the liturgical calendars used to set the mobile feasts of the Christian year no longer yielded astronomically accurate times. Because the date for the celebration of Easter was to be set for the first Sunday following the first full moon of spring, there were two periods that had to be properly synchronized with the celestial

66 Gill, *The Council of Florence*, pp. 305–48.

67 Gill, *The Council of Florence*, p. 307.

68 *Decretum pro Armenis* in *Enchiridion symbolorum, definitionum, et declarationum de rebus fidei et morum*, eds Heinrich Denzinger and Clemens Bannwart (Freiburg im Breisgau: Herder, 1908), pp. 237–43.

69 C. Ph. E. Nothaft, 'Strategic Skepticism: A Reappraisal of Nicholas of Cusa's Calendar Reform Treatise', in *Le temps des astronomes: l'astronomie et le décompte du temps de Pierre d'Ailly à Newton*, ed. by E. Mehl and N. Roudet (Paris: Les Belles Lettres, 2017), pp. 65–102.

motions. The first was the tropical year, which determined the vernal equinox; the second was the synodic month by which the lunar phases were reckoned. The Julian calendar did not accurately calibrate the parameters of either of these periods, and, for this reason, the vernal equinox had receded to an earlier day than it had had when the First Ecumenical Council of Nicaea had been convened.

The problem of the limitations of the calendar, as it was articulated in the conciliar documents, was twofold. First, because the vernal equinoxes — but even more starkly, the lunar phases — were phenomena that anyone could observe quite easily and without calculation, the fact that the Latin church's calendar no longer corresponded to sensible experience was a source of embarrassment. For the sake of their own legitimacy, it was in the best interest of the church to correct this observational error. Second, it was believed that early church councils had taught, as a matter of doctrine, that the vernal equinox should always be measured from March 21.[70] By the 1430s the date for the vernal equinox, using the Alfonsine Tables, was March 11 or March 12. Thus, from an ecclesiastical perspective, reforming the calendar would also correct a long-standing doctrinal error.

The council appointed Hermann Zoest (d. 1445), Nicholas of Cusa (1401–1464), and Thomas Strzempíński (1398–1460) to head the committee on reform, and from 1434 to 1437 these scholars deliberated over how best to solve the calendar problem. Unlike many other subjects of doctrinal dispute that arose at the council, such as the acrimonious conflict between east and west over the *filioque* clause in the creed, the members of the committee on calendar reform were remarkably open to adopting the methods of reckoning time used for other religious calendars. In fact, Zoest sought out Jewish and Orthodox authorities to discuss how the former religious group calculated the proper date for Passover and how the latter calculated the date for Easter.[71] In particular, we know that by 1434 Zoest had taken under advisement the opinions of Isidore of Kiev, who had arrived from Constantinople three years before John VIII and his delegation.[72]

Zoest, Cusa, and Strzempińksi all drew up briefs detailing their recommendations for the restoration of the calendar that would honor both physical observation and spiritual observance.[73] The final report, as described in Cusa's *De reparatione calendarii*, settled upon a seven-day shift in the date of the vernal equinox

70 Nicholas Cusanus, *Reparatio Calendarii* in *Nicolai de Cusa cardinalis … Opera*, ed. by S. Henricpetri (Basel: Ex officina Henricpetrina, 1565), p. 1160.

71 Nothaft, 'Strategic Skepticism', p. 75.

72 Nothaft, 'Strategic Skepticism', p. 84.

73 H. Zoest, *Phaselexis* in Olivier de Solan, *La réforme du calendrier aux conciles de Constance et de Bâle* (Paris: CNRS Éditions, 2016); Nicholas Cusanus, *Reparatio Calendarii*; Thomas Strzempíński, *Relacio deliberationis prime super materia kalendarii in concilio Basiliensi* in Ph. Nothaft, 'Thomas Strzempiński, Hermann Zoest, and the Initial Stages of the Calendar Reform Project Attempted at the Council of Basel (1434–1435)', *Cahiers de l'Institut du Moyen-Âge Grec et Latin* 84 (2015), pp. 164–303.

as it appeared in the calendar.[74] This would correct for the underestimated value of the tropical year used by the Julian calendar. Then the ordinal position of the year within the Metonic cycle should be shifted back by three in order to correctly adjust for the variation in the value used for the synodic period.

Finally, toward the end of *De reparatione calendarii,* Cusa wrote that, in order to return the calendar date of the vernal equinox back to the day that was used at the Council of Nicaea, namely March 21, the date of the vernal equinox found in the Alfonsine tables would have to be abandoned.[75] This was because, according to these tables, throughout the 1430s the Sun entered Aries on March 11 or 12. In either case, an adjustment of seven days would not suffice to return the equinox back to March 21.

At cross-purposes with the dates that the Alfonsine Tables generated, Cusa then invoked his acquaintance with Isidore of Kiev, and his familiarity with what he called the 'Greek Persian Tables' (tabulae Persarum Graece).[76] According to Cusa's account, when he had met Isidore, he also had the opportunity to study these tables, and he learned through them that the Greeks, too, organize their calendar using the Metonic cycle. Relying on calculations 'in Graeco libro doctissimi Isidori Abbatis ... vidi' (in the Greek book I saw belonging to the most learned Abbot Isidore), moreover, Cusa determined that the Greeks had discovered the observationally true time of the entrance of the Sun into Aries to be either March 14 or March 15, depending on the year.[77] Cusa preferred the values from the 'Greek Persian Tables', not least because the shift of seven days added to the value of the equinox using these tables would yield the desired date of March 21.

Cusa's text reports what would become the recommendation of the committee. Having consulted tables used by practitioners of Judaism and the 'Greek Persian Tables' of Isidore of Kiev, the committee ultimately rejected the Latin liturgical calendar in favor of the methods found in these two other religious traditions. At the end of the *De reparatione calendarii,* Cusa gives his argument for why they had come to this decision. Invoking the authority of the Dominican Albertus Magnus, he wrote:

> 'Dicit Albertus: Quod inimici fidei de hoc gloriantur, quoniam ex illo errore nos in aliis pariformiter errare subsumunt, hinc sacra Synodus correctioni vacare debet diligenter, quoniam ut dicit sanctus Ambrosius [...]: Licet non conveniat Christianis dies observare et lunationes superstitiose, necesse tamen est in Paschali festivitate iuxta praeceptum Lunam observare et tempus [...].'

74 Cusanus, *Reparatio Calendarii,* p. 1165.

75 Nicholas Cusanus, *Reparatio Calendarii,* p. 1166.

76 Nicholas Cusanus, *Reparatio Calendarii,* p. 1165; for more on these tables, see Bardi, *Persische Astronomie in Byzanz,* pp. 26, 55.

77 Nicholas Cusanus, *Reparatio Calendarii,* p. 1163.

> (Albertus says: Because the enemies of the faith boast about this (i.e. the Church's error), since from that error they presume us to have similarly erred in other matters, hence a sacred council ought to devote itself diligently, because, as Saint Ambrose says [...]: 'Although it is not meet for Christians to observe days and lunations superstitiously, it is yet necessary to observe the Moon and the time in the Paschal feast according to the proper rule [...].)[78]

It is clear from Cusa's conclusions here that he fully expected a convergent truth about the calendar would be found through the comparison of multiple different strategies in different religious traditions. Identifying and applying a method of calculation that would be approved by the Church's critics no less than its supporters was crucial for its success. Once again, as we had said at the beginning of this section, the role of the astronomer in such matters was to play neutral judge over all possible systems at issue, rather than serving as an interested party in defense of one over another.

By the time the committee had submitted its report, Nicholas of Cusa had already departed. In 1437 he had been called to serve as a special envoy to go to Constantinople to escort the imperial delegation back to Italy.[79] From his own recounting of it, it appears that his interest in a transcendent epistemology, went beyond calendar calculation to encompass an entire civic and spiritual philosophy. He would later memorialize these early career experiences in a Platonic dialogue that he composed shortly after the fall of Constantinople. The dialogue, *De pace fidei*, narrates events that unfold during an imaginary convocation of wise sages from all around the world regardless of faith or creed.[80] One cannot help but think about the vast Ecumenical Council of Florence when reading this text, though in the imaginary dialogue the council's judgments were heeded, unlike in the historical case. Most striking within the dialogue is the recurrence of the general refrain that, regardless of the many prophets and the many lawgivers of this world, there remains only one religion emergent out of a variety of rites.

The End of the Byzantine Epoch

As well as being reminiscent of the institutional structure of the Council of Florence itself, the lawgivers and sages in the *De pace fidei* have clear resonances with those same kinds of individuals depicted in Gemistos Plethon's *Laws*. Of course, there is no evidence that Cusa was familiar with Gemistos' sentiments,

78 Nicholas Cusanus, *Reparatio Calendarii*, p. 1163.
79 *Acta Cusana: Quellen zur Lebensgeschichte des Nikolaus von Kues*, ed. by E. Meuthen (Hamburg: Meiner, 1983), pp. 224–5, no. 334–336.
80 Nicholas Cusanus, *De Pace Fidei and Cribratio Alkorani*, ed. and trans. by J. Hopkins (Minneapolis: A. J. Banning Press, 1994), pp. 633–76.

although Cusa did likely meet Gemistos while they were both in Italy for the council. Several years later the Greek humanist Johannes Sophianos dedicated his translation of Gemistos' *De fato* into Latin to Cusa as well.[81]

Regardless of whether there is any firm connection between Cusa's thought and that of Gemistos' is beside the point, however. What is important to know is that, during this period, there was a general expectation that the old structures were falling away, and something new — and perhaps universal — was about to take their place. We see this in the writings of Gemistos who, anticipating the fall of the Orthodox church, was attempting to revive a more ancient religion that could unite all Hellenes together. We see it in Cusa's universalism as he imagined a world where all divisions of belief would ultimately resolve into religious uniformity.

One also perceives in the intellectual and political speculations of both figures a clear projection of their own social positions into the language of their reforms. For, apart from his reputation as a heterodox philosopher, Gemistos Plethon was also an expert in the law and, according to some accounts, a judge who presided over the laws of the Morean Despotate.[82] His elevation of sages and lawgivers to the highest levels of temporal authority in his astronomical tables was an endorsement of his own status within the late Byzantine hierarchy, perhaps in opposition to his ecclesiastical and monastic rivals. Cusa, too, was primarily trained as a canon lawyer whose writings on medieval political theory, such as the *De concordantia catholica*, were as influential in his own lifetime as were his works of Neoplatonic philosophy and theology.[83] Even though Cusa's later career as a bishop administering the diocese of Brixen has often been interpreted as a failure, his sermons and other activities during the period of his service show him to have been hard at work applying the principles of his reformist vision to pastoral matters.[84] In this way, his program for harmonizing different laws and religious practices through the deliberations of sages and lawgivers, as described in the *De pace fidei*, bears clear affinities to the conciliar work in which he was engaged at Basel and Florence.

Finally, even the most reactionary members of the clerical elite of Byzantium recognized that, at some root level, the world was undergoing revolutionary change. After Mehmet II entered Constantinople, he chose Gennadios Scholarios

81 P. O. Kristeller, 'A Latin Translation of Plethon's *De fato* by Johannes Sophianis dedicated to Nicholas of Cusa', in *Nicolò Cusano agli inizi del mondo moderno* (Florence: G. C. Sansoni, 1970), pp. 175–95.

82 For a discussion of the evidence for Plethon's having served officially in the capacity of a judge, see Siniossoglou, *Radical Platonism in Byzantium*, p. 8.

83 For the relationship between the *De concordantia catholica* and the *De docta ignorantia*, see J. de Guzman Miroy, *Tracing Nicholas of Cusa's Early Development: The Relationship between* De concordantia catholica *and* De docta ignorantia (Louvain: Éditions Peeters, 2009), pp. 15–45.

84 For a treatment of his episcopate as bishop of Brixen and its relationship to his earlier reforming efforts, see R. J. Serina, *Nicholas of Cusa's Brixen Sermons and Later Medieval Church Reform* (Leiden: Brill, 2016), pp. 21–52.

to be the first patriarch after the conquest.[85] Some scholars believe that the Ottoman sultan did this, knowing the kind of figure that Gennadios represented to his community. Because Gennadios was so opposed to unity with the Latin Christian church, the sultan could trust him never try to coordinate with the papacy against the Ottomans. It was the case that even Gennadios, who had been so skeptical of Gemistos' calendric work (so much so that he had Gemistos' *Laws* publicly burned), began to theorize about the rhythms and cycles of time. In his last extant work (dated to 1472), which has been characterized as a chronography, hope had passed away along with political independence.[86] The world, he concluded, was ending. The Byzantine epoch was over. Furthermore, the end of the world, he calculated was to occur 7000 years after its creation. Using the era of the tables, 5509 BCE, he concluded that the end of the world would occur sometime in 1492 (*incompleto*).[87] Of course, Gennadios was wrong when it came to his prediction of when the world would end; when it came to his recognition that the Byzantine era was over, however, he was certainly right.

Conclusion

Gemistos Plethon and Gennadios Scholarios' participation in a mutually intelligible discourse, undergirded as it was by an astronomically informed account of sacred history, reveals some important insights about the categories of science and Orthodoxy in the late medieval Byzantine context. These categories do not map neatly onto Western notions of science and religion as they developed in the Early Modern period, and as they have been documented in recent histories of science and religion.[88] Some aspects of these differences are shared across multiple religious communities of the pre-modern world; others are particularly salient to medieval Orthodoxy.

For one, the writings of figures like Gemistos make clear that timekeeping and calendric reform were a branch of astronomical practice that was constitutive of the ceremonies, rites, and duties of Orthodox Christianity. In this way, astronomical principles served a regulative function within the community of believers who were subject to the rule of the imperial dynasty itself. This kind

85 Kritovoulos, *History of Mehmed the Conqueror*, trans. by Ch. T. Riggs (Princeton: Princeton University Press, 1954), pp. 94–95.

86 M. Jugie, L. Petit, et X. Siderides, *Oeuvres complètes de Georges Scholarios*, Tome 3, p. XXI.

87 C. Mango, *Byzantium: The Empire of New Rome* (London: Phoenix, 1980), p. 213.

88 A large literature has emerged in recent years interrogating the boundaries between science and religion, and the historically contingent nature of the terms as they are deployed in the contemporary West. See, for instance, Ch. Taylor, *A Secular Age* (Cambridge, MA: Harvard University Press, 2007); D. Livingstone, 'Which Science? Whose Religion?' in *Science and Religion Around the World*, ed. by J. H. Brooke and R. L. Numbers (Oxford: Oxford University Press, 2011); P. Harrison, *The Territories of Science and Religion* (Chicago: University of Chicago Press, 2015).

of religious expression was neither confessional nor strictly doctrinal, as the later Protestant sects of post-Reformation Europe would come to define themselves.[89] Nor was it understood in terms of a *habitus* or an internal spiritual virtue as Peter Harrison has argued pre-modern Latin notions of religion were understood. Instead, ceremonies and rites defined a community outwardly through actions that were guided by collective imperatives and commands.[90] In this way, the best lexicon to rely upon to express this kind of relationship between science and religion comes from the domains of law and jurisprudence.

This juridical characterization of the religious forms could be extended to other traditions of the late Middle Ages, such as many Muslim and Jewish communities.[91] But there was a way in which Gemistos Plethon's emphasis on sages and lawgivers drew particularly on Orthodox Byzantine notions of order and hierarchy, as well. His book, *The Laws*, served as an instrument of governance for the despot of Morea, or even the emperor himself, as this body of laws would have defined the contours of religious expression of communities living within the imperial domain. Thus, the reforms described by Gemistos were intended to shape the already existing Orthodox society, and appealed to the emperor as its chief exemplar. Since at least the time of Emperor Justinian I, the imperial role within Orthodox Christianity was that of the vicegerent of Christ, the seal of justice within the community, and the giver and reformer of laws. Justinian even presented himself as the *νόμος ἔμψυχος*, (the living manifestation of the law), and this title remained relevant for successive dynasties as well.[92]

Gemistos Plethon's own status as a jurist in the service of the emperor and his family meant that, by extension, he too participated in the divine mandate of the imperial law. In fact, it is probably because Gemistos remained a loyal servant to the Palaiologan family throughout his life that labels of 'heresy' or 'paganism' have been contested even in the secondary literature today. Whatever the content of his beliefs as documented in the heterodox program found in his tables of time and his legal opinions, the fact that they were granted the imprimatur of the emperor's approval shielded him from accusations of apostasy until after his death and the collapse of the empire.

Finally, the dual capacity of astronomer-historians to serve as both members of religious communities on the one hand and judges over these communities on the other, resulted in the development of a sort of science of religion *avant la lettre*. A judge or a lawgiver who set down and established the principles

89 C. Calhoun, M. Juergensmeyer, and J. Van Antwerpen, eds, 'Introduction', in *Rethinking Secularism* (Oxford: Oxford University Press, 2011), pp. 14–20.

90 Harrison, *The Territories of Science and Religion*, pp. 91–97.

91 For a discussion of the tight–knit connection between religion and law in certain Islamic and Jewish traditions, see B. Lewis, 'Introduction', in I. Goldziher, *Introduction to Islamic Theology and Law* (Princeton: Princeton University Press, 1979), pp. vii–xiii.

92 H. Hunger, 'State and Society in Byzantium', *Proceedings of the Royal Irish Academy: Archaeology, Culture, History, Literature* 82c (1982), pp. 197–209.

guiding a particular religious tradition necessarily had to work from outside of the tradition itself, thus relying on principles and considerations that were not immanent to the system that they created and regulated. In the case of Gemistos — but also Nicholas of Cusa, Gennadios Scholarios, and others — the methods of astronomical timekeeping served in this capacity to both contain and explain the rise and mutation of religious laws. It could contain these laws through the numerical measurement of time, starting at the so-called radix of a historical era, and tracing the era to its final culmination or end. It could explain the rise and mutation of these laws by defining their unfolding as being a result of either divine providence, fate, or the naturalistic life span of a human community. This way of approaching religious difference and change, though it arose at many different times and in different traditions, was particularly well suited to the proclivities of the Orthodox emperor who saw himself as the living manifestation of the law, its creator, and its guarantor.

OLGA ČADAJEVA

Cosmology in Russian Thought on the Eve of Modernity

▼ **ABSTRACT** This paper focuses on the seventeenth-century Russian, Ukrainian and Belarusian (Ruthenian) texts reflecting cosmology. It addresses the following problems: the background of the Muscovite readership in cosmology; translated sources; the traditionalists' attempts to grope for an adequate image of the universe based on acknowledged Orthodox sources; and the controversies around certain cosmological aspects. The discussed texts illustrate the slow and gradual transfer of Western knowledge in seventeenth-century Russia. The essay aims at demonstrating how particular aspects of the worldview, such as cosmology, provide an insight into the cultural crisis and processes of change that Russian society was going through in the seventeenth century.

Introduction

The seventeenth century was a remarkable period of Russian history. It saw numerous internal crises, along with gradual preparation for westernization with a cautious start of transfer and absorption of Western cultural patterns and knowledge. The eve of modernity started with a severe political crisis, known as the Time of Troubles (1598–1613/18), accompanied by the invasion of Polish-Lithuanian troops. As a result, the first decades of the century were marked by the policy of isolationism, which came definitely to an end with the reign of Alexei Mikhailovich (1645–1676). Although tsar Alexei acquired the moniker *Tishayshy*, which means 'most quiet', his reign could hardly be referred to as 'the most peaceful'. It saw wars, revolts, and Russian Orthodox Church schism (*raskol*). At the same time, the number of foreigners living in Muscovy increased, numerous

Olga Čadajeva • Palacký University

Orthodox Christianity and the Study of Nature, ed. by Kostas Tampakis and Ronald Numbers, Science and Orthodox Christianity, 4 (Turnhout: Brepols, 2024), pp. 77–103
BREPOLS PUBLISHERS 10.1484/M.SOC-EB.5.141381

books were translated into Russian, and the first projects of higher education establishments appeared.[1]

However, the intellectual milieu of Muscovy had yet to approach the advances of European thought. Russian culture had almost no background in antiquity, and for many centuries it had been mostly relying on Byzantine sources in building the worldview. The reception of Byzantine knowledge was limited not only by the distance, but also by the necessity of translation into Church Slavonic. For such a task, usually religious sources were almost solely selected and transferred through Bulgaria and Serbia. The contacts with the Catholic West, on the other hand, were restricted due to the religious antagonism with *Pax Latina*, along with great geographical distances. Having a rather selective intellectual background in actual Byzantine knowledge or its creative perception of antiquity, it was almost impossible for Muscovy to build its own strategies in the reception of European thought. The full-fledged inclusion of Russia into the Western scientific discourse starts only as late as in the eighteenth century prompted by the secularization after the reforms of Peter the Great and the establishment of educational and scientific institutions.

In the seventeenth century, while Europe was involved in a heated debate about the Copernican system, it could seem that readers in Muscovy were still largely relying on the image of the world suggested solely by the texts offering the religious interpretation of the universe.[2] The sources available for the Muscovite audience in the early modern period still consisted of the works of Church Fathers and their later commentators, apocryphal texts, translated astrological and divinatory literature,[3] and a few translated historical or geographical works of Western provenance, often outdated at the moment of their translation.

Cosmological Background of Muscovites: Two Very Different Traditions

The cosmological models known in Muscovy at the beginning of the seventeenth century were related to four main groups of sources. The first group presents the geocentric model. Its Christian adaptation derives from the Cappadocian tradition of exegesis, namely from Basil of Caesarea and George of Pisidia, whose po-

---

1 A. Vucinich, *Science in Russian Culture: a History to 1860* (Stanford: Stanford University Press, 1963), pp. 16–37.

2 E. Nicolaidis, *Science and Eastern Orthodoxy: From the Greek Fathers to the Age of Globalization* (Baltimore, MD: Johns Hopkins University Press, 2011), pp. 140–41; N. Chrissidis, *An Academy at the Court of the Tsars: Greek Scholars and Jesuit Education in Early Modern Russia* (DeKalb: Northen Illinois University Press, 2016), p. 139.

3 W. F. Ryan, *The Bathhouse at Midnight: An Historical Survey of Magic and Divination in Russia* (University Park: Penn State University Press, 1999), pp. 394–403.

etic *Hexaemeron* or *Cosmourgia* was translated into Old Church Slavonic in 1385.[4] The Cappadocian tradition preferred allegorical exegesis to the literal one, which enabled greater interpretational liberty. Prominent Church Fathers, such as Basil of Caesarea and Gregory of Nyssa, were well-educated in Greek and Hellenistic culture, thus they incorporated a system of the Aristotelian-Ptolemaic world into their Christian teachings. Their somewhat simplified notion of the Aristotelian-Ptolemaic model facilitated a geocentric spherical universe with seven classical planets or seven luminaries, and spherical Earth. Cappadocian thinkers did not focus on the mathematical apparatus of Ptolemaic corrections, they did not reflect on the concepts of eccentric circles, epicycles, and equant to explain the apparent retrograde motion of the planets. They probably presupposed the knowledge of these concepts in their educated audience from Hellenistic sources. Russian readers, however, were not familiar with antiquity, and therefore they were deprived of a solid background in Aristotelian natural philosophy and Greek mathematics.

Moreover, although Basil of Caesarea was highly regarded in Rus', the full translation of Basil's *Hexameron* in Church Slavonic appeared in Moscow as late as in 1656 and was printed in 1665. Before that, the text was mostly known through excerpts in the *Hexameron* by John the Exarch, a Bulgarian scholar of the ninth and early tenth century. John the Exarch was heavily influenced by Basil of Caesarea and John Damascene. At the same time, he did not omit the thinkers representing a different tradition, namely Severian of Gabala and Theodoret of Cyrus, and included them in his compilatory *Hexameron*.[5] John the Exarch's *Hexameron* could be probably labelled as the richest compendium of the theological and rationalistic Christian worldview and information on ancient philosophy known in Rus'. This Bulgarian scholar also translated the works by John of Damascus into Old Church Slavonic. His main text, *An Exact Exposition of the Orthodox Faith* was rather well-known, and it was found in Russian manuscripts of the twelfth and thirteenth centuries. John Damascene preferred the geocentric model and discussed cosmology and even certain elements of practical astronomy, which, however, contained some serious errors.[6] Subsequently, various compilations and adaptations were created in Church Slavonic based on Damascene's and John the Exarch's texts. They also reflected the geocentric picture of the world but often included peculiar apocryphal elements.[7]

---

4 V. Mil'kov, and S. Poliansky, *Космологические произведения в книжности Древней Руси. Ч. I. Тексты геоцентрической традиции* (Saint Petersburg: Mir", 2008), p. 19. The Cappadocian Fathers, in turn, belonged to the school of Alexandria: Nicolaidis, *Science and Eastern Orthodoxy*, p. 24. More on Cappadocian tradition see H. Inglebert, 'Inner and outer knowledge', in *A Companion to Byzantine Science*, ed. by S. Lazaris, (Leiden: Brill, 2020), p. 46.

5 For on this *Hexameron* and its sources cf.: G. Barankova, and V. Mil'kov, *Шестоднев Иоанна экзарха Болгарского* (Saint Petersburg: Aleteiya, 2001). Mil'kov, *Космологические произведения в книжности Древней Руси. Ч. I.*, pp. 110–124; L. Sels, 'John the Exarch and his Sources: New Sources of the Sixth Book of the Šestodnev', *Slavica Gandensia*, 35 (2008), 123–49.

6 Nicolaidis, *Science and Eastern Orthodoxy*, p. 46.

7 Mil'kov, *Космологические произведения в книжности Древней Руси. Ч. I.*, pp. 24–35.

Some scholars, however, have been arguing that another model was prevailing in Rus' and then in Muscovy.[8] That was the model of the flat Earth, suggested by the school of Antioch, whose adherents insisted on the literal interpretation of the Scripture.[9] Its main proponent was Cosmas Indicopleustes, or Constantine of Antioch, who was probably a spice merchant, a widely travelled man in the Oikoumene of the sixth century.[10] His model was a perfect illustration of the idea of the cosmos acceptable for the Antiochene school. In this system, the world takes shape of a tabernacle, topped by a vault, with the flat rectangular Earth at the base of the whole construction.[11] In his treatise *Christian Topography*, Indicopleustes adheres to the principle of polar ontology. On one pole, he places the noumenal sphere, close to God, and on the other pole, the natural world, separated from the angelic sphere by a firmament. The firmament in his scheme is understood as the upper boundary of the material world, above which the first heaven, created together with the Earth on the first day, is located. The first heaven has the special ontological qualities of a non-material creation. It is essentially a noumenal boundary above which is God in the eternal infinity of superspace.[12] The firmament is surmounted by celestial waters to prevent it from being melted by the heat of the luminaries and to reflect the rays to illuminate the natural world. The motion of celestial bodies was explained by angelic forces, and the day and the night were attributed to an elevation in the northwest part, where the luminaries were hiding behind.

In Muscovy, the *Christian Topography* by Indicopleustes was one of the most widely spread texts consistently describing the image of the universe even as late as in the sixteenth and seventeenth centuries. At least that is what appears if we look into the considerable number of cosmographies still preserved in Russian archives.[13] However, the *Christian Topography* was by far not the only

8 B. Raikov, *Очерки по истории гелиоцентрического мировоззрения в России* (Moscow: Izdatel'stvo Akademii Nauk SSSR, 1947), pp. 11–45; Nicolaidis, *Science and Eastern Orthodoxy*, p. 141. More on scientific attitude and available sources in Early Rus' and Muscovy see in Vucinich, *Science in Russian Culture*, pp. 3–37.

9 Mil'kov, *Космологические произведения в книжности Древней Руси. Ч. I.*, p. 14; Nicolaidis, *Science and Eastern Orthodoxy*, pp. 24–39.

10 On Cosmas and his *Cristian Topography* see: V. Mil'kov, and S. Poliansky, *Космологические произведения в книжности Древней Руси. Ч. II. Тексты плоскостно-комарной и других космологических традиций*, (Saint Petersburg: Mir", 2009), pp. 8–10; M. Kominko, *The World of Kosmas: Illustrated Byzantine Codices of the Christian Topography* (Cambridge: Cambridge University Press, 2013); Inglebert, 'Inner and outer knowledge', pp. 42–48; Telelis, Ioannis, 'Meteorology and Physics in Byzantium', p. 190; A. L. Caudano, 'Astronomy and Astrology', p. 204, P. Martín, Pérez, and C. Andreotti, 'Geography', p. 236, all in *A Companion to Byzantine Science*, ed. S. Lazaris, (Leiden: Brill), 2020.

11 Nicolaidis, *Science and Eastern Orthodoxy*, p. 28.

12 Mil'kov, *Космологические произведения в книжности Древней Руси. Ч. II.*, p. 7.

13 E. Piotrovskaia, *'Христианская Топография Козьмы Индикоплова' в древнерусской письменной традиции* (Saint Petersburg: Dmitry Bulanin, 2004).

text presenting the flat Earth model. Among those sources, we can find the *Hexameron* of Severianos of Gabala. This text was first known within excerpts included in the 'geocentric' *Hexameron* by John the Exarch. Later, around the fifteenth century, the first full separate manuscripts with his text appeared.[14] The Antiochene tradition was particularly sound in one of the most read and most widely spread texts, i.e. *Chronographs,* which described the history of the world from its creation. Typically, the *Chronographs* were preceded by a 'cosmological' hexameral part and gave the basic notion of the image of the world.[15] Another influential Slavonic text linked with the flat Earth tradition was *Paleia Tolkovaia,* or *Paleia Interpretata.* It presents a 'hybrid' model which combines elements of the geocentric system and that of the flat Earth. *Paleia* influenced several shortened versions of smaller *Hexamerons* known in Russian literary culture and also served as a source for religious thinkers even in the seventeenth century.[16]

The third massive group of cosmological sources available for Early Rus' and then Muscovite readers consisted of various apocrypha or pseudepigrapha.[17] They could either be loosely linked with the geocentric tradition, often reproducing an egg-shaped model, attributing it to John of Damascus,[18] presenting a variant of the flat Earth concept, or combining both of them, also adding pantheistic elements. Cosmological apocrypha were frequently related to the genre of [dream] vision, visio, e.g. *The Second Book of Enoch* (*Slavonic Enoch*), *Ascension of Isaiah* or *The Dream of Paul the Apostle.* Another type was presented by the genre of erotapokrisis, or 'questions and answers', where the disciple typically asked the teacher various questions, including the ones on cosmology. The most prominent feature of apocrypha and pseudoepigrapha was pictorial imagery and mythological motifs.

The fourth set of cosmological sources included various translated texts of the Western provenance, mostly not devoted to cosmology itself, but referring to it to different extents. Among such texts we can find e.g. *Lucidarius,* the first German-language medieval summa dated to the twelfth century and translated into Russian probably at the beginning of the sixteenth century, whose apocryphal nature

14 Mil'kov, *Космологические произведения в книжности Древней Руси. Ч. II.,* pp. 10–14, 157–66.

15 O. Tvorogov, *Древнерусские хронографы* (Leningrad: Nauka, 1975), pp. 3–5; Bulanin, Dmitry 'Утраченный рукописный источник русского хронографа (опыт реконструкции состава)', *Rossica Antiqua,* 1/2 (2016), 99–136.

16 Mil'kov, *Космологические произведения в книжности Древней Руси. Ч. II.,* pp. 13–15.

17 For studies on Slavic apocyphal literature and pseudoepigraphia see e.g. V. Mil'kov, *Древнерусские апокрифы* (Saint Petersburg: Izdatel'stvo khristianskogo gumanitarnogo instituta, 1999); Mil'kov, *Космологические произведения в книжности Древней Руси. Ч. II.,* pp. 429–442; A. Orlov, *Selected Studies in the Slavonic Pseudepigrapha* (Leiden: Brill, 2009); F. Badalanova Geller, 'Heavenly Writings: Celestial Cosmography in the Book of the Secrets of Enoch', *Starobulgarska Literatura,* 45/46 (2012), 197–224.

18 Mil'kov, *Космологические произведения в книжности Древней Руси. Ч. I. Тексты геоцентрической традиции,* pp. 26–27.

makes it closer to the third group.[19] The growing interest in geography and history in the sixteenth century resulted in the translation of cosmographies and historical books, which sometimes contained very brief insights into cosmology, mostly with austere descriptions of the geocentric model.[20] These included e.g. *Universal Chronicle* (*Kronika wszytkiego świata*, 1551–64) by a Polish historian Marcin Bielski, translated in 1584, and later it was *Theatrum Orbis Terrarum* (1570) by Abraham Ortelius.

In the seventeenth century, the number of translations from Latin and Polish grew and contacts with Europe gradually intensified, which broadened the perspectives of Muscovites. Moreover, there were the first individual attempts at the reflection on cosmological issues, as well as the first controversies, disputes, and debates related to understanding the universe. And while they could be considered outdated from the perspective of European scientific advances, they reflected important processes in the Russian cultural milieu.

Ukrainian Perspective: Kyrylo Stavrovetsky-Tranquillon

A significant number of the seventeenth-century advances and controversies were rooted in the cultural transfer and clash between intellectuals from Ruthenia (modern-day Ukraine and Belarus) and Muscovite Orthodoxy.[21] Ruthenian scholars maintained much closer contact with Western thought thanks to the fact that they were living on the territories of the Polish-Lithuanian Commonwealth. While trying to preserve and define their Orthodox identity, they incorporated certain concepts from the Catholic or even Protestant provenance into their worldview, had a better background in antiquity, and were not limited to reading texts in Old Church Slavonic or Russian. This process intensified with the establishment of the Kyiv-Mohyla Collegium (Kyiv Ecclesiastical Academy) in 1632. The focus of Ruthenian intellectuals was primarily religious; therefore, they did

19 Raikov, *Очерки по истории гелиоцентрического мировоззрения*, p. 30.

20 A. Sobolevsky, *Переводная литература Московской Руси XIV–XVII веков. Библиографические материалы* (Saint Petersburg: Tipografiya Imperatorskoi Akademii Nauk, 1903); Raikov, *Очерки по истории гелиоцентрического мировоззрения*, pp. 30, 46; T. Matasova, 'Первая книга "Географии" Помпония Мелы в древнерусском переводе: о рецепции античного наследия в русской культуре XV–XVI вв.', *Аристей: Вестник классической филологии и античной истории*, 9 (2014), 310–43.

21 See e.g. L. R. Lewitter, 'Poland, the Ukraine and Russia in the 17th Century', *The Slavonic and East European Review*, 27.68 (1948), 157–71; 28.69 (1949), P. *Bushkovitch, Religion and Society in Russia: The Sixteenth and Seventeenth Centuries (New York: Oxford University Press, 1992)*, pp. 7–8; S. Plokhy, *Ukraine and Russia: Representations of the Past* (Toronto: University of Toronto Press, 2008), p. 294; M. *Kiseleva, Интеллектуальный выбор России второй половины XVII — начала XVIII века: от древнерусской книжности к европейской учености (Moscow: Progress-Traditsiia, 2011); P. Bushkovitch, Religion and Society in Russia: The Sixteenth and Seventeenth Centuries (New York: Oxford University Press, 1992), pp. 7–8.*

not aim to bring the advances of European scientific thought to the Orthodox soil. On the other hand, their activities propelled the discourse of self-identification in Russian Orthodoxy. Even though cosmology, or science in general, was not at the forefront of this discourse, the mere introduction of these topics to the cultural milieu was a significant step forward.

Among the first East Slavonic authors of the seventeenth century, who attempted to create an original reflection on cosmological matters, was Kyrylo Stavrovetsky-Tranquillon (died 1646).[22] He was a Ukrainian preacher, theologian, and poet, who published three major works — a theological compendium *The Mirror of Theology* (*Zertsalo bogosloviia*, Pochaiv, 1618), a collection of sermons *Didactic Gospel* (*Evangelie uchitel'noe*, Rokhmanovo, 1619) and a poetic cycle *Priceless Pearl* (*Perlo mnogotsennoe*, Chernihiv, 1646). The first two works came under severe criticism both in Kyiv and then in Moscow. Consequently, they were condemned and prohibited.[23] As a result, he decided to adopt the Union of Brest in 1626, and from 1628 to his death he served as an archimandrite of Yelets Dormition Monastery in Chernihiv. His first work *Mirror of Theology* included a part devoted to the description of the universe, which was only loosely connected with the hexameral genre and was based rather on the perception of the world through the main four elements. This narrative, in turn, was most likely inspired by *An Exact Exposition of the Orthodox Faith* by John of Damascus.[24]

Stavrovetsky's cosmological beliefs were presented in the second part of the book entitled 'On the four-part world'. According to Stavrovetsky, all the components of creation are classified in a four-part scheme, where the first level hosts the invisible world (angels), the second is the visible world (superlunary and sublunary worlds, the four primary elements), the microcosm (a human) and an 'evil world' (sin and sinners). The description of the universe in Stavrovetsky's treatise is dominated by the idea of the harmonious co-existence of various elements and components of being. The visible world got its name — *Mir*, which means both 'world' and 'peace' — because even though it contains various elements of different substances, they 'live in peace or in quiet until the time measured

22 The life and work of Kyrylo Tranquillion Stavrovetsky was thoroughly studied by Sergei Maslov in the beginning of the twentieth century, the study, however, was published almost eight decades later, see: S. Maslov, *Кирилл Транквиллион-Ставровецкий и его литературная деятельность: Опыт историко-литературной монографии* (Kyiv: Naukova dumka, 1984). Lately, the interest towards Kyrylo Stavrovetsky has been on its rise, see e.g. the collective study: B. Krysa, and D. Syroyid, eds, *Кирило Транквіліон Ставровецький — проповідник Слова Божого* (Lviv: Vidavnitstvo Ukrains'kogo Katolits'kogo Universitetu, 2017).

23 The trial is analysed in: A. Bulychev, *История одной политической кампании XVII века* (Moscow: Iazyki slavyanskoi kul'tury, 2004).

24 Maslov, *Кирилл Транквиллион-Ставровецкий и его литературная деятельность*, pp. 139–56; O. Chadaeva, 'Visions of universe in theological works of Ruthenian authors in the seventeenth century: K. T. Stavrovetsky and S. Polotsky', *Studia Comeniana et historica*, 48.99–100 (2018), p. 58.

for them by God'.[25] The opposite elements are interrelated, and while water is connected to the earth, the air is related to the heavenly fire. The idea of harmony also applies to the stars, which follow the given order, and are complementary to one another and the main celestial object, i.e. the Sun. Thus, they serve their Creator in peace, not harming one another.[26]

Stavrovetsky did not particularly focus on heavenly matters, most of his description of the world is devoted to the reflection of the four primary elements. The model presented by Stavrovetsky was geocentric, with the spherical Earth in the centre of the universe, equidistant from the borders of the firmament, being founded 'in the middle of the heaven as a spot and maintained by the unseen power of God'.[27] In a very poetic manner, Stavrovetsky compared the world to a luxurious palace, designed exclusively for a human. However, the author insisted that a man, despite being the king over nature, is not equal to divine Providence, and therefore there is a boundary separating him from understanding the fullness of Providence and many things remain hidden from human perception.[28]

The image of the universe presented by Stavrovetsky fully remained within the patristic tradition. It was mostly referring to John of Damascus and Basil of Cesarea, some similarities can be found with *Paleia Interpretata*. However, the major novelty of the text was its language, since it was written in 'Prosta Mova', or 'simple speech', which referred to the vernacular form of Ukrainian. In contrast with older similar texts, Stavrovetsky primarily focused on a human, his central position in the universe and celebrated harmony as a main property of the physical world.

Stavrovetsky's second book, the collection of sermons entitled *Didactic Gospel*, also included certain reflections on the universe, e.g. he mentioned the seven levels of the physical world along with another theory of nine levels,[29] or again admired the divine wisdom in the creation of the world that serves as a perfect home for a human.[30]

When Moscow editors hegumen Ilia and Ivan Nasedka prepared a critical review of the *Didactic Gospel* in 1627, they did not pay particular attention to Stavrovetsky's description of the universe and rather focused on other matters that could cast a 'Latin' shadow on the work.[31] Another critical review, this time of the *Mirror of Theology*, came several decades later. In 1679 it was produced by a

---

25 Stavrovetsky-Tranquillon, *Зерцало богословия* (Pochaiv, 1618) part 2, fol. 1v. (Translation is mine).

26 Stavrovetsky-Tranquillon, *Зерцало богословия*, part 2, fol. 1v.

27 Stavrovetsky-Tranquillon, *Зерцало богословия*, part 2, fol. 16r. Here Kyrylo Tranqillion refers to Basil of Cesarea.

28 Stavrovetsky-Tranquillon, *Зерцало богословия*, part 2, fol. 18v.

29 Stavrovetsky-Tranquillon, *Евангелие учительное* (Rokhmanovo, 1619), part 2, fol. 30v.

30 Stavrovetsky-Tranquillon, *Евангелие учительное*, part 1, fols 259r–259v, part 2, fol. 4v.

31 Maslov, *Кирилл Транквиллион-Ставровецкий и его литературная деятельность*, p. 173; A. G[olubtso]v, 'Судьба «Евангелия учительного» Кирилла Транквиллиона-Ставровецкого', *Чтения в обществе любителей духовного просвещения*, 28.4 (1890), p. 554.

Belarusian scholar Simeon Polotsky at the order of Tsar Feodor. When analysing the cosmological part, he pointed out some controversial aspects, such as inaccuracies in the order of creation and the overly harmonious nature of the physical world. According to Polotsky, who followed the scholastic adaptation of the Aristotelian natural philosophy, the sublunary world could not be harmonious, since it is corruptible. Hence Polotsky criticised the notion of peaceful coexistence of the four elements, which were, according to him, incorrectly described. He also questioned the notion of waters placed above the firmament in order to prevent it from being burnt.[32] The treatise was judged as a 'very harmful' and poorly written work.[33] Interestingly, some of the aspects pointed out by the reviewer such as the description of the firmament would not seem troubling for a Muscovite reader, as they were fully in line with accepted interpretations from the Church Fathers.

Simeon Polotsky, whose views will be discussed later, adhered to the scholastic interpretation of the geocentric system, which impacted his judgement. The criticism, however, did not prevent the dissemination of Stavrovetsky's ideas and books, particularly among Old Believers. Moreover, references to Stavrovetsky could be even found in early eighteenth-century compilatory works on cosmology next to the names of Tycho Brahe and Galileo.[34]

The First Disputation on Cosmology: Lavrentii Zyzanii and His Great Catechism

In the same year that Stavrovetsky's works were placed under a ban, the problematic description of the world became the subject of controversy in the debate between a Ukrainian enlightener Lavrentii Zyzanii (around 1560s–after 1634) and the Moscow editors hegumen Ilia and Grigory Onisimov.[35] Zyzanii brought his *Great Catechism* to Moscow in 1626, hoping to publish it with funding from the Moscow Patriarchate. However, his work seemed suspicious to Patriarch Filaret, who ordered the editors of Pechatnyi dvor (Print Yard) to review and edit the text. The editors found numerous problematic areas with a distinct 'Western' theological influence, including the relations between the hypostases of the Trinity and their particular features, the sacred sufferings of Jesus Christ,

32 Moscow, State Historical Museum, MS Sinodal'noe sobranie 130, Collection of materials from the archives of Simeon Polotsky 130, fols 246^{r}–248^{r}.

33 Moscow, State Historical Museum, MS Sinodal'noe sobranie 130, fol. 251^{v}.

34 Moscow, Russian State Library, MS Belyaev (29) 45, The Celestial Motion (Begi nebesnyia, an anonymous compilation), first quarter of the eighteenth century, fol. 5^{r}.

35 For the whole case in more detail see: T. Oparina, *Иван Наседка и полемическое богословие киевской митрополии* (Novosibirsk: Nauka, 1998); M. Korzo, *Украинская и белорусская катехетическая традиция конца XVI–XVIII вв.: становление, эволюция и проблема заимствований* (Moscow: Kanon+, 2007).

the afterlife of the soul, and the concept of the 'first hell', i.e. the purgatory, the doctrine of repentance, the sacrament of Baptism and the free will, etc.

After the editors reviewed the text, there were three debates between the author and the editors,[36] and Zyzanii consequently accepted almost all the changes. The third debate included the discussion over cosmological issues. The author noticed that the original text related to the description of the universe was removed and replaced by some strange passages. He consequently inquired about the reasons for such editing. The editors responded that it was due to his use of 'the book of astrology', and they did so by direct order of Patriarch Filaret. Unfortunately, the original Zyzanii's manuscript did not survive, therefore we can judge the content of the problematic cosmological concepts only based on the disputation.[37] The editors mentioned 'heavenly circles and planets, and the zodiac, and the solar eclipse, thunder and lightning, and the crush and clash and Perun, comets and other stars'.[38] It is noteworthy that such an explanation surprised the author, who could not even have thought of any controversies that might have been related to such an issue. According to his words, he had not written anything about the influence of the stars on the human life and happiness or the guidelines given by the stars (in the astrological sense), he wrote only 'for the sake of knowledge' that a human is a divine creature, same as all beings created by God.[39] However, for the editors, the main error of Zyzanii's interpretation of the universe was in 'choosing false speech and stars' names from the book of astrology'. After the author cautiously asked about the proper way to speak about cosmological matters, the editors first referred to the book of Genesis, where Moses had not provided any 'animate and animal names'. As to the motion of celestial objects, they should have been solely explained by the force of angels, similarly to meteorological phenomena.

It would be an overstatement to claim that the editors of the Moscow Print Yard were not familiar with the ancient nomination of celestial bodies or at least certain elements of the geocentric system, which, most likely, had been presented in Zyzanii's treatise. It could be possible that Zyzanii used Latinisms, which automatically cast a shadow of suspicion on his text for the editors, who were used to Grecisms in the nomination of celestial bodies, e.g. Aphrodite for Venus, Ares (Arris) for Mars, etc. The rejectionist attitude of the editors signified an extreme conservative turn to the most rooted tradition that derived from the Antioch school, especially since it had been supported by the most respected Russian thinkers of the preceding centuries. In particular, we can refer to Joseph Volotsky

---

36 First published in: 'Прение литовского протопопа Лаврентия Зизания с игум. Илиею и справщиком Григорием по поводу исправления составленного Лаврентием Катехизиса', in *Летописи русской литературы и древностей, изданные Н. С. Тихонравовым*, vol. 2, part 2, ed. Nikolai Tikhonravov (Moscow 1859), pp. 80–100.

37 Korzo, *Украинская и белорусская катехетическая традиция*, pp. 296–304.

38 'Прение литовского протопопа Лаврентия Зизания', p. 94.

39 'Прение литовского протопопа Лаврентия Зизания', p. 94.

(1439/40–1515), who in his *Prosvetitel'* (*The Enlightener*) laid the foundation for the vague description of the universe, which, at the same time, is closest to the diction of the Scripture.[40] Joseph Volotsky was the teacher and an uncle of Dosifei Toporkov, a leading figure in the creation of the *Chronograph* of 1512, where we can find precisely the cosmological picture offered by the Moscow Print Yard editors as a normative one.[41] The attention to nomination of the heavenly bodies expressed by the Muscovite editors reveals the linguistic problem, the opposition of the profane and sacred language, which became topical with the correction of divine service books at the end of the 1640s.[42]

Apparently, cosmological aspects were not crucial for the author. Zyzanii consequently accepted the changes,[43] which, however, did not help the book find its target reader. Almost all printed books were ordered to be eliminated, and only a small number of copies from the first issue survived. However, the *Catechism* later became popular among Old Believers.[44]

Translations: Old and New Texts

During the reign of Mikhail Fedorovich (1613–1645), the attention of the Muscovite audience to cosmological problems, on the one hand, was not particularly strong, since other problems absorbed the focus of intellectual activity. Nevertheless, it was still present, and certain attempts to collect and compile existing knowledge were also made. Muscovite readers were particularly interested in geography.[45] Their inquisitiveness became even more prominent in the seventeenth century, when the contacts with the West increased, and, at the same time, Muscovy was involved in the territorial expansion to the East. The most active flow of translations was registered in Posolsky prikaz (the Ambassadorial Office). One of their endeavours was a translation of *Theatrum Orbis Terrarum*, a cosmography by Abraham Ortelius,[46] which contained a brief description of the geocentric system.

40 M. Shpakovsky, 'Отношение Иосифа Волоцкого к космологической проблематике', *Древняя Русь. Вопросы медиевистики*, 42.4 (2020), p. 129.

41 Shpakovsky, 'Отношение Иосифа Волоцкого', p. 120.

42 B. Uspensky, 'Раскол и культурный конфликт XVII в.', in Boris Uspensky, *Избранные труды. Т. 1. Семиотика истории. Семиотика культуры* (Moscow: Iayzki russkoi kul'tury 1994), p. 340. However, the roots of this discourse can be traced up to the end of 1610s, see G. Florovsky, *Ways of Russian Theology: Part 1*, in *The Collected Works of Georges Florovsky*, vol. 5 (Vaduz: Büchervertriebsanst, 1979), p. 90.

43 On the content of the cosmological part in the edited Catechism see O. Čhadaeva, 'Reception of Western cosmological models in the seventeenth-century Russia', *Studia Comeniana et Historica*, 101–102 (2019), 195–213, on p. 200.

44 Korzo, *Украинская и белорусская катехетическая традиция*, p. 298.

45 Sobolevsky, *Переводная литература Московской Руси XIV–XVII веков*, p. 46.

46 N. Kazakova, 'Русский перевод XVII в. "THEATRUM ORBIS TERRARUM" А. Ортелия', *Вспомогательные исторические дисциплины*, 18 (1987), 130.

This cosmography was very popular and was sometimes added as a supplement to the text of the *Chronograph*, the most read and spread book on history at that time.

In 1637, the translators of the Posolsky prikaz Bogdan Lykov (Lyzlov) and a detained ambassador of the Holy Roman Empire to Persia Ivan (Adam) Dorn, completed an ambitious translation project, which was *Atlas Sive Cosmographicae Meditationes de Fabrica Mundi et Fabricati Figura* by Gerhard Mercator. The atlas was published in 1595 and at the time of translation into Russian it was already outdated. Nevertheless, the text was popular in Russia and survived in several manuscripts. The translation was vastly edited, with numerous reductions and expansions. The most significant change, however, was related to cosmology. While the original text contained an extensive cosmological chapter entitled 'De mundi creatione ac fabrica luber', the translators completely omitted this controversial Neoplatonist treatise[47] and replaced the initial chapter with a compiled introduction.[48] Most likely the replacement was driven by the controversial nature of Mercator's interpretation of the divine matters, which even resulted in its inclusion into the *Index librorum prohibitorum* in the Catholic provenance in 1603.[49] Therefore, the translators could have logically decided to remove it in order not to threaten the whole project. The emphasis in the introduction they added is on the inquisitive nature of a human being regarding the sublunary realm, not the heavenly matters. The text reads:

> 'Every human is naturally inclined by the reason granted to him by God to know various things that are placed *below* the heaven and that he can see with his own eyes, and to search out for the [things] he has heard of, and to invent and collect different [things] together, and to present it in reality for all the people for a better understanding'.[50]

This case demonstrates the cautious attitude of Muscovite learned men to any cosmological texts of Western provenance, regardless of their content, if any divine or philosophical matters were discussed.

With the reign of Alexei Mikhailovich, the amount of translated texts sharply increased, particularly thanks to the activities of Kyivan monks, sent to Muscovy by Sylvester Kosiv, the metropolitan of Kyiv, Galicia and all Ruthenia, at the

---

47 For more on Mercator's Atlas see e.g.: N. Crane, *Mercator: the man who mapped the planet* (London: Orion Books Ltd, 2003); A. Mosley, 'The Cosmographer's Role in the Sixteenth Century: A Preliminary Study', *Archives internationales d'histoire des sciences*, 59 (2009), 423–39; A. Ramachandran, *The Worldmakers: Global Imagining in Early Modern Europe* (Chicago: The University of Chicago Press, 2015), p. 67.

48 Moscow, Russian State Library, MS MDA (173.1) 101, fol. 1r. A fragment was published in A. Popov, *Обзор хронографов русской редакции. Вып. II* (Moscow: Tipogragiya A. I. Mamontova i K°, 1869), pp. 190–91.

49 Raikov, *Очерки по истории гелиоцентрического мировоззрения*, 52.

50 The translation is mine, italics and brackets supplied. Published in: Popov, *Обзор хронографов русской редакции. Вып. II*, pp. 190–91.

request of tsar.[51] The Ukrainian scholars were mostly preoccupied with translating works of Church Fathers, compiling a Latin-Slavonic lexicon, and they were involved in the process of liturgical books' editing. However, particularly the translation activities of Epifanii Slavinetsky (?–1675) deserve attention in the cosmological context. In the second half of the 1650s, Slavinetsky, together with another Kyivan translator, Arsenii Satanovsky, and Isaiia Chudovsky, who, apparently, was also from Ukraine, completed a translation of *Theatrum Orbis Terrarum, sive, Atlas Novus* by Willem Blaeu (1635). The original atlas was brought to Moscow in 1645.[52] The translation was probably ordered by Patriarch Nikon, who owned the original, and was never printed. The fact that it was the patriarch who was the commissioner of the translation, should not be surprising. His attitude to education and knowledge was debatable, but it was he who undertook an attempt to purge the church of any mystical and magical element.[53]

The introductory part of the *Atlas*, translated as *Cosmography*, contains the general information on cosmography, astronomy, and geography, and presents two conceptions of the centre of the world, with the description of the geocentric model with planets and their orbiting cycles, and a short explanation of the heliocentric model. This part was later detached from the main text and distributed in a limited number of separate manuscripts and collections of mixed content in the second half of the seventeenth century.[54] Here we find the first known mention of the heliocentric system in the Russian language.[55] In fact, the text presents two models, the geocentric and the heliocentric one, and although the author considers them equal, he prefers the heliocentric model, referring to the most accomplished mathematicians of the time.[56]

In contrast with Mercator's atlas, the cosmological introduction of the *Atlas Novus* was fully translated. However, these two texts are very different in their nature and character. While Mercator's introduction was deeply philosophical, the text of Blaeu is austere, factual, technical, and scientific, devoid of remarks on theology. As a result, the text did not attract much attention from the Muscovite audience and did not cause any criticism. It is quite possible that its brief and technical character with no theological references was not giving the idea of the potential threats to the Orthodox worldview. On the one hand, there was still no consensus on the only one 'proper' and 'right' image of the world, since the

---

51 A. Panchenko, '*Епифаний Славинецкий*', in *Словарь книжников и книжности Древней Руси. Вып. 3 (XVII в.), Ч. 1, А–З,* (Saint Petersburg: Dmitry Bulanin, 1992), p. 310.

52 N. Nikolenkova, 'Русская географическая терминология во «Ввождении в космографию». Лингвистический аспект', in *Историческая география*, vol. 3, ed. by I. Konovalova (Moscow: Akvilon, 2016), pp. 117–18; 'Переводческая деятельность в московском Чудовом монастыре: новые данные о кружке Епифания Славинецкого', *Вестник Московского университета. Серия 9. Филология*, 6 (2018), 77–91.

53 Nicolaidis, *Science and Eastern Orthodoxy*, p. 141.

54 Nikolenkova, 'Русская географическая терминология', p. 121.

55 Raikov, *Очерки по истории гелиоцентрического мировоззрения*, p. 117.

56 Moscow, State Historical Museum, MS Sinodal'noe sobranie 19, fol. 2^{r}.

Aristotelian-Ptolemaic model had coexisted with the notion of the flat Earth for many centuries, not causing specific debates or controversies, except for the aforementioned specific episode with Zyzanii's *Catechism*. The situation when the particular interpretation was rather a matter of choice might have led to a certain relativism in the perception of the universe. On the other hand, at the end of the 1650s and in the 1660s, Muscovy was facing a deep cultural crisis which initiated the Church schism, along with the conflict of the patriarch with the tsar. Therefore, the primary intellectual focus was definitely directed towards other matters than the image of the universe. Nevertheless, it would be an error to claim that the text went completely unnoticed by the readers. It can be found in several separate manuscripts detached from the descriptions of the countries, which means that it found some little audience.

Another significant translation project which contained a description of the heliocentric model was undertaken with *Selenographia* (1647) of Johannes Hevelius possibly by S. Chizhinsky during his service in Posolsky prikaz in 1678–1681.[57] This text, however, is known only in one copy, which leads to the conclusion that it went unnoticed by the Muscovite readership.

New Compilations

In the second quarter of the seventeenth century there was a certain interest in finding the acceptable picture of the universe within available and approved Orthodox sources. One of the most prominent and odd examples was a new compilation, added to the *Chronograph* in the 1640s.[58] The text of the *Chronograph* had already undergone numerous changes, e.g. in the redaction of 1617, information about America was added to the description of the Earth.[59] However, the new compilation was much bolder in its content. This text, which among other additional textual units preceded the main body of the *Chronograph*, is also known in separate manuscripts of later years. It is entitled *On the creation of the seven days and the whole universe* (*O sotvorenii sedmi den' i vsego mira*) and consists of a bizarre mixture of apocryphal and canonical texts of different origins. They included the *Book of Enoch*, the *Dispute between Panagiotes and the Azymites*, *Μαργαρῖται*

57 R. Cicėnienė, 'Rankraštinė Jano Hevelijaus Selenografija Vilniuje', *Knygotyra*, 72 (2019), 34–61.

58 This text definitely deserves attention, but it has been neglected by scholars so far. It was reflected in the nineteenth-century classical work on *Chronographs* by A. Popov (Popov, *Обзор хронографов русской редакции. Вып. II*, pp. 154–69), and briefly mentioned by some researchers, including N. Savel'eva, 'Космологическая компиляция «Слово о Святой Троице» и ее бытование в составе древнерусских сборников', *Труды Отдела древнерусской литературы*, 58 (2007), p. 571, and D. Bulanin, 'Утраченный рукописный источник русского хронографа (опыт реконструкции состава)', *Rossica Antiqua*, 1/2 (2016), 122.

59 A. Popov, *Обзор хронографов русской редакции. Вып. I.* (Moscow: Tipogragiya Lazarevskogo instituta, 1866), p. 93; Popov, *Обзор хронографов русской редакции. Вып. II*, p. 97.

(*Pearls*) attributed to John Chrysostom, *Life of Andrew the Fool*, along with *Paleia*, *Christian Topography* and Basil's *Hexameron*, and, surprisingly, the *Lucidarius*, an anonymous medieval German-language summa of the twelfth century, which was translated into Russian in the sixteenth century. The text does not provide a coherent picture of the universe, rather, it offers various points of view as well as an opportunity to choose whether the presented information is appropriate or not. There is even a warning that the text should only be read in case the reader wishes to do so.[60] Another specific feature of the text is that it quite carefully lists references to the sources, e.g. references to *Lucidarius* are marked as 'Hellenic and foreign'. Whenever the compiler disagrees with the information, he does not hesitate to express it, particularly regarding the statement that there were two initial principles in the world, one of which ascended to heaven, while the other one remained on the Earth.[61]

The reader is offered, among others, three theories on the number of heaven (a single one, three, and seven levels).[62] The information collected in the compilation is scattered and contradictory, therefore it is even impossible to classify the presented model(s) as a geocentric or flat Earth system. The information drawn from texts with an uncertain attitude to the structure of the universe or the flat Earth tradition prevails in the compilation. However, the geocentric view is also quite firmly outlined. The readers are invited to put together their own image of the universe based on reflection of the compiled fragments.

This cosmological compilation quite often appears as an addition to *Chronographs* even in the eighteenth century, as well as in separate manuscripts, and could be found in Old Believers' libraries. The reasons for its creation and dissemination could lie in the need to finally collect all the cosmological information of the domestic provenance, as opposed to alien newly translated texts or emerging works that were not tested by time and therefore caused suspicion. On the other hand, this text clearly demonstrates that there was absolutely no coherent and unified picture of the world shared by Russian Orthodox believers in the seventeenth century. Without solid theological, philosophical, or scientific foundations, Muscovites could only collect a mosaic of 'possible' models and attempt to solve this puzzle themselves.

At the same time, other attempts at approaching the cosmological matters were made in the second quarter of the seventeenth century. According to recent research, the *Hexameron* that has been traditionally attributed to a Graecophile bishop Athanasius of Kholmogory (1641–1702) and dated to the late 1660s,[63]

---

60 Saint Petersburg, Russian National Library, F. IV. 168, fol. 8^{r}.

61 Saint Petersburg, Russian National Library, F. IV. 168, fol. 7^{r}.

62 Saint Petersburg, Russian National Library, F. IV. 168, fols 13^{r}–15^{r}.

63 T. Panich, '"Шестоднев" Афанасия Холмогорского', in *Христианство и церковь в России феодального периода* (Novosibirsk: Akademia Nauk SSSR, 1989), pp. 136–93; T. Panich, *Литературное творчество Афанасия Холмогорского. «Естественнонаучные» сочинения* (Novosibirsk: Sibirskii khronograf, 1996).

in fact, could have appeared earlier, possibly in the 1640s.[64] In this text, the author, or compiler, made an attempt to create an 'updated' and more coherent version of *Hexameron*. Apart from relying on the Scripture, the author was using direct and indirect references to Basil the Great, Gregory of Nazianzus, John Chrysostom, John of Damascus, (Pseudo)Dionysius the Areopagite, Gregory of Nyssa, Explanatory Gospel as well as Plato and Aristotle. This *Hexameron* presents a geocentric version of the universe, and the idea of the spherical Earth is consistently emphasised in the text. The Earth is placed inside two circles — the physical heaven and the divine one, separated from the latter by the firmament and consisting of three levels. The seven classical luminaries are placed within the physical heaven in separate zones (belts) that are 'fine, like smoke', with reference to the interpretation of Basil of Caesarea and John of Damascus.[65]

If compared to the most influential predecessors, this seventeenth-century *Hexameron* is more concise and, perhaps, more logical than John the Exarch's work, since it does not contain extensive excerptions from the proponents of the flat-Earth model. On the contrary, it interprets even certain metaphors, typically used by the followers of the Antiochene tradition, such as the world as a skull, with the brain being the invisible heaven, within the geocentric framework.[66] Compared to John the Exarch's *Hexameron*, there are no stories of animals, strange and miraculous creatures, as, e.g., alkyon or turtle-dove. As a result, the text is less entertaining.[67] The author of this compilation also avoided any numerical information about the measures of the Earth, Moon, and Sun or the lunar cycle, which is present in John the Exarch's text.[68] However, there are a lot of similarities, including the attitude to astrology. While the 'signs' of the luminaries can be interpreted either theologically, or meteorologically, the author of the text confronts astrologers who try to indicate the character properties of a human by the zodiac sign.[69]

This *Hexameron* is strongly influenced by the apophatic tradition of theology, which is explicitly demonstrated in the introduction to the text.[70] The author, while using logical reasoning when commentating on the Scripture, still insists on the limitations of human knowledge and the cognoscibility of the world.[71] The image of the universe this text produces is ambiguous, it is dynamic and static at the same time. On the one hand, the world is in constant motion, the

64 T. Anisimova, 'Шестоднев второй четверти XVII в., приписываемый Афанасию Холмогорскому', in *Тринадцатые Загребинские чтения: Сборник статей по итогам международной научной конференции, Санкт-Петербург, 03–04 октября 2018 года* (Saint Petersburg: Российская национальная библиотека, 2019), pp. 11–34.

65 Panich, '"Шестоднев" Афанасия Холмогорского', p. 165.

66 Panich, '"Шестоднев" Афанасия Холмогорского', p. 158.

67 Panich, *Литературное творчество Афанасия Холмогорского*, p. 72.

68 Cf. Barankova, and Mil'kov, *Шестоднев Иоанна экзарха Болгарского*, pp. 727–29.

69 Panich, '"Шестоднев" Афанасия Холмогорского', p. 170.

70 Panich, '"Шестоднев" Афанасия Холмогорского', p. 141.

71 Panich, '"Шестоднев" Афанасия Холмогорского', p. 138, 157.

heaven and stars rotate and do not change their places, the seasons alternate, the moon decreases and increases. The weather changes and plants are born and grow, similarly to animals and humans. However, all of these changes are cyclical, and the only truly dynamic element is the incarnation of Christ and the Second Coming.[72]

Whether Athanasius of Kholmogory authored this text or not, remains a debatable issue. It is not surprising, however, that this intellectual bishop owned this text, as his interest in that matter was obvious. His library contained European astronomical calendars, globes, compasses and telescopes.[73] Along with that, he was known as a Graecophile, and the *Hexameron* presents a picture of the world that almost exclusively relies on Greek sources.

The two aforementioned compilations signify the importance of the cosmological discourse for Muscovite authors, as well as attempts to either reconstruct the 'standardized' picture of the universe, or collect all available sources that can be referred to for these purposes. This can be interpreted as a reaction to the feeling of the growing 'Western' influence and the need to find a foundation for the image of the world on approved and 'localized' material.

Cosmology and Theology: Simeon Polotsky

The issue of cosmology initiated a discussion only in the case when it was related to some dogmatic matters, certain perceptions of the world related to theology, not natural philosophy or science per se. The reaction of the Russian intellectual environment to the texts where cosmology was related to theology was much more critical than in the case of translated works with no reference to the religious understanding of the world. Such texts could be produced by Ruthenian scholars, whose background was much more closely tied with the Western scholastic tradition.

Perhaps the most influential intellectual in pre-Petrine Russia was Simeon Polotsky (1629–1680), who relocated to Moscow in 1663, having received his education at the Kyiv Ecclesiastical Academy and probably at the Jesuit college of Wilno.[74] This Belarusian scholar is most known as a poet, author of sermons, educator, publicist and translator, however, he also had theological ambitions.[75] In

72 Panich, '"Шестоднев" Афанасия Холмогорского', p. 193.

73 Vucinich, *Science in Russian Culture*, p. 21.

74 More on Polotsky in English see e.g.: A. Kahn, M. Lipovetsky, I. Reyfman, and S. Sandler, *A History of Russian Literature*, (Oxford: Oxford University Press, 2018), 125, 146, 150–2, 154–8, 160–8, 181, 205, 213.

75 G. M. Hamburg, *Russia's Path toward Enlightenment: Faith, Politics, and Reason, 1500–1801* (New Haven: Conn, 2016).

1670 he completed his major theological treatise, *Venets very*, or *Crown of Faith*,[76] which was allegedly intended for the Tsar's children as a theological study guide.[77] For this treatise, Polotsky used mostly Latin sources, mainly *Hortus Pastorum* of Jaques Marchant[78] and *Compendium Theologicae Veritatis* of Hugh Ripelin of Strasbourg.[79] The latter work served as the main source for a detailed description of the classical scholastic adaptation of the geocentric model with eleven heavenly spheres in Polotsky's compendium. These levels included the empyrean heaven, the crystalline heaven with the Primum Mobile, the crystalline watery level, and the firmament, which consisted of the stellar sphere and seven planetary celestial orbs. Apart from Hugh Ripelin, Polotsky also used Pliny's *Naturalis Historia* and some other Latin sources.[80] In his description, Polotsky focused on the differences between the sublunary and heavenly realms, where the latter was described as incorruptible by the Grace of God, while the former was corruptible since it consisted of the four elements.[81] What was particularly unusual for the Orthodox provenance, is that Polotsky provided various accounts of distances between the Earth and individual planets and the size of the celestial objects. Along with reasoning about the nature of heavenly objects, he also tried to explain the celestial motion and briefly mentioned the retrograde motion. However, his explanation might have seemed slightly confusing for an unprepared reader.[82] The extensive cosmological part in Polotsky's work is not surprising, since he was known for his interest in natural philosophy and possibly astrology, and his personal library contained numerous Latin works on the matter.[83]

For a conservative Catholic reader, Polotsky's description of the universe would seem uncontroversial at its core, but certainly outdated and plain, especially as regards the phantasmal and contradictory distances between the Earth and other luminaries given in the treatise. However, the readership in Muscovy

76 M. Korzo, *Нравственное богословие Симеона Полоцкого: освоение католической традиции московскими книжниками второй половины XVII века* (Moscow: IFRAN, 2011), p. 13.

77 Korzo, *Нравственное богословие Симеона Полоцкого*, p. 137.

78 Korzo, *Нравственное богословие Симеона Полоцкого*, p. 20.

79 O. Chadaeva, 'Космос как теологическая проблема: к вопросу о «Венце веры кафолическия» Симеона Полоцкого', *Вопросы философии*, 4 (2017), 139–149; O. Chadaeva, *Two Authors, Two Universes: The Cosmological Models in the Works of Simeon Polotskii and Archpriest Avvakum Petrov*, *Russian Literature*, 99 (2018), 1–37.

80 Chadaeva, *Two Authors, Two Universes*, pp. 9–12.

81 Moscow, State Historical Museum, MS Sinodal'noe sobranie 286, fols 29ʳ–37ʳ; Moscow, Russian State Library, MS MDA (173.1) 66, fols 15ʳ–19ʳ.

82 Cf: 'The lower stars are called planets, of wondering [stars], since they sometimes go backwards to their own motion and move opposite to the firmament. The motion that they have from the flow of the Primum Mobile is straightforward. [But] their own motion is oblique.' Moscow, State Historical Museum, MS Sinodal'noe sobranie 286, fol. 35ᵛ. Translation is mine, brackets supplied.

83 A. Hippisley, and E. V. Luk´janova, *Simeon Polockij's Library: A Catalogue* (Köln: Böhlau 2005) pp. 167, 170, 175.

reacted to Polotsky's work with severe criticism. Polotsky's main opponent was a Graecophile translator, editor and writer Evfimii Chudovsky (?–1705), who was a student of Epifanii Slavinetsky. He had already criticized other Polotsky's works[84] and attempted to perform a critical analysis of the *Crown of Faith*. Having scrupulously listed all the statements he found heretic (i.e. Latin), errors, and suspicious and incorrect passages,[85] Evfimii concluded that the work is a mere 'delusion and heretical deceit'.[86] Apparently, he was infuriated by an attempt to include the cosmological part with particular measures of distances between the seven luminaries and the size of the Earth and other celestial bodies. It is noteworthy that it was the cosmological part of the treatise that came under the most severe criticism. Evfimii particularly pointed out the lack of references to the Church Fathers and brought up the question of whether the celestial problems could be cognizable in principle, measured, and expressed in numbers. According to Evfimii, 'the holy fathers […] Basil the Great and Saint John Chrysostom […] did not tell the measures and distances of heavenly orbs: moreover, [they] did not tell the size of the Earth itself', and '[…] the size of heaven exceeds the measure of the human mind'.[87] Later, in 1689–90, Evfimii's reasoning was used at a Moscow Church Council, when Polotsky's works were posthumously condemned along with works of other Ruthenian authors, including the aforementioned Tranquillion-Stavrovetsky. In the text of condemnation, which was officially attributed to Patriarch Joachim, but, in fact, was written by Evfimii Chudovsky,[88] a special emphasis was again made on the 'heretic' cosmological part with numbers and measures, whose source was unclear.[89]

On the one hand, the whole episode could be explained by a certain personal animosity between the two authors, one of which, according to scholars, belonged to a so-called Graecophile party, while the other one represented Latinophiles.[90] Thus, the cosmological issue could have served as a mere pretext for the attack. Indeed, cosmology was an easy target, since, as was stated above, there was no decisive and canonical approach to tackling cosmological matters. Therefore, it can be similar to the case of Zyzanii, when the choice of 'errors' was arbitrary, and cosmology merely came in handy for the purpose of criticism.

84 Korzo, *Нравственное богословие Симеона Полоцкого*, p. 140.

85 Moscow, State Historical Museum, MS Sinodal'noe sobranie 396, fols 20^{r}–27^{v}.

86 Moscow, State Historical Museum, MS Sinodal'noe sobranie 396, fol. 27^{v}.

87 Moscow, State Historical Museum, MS Sinodal'noe sobranie 396, fol. 27^{v}. The translation is mine, brackets supplied.

88 T. Panich, 'Евфимий Чудовский как интеллектуальный лидер сообщества церковных писателей второй половины XVII в.', *Гуманитарные науки в Сибири*, 25.3 (2018), 58–63.

89 E. Inok and Patriarkh Ioakim, *Остен. Памятник русской духовной письменности XVII века* (Saint Petersburg: Obshchestvo pamiati igumen'i Taisii, 2006), p. 105.

90 M. Kiseleva, *Интеллектуальный выбор России, p. 63. This division, however, was recently questioned. Cf. A. Bogdanov, '«Грекофилы» и «латинствующие» XVII в.: тексты в контексте', Genesis: исторические исследования, 4 (2021), 1–46.*

There were many more aspects of the treatise that could have been subjected to criticism. The choice of the cosmological problem as the main target for scrutiny seems to demonstrate that it resonated with a certain intellectual focus of the period. Polotsky included the description of the universe in a theological treatise, and therefore needed more theological foundation and references to patristics. As was pointed out above, both Antiochene and Cappadocian traditions were present in the Russian cultural milieu, and there was no resolute answer to the question about the limits of human cognition or method of interpreting the Scripture with regard to the image of the universe. Eastern Christian cosmology was balancing between two tendencies, when some theologians rejected the attempts of 'natural theology', while other thinkers were inclined to understand cosmology as a part of theology.[91] However, it seems that the ideas of the Hesychast movement, which emphasized the weakness of human reason, deeply influenced the complicated approach to science in Russian society.[92] Therefore, such a scientific and technical text as the aforementioned *Atlas*, which did not consider any theological aspects, was not subjected to criticism and was barely noticed. At the same time, Polotsky's *Crown of Faith* clearly contradicted the Russian theological tradition, which preferred metaphorical imagery to concrete numbers and avoided resolute answers to cosmological questions.

Despite all controversies, however, Ruthenian scholars played a significant role in a gradual drift towards accepting Western intellectual models. With the rise of baroque culture in Muscovy, cosmological (and astrological) motifs and imagery penetrated poetry,[93] homiletics,[94] and fine arts.[95]

---

91 T. Obolevitch, *Faith and Science in Russian Religious Thought* (Oxford: Oxford University Press, 2019), p. 10.

92 Nicolaidis, *Science and Eastern Orthodoxy*, pp. 104, 141.

93 See e.g. A. Robinson, 'Симеон Полоцкий — астролог', in *Проблемы изучения культурного наследия*, ed. by G. Stepanov (Moscow: Nauka, 1985); L. Zvonareva, 'Натурфилософские представления Симеона Полоцкого', in *Естественнонаучные представления Древней Руси*, ed. by Rem Simonov (Moscow: Nauka, 1988), pp. 228–46; A. Bogdanov, 'Естественнонаучные представления в стихах Кариона Истомина', in *Естественнонаучные представления Древней Руси*, ed. by R. Simonov (Moscow: Nauka, 1988), pp. 260–79; L. Sazonova, *Поэзия русского барокко (вторая половина XVII — начало XVIII в.)* (Moscow: Nauka, 1988); L. Sazonova, *Литературная культура России: Раннее Новое время* (Moscow: Iazyki slavianskikh kul'tur, 2006); H. Ram, *The Imperial Sublime: The Russian Poetics of Empire* (Madison: University of Wisconsin Press, 2006).

94 A. Eleonskaya, *Русская ораторская проза в литературном процессе XVII века*, (Moscow: Nauka, 1990); Iu. Zvezdina, 'Космос в поздних памятниках православной культуры второй половины XVII–XVIII в.', in *VIII научные чтения памяти Ирины Петровны Болотцевой*, ed. by O. Kuznetsova (Yaroslavl: Iaroslavsky khudozhestvennyi muzei, 2004), pp. 3–17; O. Čadajeva, 'Cosmological, Astronomical and Astrological Elements in Sermons of Seventeenth-Century Ruthenian Authors', *Teorie Vedy/ Theory of Science*, 42.1 (2020), 141–71.

95 Iu. Zvezdina, 'Зодіак у виданнях Києво-Печерської лаври другої половини XVII століття', in *Українське небо 2. Студії над історією астрономії в Україні*, ed. by O. Petruk, (Lviv: Institut prikladnikh problem mekhaniki i matematiki im. Ia.S. Pidstigacha NAN Ukraini,

Radical Simplicity: Archpriest Avvakum and Old Believers

'Westernized' Ukrainian and Belarusian scholars were not the only ones who were interested in reflecting on cosmological issues in the second half of the seventeenth century. Another description of the universe comes from their most prominent ideological enemy, protopope (archpriest) Avvakum (1620–1682). Avvakum opposed Patriarch Nikon's reforms of the Russian Orthodox Church and criticized any sign of Western influence. In his vivid and lively texts, he attacked 'almanac-mongers' and astrologers, and his outlook was clearly anti-scientific. This prolific author, however, attempted to cover the most fundamental topics, and therefore he also dealt with the image of the universe in his writings.

Cosmological imagery is present in many of Avvakum's works, among which the most celebrated is his *Life of the Archpriest Avvakum, by Himself* (1672–76).[96] One of his works entitled *Sniskanie i sobranie o Bozhestve i o tvari* (*An Inquiry and Collection about God and His Creation*) and dated 1672, most fully reveals Avvakum's attitude to the question of creation and his perception of the physical world. Interestingly, the incentive for writing this text was the spread of an apocryphal *Discourse of Three Saints* (*Beseda trekh sviatitelei*) in the newly forming Old Believer community. Having noticed that, Avvakum, who could have been considered the leader of the community, felt the need to refute the misconceptions of the world that the text contained.[97]

Avvakum first focuses on theological misconceptions that literally infuriated him, i.e. on the idea of three chambers, where the Trinity had resided before the creation of the world. For Avvakum, any idea of limitations when applied to God was the worst kind of heresy. Avvakum further disproves fantastic ideas related to cosmology. According to the *Discourse*, the Sun was made of the Holy Robe, the Moon was a mixture of air and the Throne of God. The Earth, which had been taken from the sea foam, was placed on three large and thirty-three small whales, while the sea was situated on columns.[98] This leads Avvakum to explain in more detail the 'correct' image of the world. Avvakum was heavily influenced by the

---

2016), pp. 176–186; E. Orlova, 'Космологические мотивы в русской иконописи рубежа Позднего Средневековья и Нового Времени: художественно-эстетические и религиозно-философские вопросы', *Древняя Русь. Вопросы медиевистики*, 3 (2015), 90–91.

96 For more on Avvakum's texts with cosmological aspects see: Chadaeva, *Two Authors, Two Universes*, p. 14.

97 Avvakum, *Снискание и собрание о Божестве и о твари и како созда Бог человека*, Пустозерский сборник: Автографы сочинений Аввакума и Епифания, ed. by N. Demkova, (Leningrad: Akadeniia nauk SSSR), pp. 92–99.

98 For the imagery in the *Discourse of Three Saints* see 'Беседа трех святителей', in *Памятники старинной русской литературы, издаваемые Графом Григорием Кушелевым-Безбородко*, vol. 3 ed. by Aleksandr Pypin (Saint Petersburg: Tipografiya Kulisha, 1862), pp. 169–184; Badalanova Geller, Florentine, '"These Blasphemous Rustic Scriptures" (Indigenous Apocryphal Heritage of Slavia Orthodoxa)', in *Triantaphyllo: In Honorem Hristo Trendafilov* (Shumen: Shumenski universitet 'Konstantin Preslavski', 2013), pp. 66–106.

Hexameron of John the Exarch and *Tolkovaia Paleia*, the texts that he frequently mentioned.[99] However, in *Sniskanie* he most often referred to *Chronograph*, thus demonstrating the authority of this text in the conservative part of the Russian society. As has been mentioned before, the text of *Chronograph* was related to the Antiochene cosmology, as well as most of the texts that were preceding the main body of the text.

The world described by Avvakum is very simple. It includes only a few levels, with the first one being the fiery heaven of heavens with the Throne of God, the firmament, which forms the border between the divine realm and the physical world. The watery visible heaven, or the sky, is not divided into several levels. The firmament is a thick watery structure that does not allow the waters to come down from above. The stars and planets are akin to the fiery Primary Heaven and serve as lighting instruments and decorations, the Earth is akin to water and lies on the water. The only curious element here is the image of hell, which is located *above* the firmament and the zodiac is located underneath. It should be also noted that Avvakum's text contains inaccuracies and errors. To name a few, there was a confusing order of the creation, which did not particularly correspond with the Scripture, and mistakes in the order, number, and names of the luminaries.[100]

The simplistic image of the world presented by Avvakum, on the one hand, fits completely within the Antiochene tradition and follows literal exegesis of the Scripture in the context of cosmology. However, it is not clear whether Avvakum believed in the model of the flat Earth, the semi-spherical universe, or the geocentric model, since it is not stated clearly in his text and some indicators can be interpreted ambiguously. In his other writings, e.g. in the treatise '*How to Live in Faith*' or in one of his letters to his followers, he extensively refers to John the Exarch's *Hexameron*, which advocates the geocentric worldview.[101] Moreover, the reasoning Avvakum used combines references to the most authoritative sources and an appeal to common sense, demonstrating a hidden or even subconscious attempt at rationalization. For example, while explaining why the *Discourse* is heretical, Avvakum does not only refer to Scripture, but uses rational arguments: '[He] took earth from water, not from foam, thus the earth is heavy, and foam

99 N. Sarafanova (Demkova), 'Произведения древнерусской письменности в сочинениях Аввакума', *Труды отдела древнерусской литературы*, 18 (1962), 329–40.

100 Avvakum, *Снискание и собрание*, p. 100. Analysed e.g. in O. Chadaeva, 'Потерянный рай протопопа Аввакума: сочинение "Снискание и собрание о Божестве и о твари" как интерпретация библейской истории', *Slověne= Словѣне. International Journal of Slavic Studies*, 8. 1, (2019), 188–222.

101 Avvakum, Как жить в вере, ed. by Sergey Platonov, Памятники истории старообрядчества XVII в., Русская историческая библиотека, издаваемая Археографической комиссией, 39 (Leningrad: Izdanie Akademii Nauk SSSR, 1927), cols 549–550; Avvakum, *Послание Симеону, Ксении Ивановне и Александре Григорьевне*, ed. by. Nikolay Gudzy, Житие протопопа Аввакума им самим написанное и другие его сочинения (Moscow: Goslitizdat, 1960), pp. 266–73.

is a light thing; the nature of water is heavy, so the earth is akin to water.'[102] Likewise, he does not adopt the notion of angelic forces governing the celestial or weather phenomena. Avvakum inclines to either rational explanation or attributes everything that happens in the physical world solely to the will of God.

Avvakum is sometimes referred to as the most radical retrograde rejecting science and scientific approach.[103] Indeed, in *Sniskanie* he also attacked 'almanac-mongers', aiming both at superstition (the zodiac readers) and those who attempt at explaining physical reality by numbers and reasoning. However, it is not the anti-scientific outlook that drove his reasoning. Avvakum deliberately chose the simplest explanation for natural phenomena, opposing it to the pride of the stargazers, superstition, and overly complicated and artificially constructed concepts.[104]

The Old Believer community, which was beginning to form, also needed to define its worldview, as well as to formulate ideas about the physical world and integrate them into their comprehensive picture of the world. Hence their leaders were forced to select the appropriate sources and reflect their ideas. Therefore, it is not surprising that this community later adhered to the aforementioned texts, i.e. the *Mirror of Theology*, Zyzanii's edited *Chatechism*, and even the hexameral phantasmal compilation from *Chronograph* discussed above. These texts were owned, copied and distributed by Old Believers in the eighteenth century. Although these texts were very different, they shared one thing in common: they were either banned, disapproved of, or frowned upon by the official Russian Orthodox Church.

Introducing Natural Philosophy: A Foreign Take

While the conservative part of the Russian society was looking for authoritative sources in the Old Church Slavonic provenance, the first attempts at establishing the curriculum in natural philosophy in the Western fashion were made in Moscow. In 1687, brothers Ioannikios and Sophronos Leichoudes founded the first higher education establishment in Russia, Slavic Greek Latin Academy, and introduced Aristotelian natural philosophy to Russian students. They based their educational programs on the Jesuit model, which also included natural philosophy.[105] The course of natural philosophy in the Academy started around 1691 and was apparently based on the Aristotelian physics with the commentaries of Nikolaos Koursoulas, a Western-educated Greek of the seventeenth century.

102 Avvakum, *Снискание и собрание*, p. 98. Translation is mine.

103 L. Heretz, *Russia on the Eve of Modernity: Popular Religion and Traditional Culture under the Last Tsars* (Cambridge: Cambridge University Press, 2008), p. 53.

104 Avvakum, *Снискание и собрание*, p. 110.

105 Their activities were profoundly studied by Nikolaos Chrissidis, in Chrissidis, *An Academy at the Court of the Tsars*.

Another probable source for the course was an anonymous work *De Mundo*. The treatises copied by Ioannikios presented the six models known by the middle of the seventeenth century in Europe, i.e. the Ptolemaic, Platonic, Egyptian, Copernican, Tychonic, and semi-Tychonic systems.[106] The last one was preferred in *De Mundo*, while Koursoulas vacillated between the Ptolemaic and the semi-Tychonic systems.[107] Unfortunately, we have no records of the reaction to the course or how well it was received. However, there were probably no debates or controversies on that issue, since those would have certainly been reflected otherwise. Nevertheless, since there was no translation of the texts owned by Ioannikios and used for the curriculum, the immediate dissemination of these ideas should not be overestimated. On the other hand, the graduates of the Academy, such as e.g. Fedor Polikarpov, Leontii Magnitsky,[108] Petr Postnikov, Stepan Krasheninnikov, Mikhail Lomonosov, and Dmitrii Vinogradov later significantly contributed to the dissemination of knowledge in Petrine and post-Petrine Russia.

Astrology also acquired a more influential position in the seventeenth century. Despite the critical attitude rooted in the beginning of the sixteenth century and soundly expressed by Maximus the Greek and Philotheus of Pskov,[109] it found its way through translations of astrological texts and foreign physicians serving at the Russian court, such as Arthur Dee (1579–1651), Wendelin Sybelist (1597–1677), Andreas Engelhardt (?–1683). Engelhardt should be noted in the context of astrology, since two of his prognostic letters to Alexei Mikhailovich written at the tsar's order were preserved.[110] The tsar himself displayed a decent interest in astrology and astronomy, and a map depicting the distribution and movement of stars was painted on the ceiling of the dining room in the Kremlin. Later a similar map appeared in the Kolomensky Palace, images of the Zodiac embellished the ceilings in the chambers of princesses Sophia Alekseevna, Tatiana Mihkailovna and Maria Alekseevna, as well as Moscow nobility and even churches.[111] Despite the negative attitude towards astrology expressed by the Church, it gained the court's endorsement. The astrological charts and predictions served to glorify the tsar's family and courtly life.[112] Astrological imagery was abundant in the works of

---

106 Chrissidis, *An Academy at the Court of the Tsars*, p. 146.

107 Chrissidis, *An Academy at the Court of the Tsars*, p. 156.

108 His education in the academy is a hypothesis that is questioned.

109 R. Simonov, *Русская астрологическая книжность, XI — первая четверть XVIII века* (Moscow: URSS, 2019), pp. 116–165; O. Akopyan, 'Ptolemy, Fortune, and Politics: A Case of the Reception of Western Scholarship in Early Modern Russia' in *Fate and Fortune in European Thought, ca. 1400–1650*, ed. Ovanes Akopyan (Leiden; Boston: Brill), p. 256.

110 A. Bogdanov, and R. Simonov, Прогностические письма доктора Андреаса Энгельгардта царю Алексею Михайловичу', in *Естественнонаучные представления Древней Руси*, ed. by Rem Simonov (Moscow: Nauka, 1988), pp. 260–79.

111 D. Sviatsky, *Астрономия Древней Руси* (Moscow: Russkaya panorama, 2007), p. 429.

112 O. Akopyan, 'Ptolemy, Fortune, and Politics', p. 221.

court poets. The tradition was established by Simeon Polotsky and continued by Silvestr Medvedev and Karion Istomin.

The borderline between astrology and astronomy was blurred, thus suspicious attitude to the latter was not without reason. However, first texts explaining the difference between the two disciplines appeared in the 1670s. The notable example is a compilative translation of several works undertaken by Nikolai Spathari (1636–1708), a Moldavian-born writer, translator, diplomat and traveller. His work *A Selected Brief Book on the Nine Muses and Seven Liberal Arts* (*Kniga izbrannaia vkrattse o deviatikh musakh i o sedmi svobodnykh khudoyhestvakh*) was based on int.al. *Iconologia* by Cesare Ripa and Vincenzo Cartari's *Le Imagini con la sposizione dei dei de gli antichi* (*The Images of the Gods of the Ancients and their Explanations*).[113] In this text, astrology is presented as a discipline that teaches 'about the perfection of stars and their usefulness', while astronomy explains the notion of stars.[114] Astronomy is seen here as a difficult, but necessary tool, particularly for agriculture, navigation and measuring time. At the same time, astrology is glorified as the jewel in the crown of cognition, and several serious reasons to study it are given in this brief text. Certain astrological, astronomical and even hermetic information can be found in other texts, compiled or translated by Spathari, e.g. in *Arithmologion* and unfinished *Hieroglyphic Book* (*Kniga ieroglifiiskaia*).

Spathari's works did not achieve large dissemination in Muscovy, in contrast with another translator-compiler, Jan Andrzej Białobocki (?–after 1712), a Polish Calvinist convert to Orthodoxy. Białobocki arrived in Moscow in 1681 and served as a translator, teacher, and interpreter while participating in theological debates and conflicts.[115] Among his translation, one book stands out as the text which became widespread in the eighteenth century. It was *The Great and Amazing Teaching of Ramon Llull* (*Velikaia i predivnaia nauka Raimunda Liulliya*). This text is not a direct translation of Llull's *Ars Magna* or his other works, but rather a compilation and a re-written version.[116] Its fifth part contains another picture of the universe that follows Llull's hierarchical model. In this model, heaven comes third after God and angels. It is followed by humankind, thinking and feeling creatures, the natural world (flora), and the elemental and instrumental world. According

113 O. Belobrova, 'К изучению «Книги избранной вкратце о девятих мусах и о седмих свободных художествах» Николая Спафария', *Труды отдела древнерусской литературы*, 30 (1976), 307–17.

114 O. Belobrova, *Николай Спафарий. Эстетические трактаты* (Leningrad: Nauka 1978), 45.

115 More on Białobocki see: L. Suchanek, 'Jan Andrzej Bialobocki i Barok Rosyjski', in *Polacy w życiu kulturalnym Rosji*, ed. by R. Łużny (Wrocław: Zakład Narodowy im. Ossolińskich, 1986), pp. 9–20, V. Kul'matov, 'Луллианские произведения А.Х. Белобоцкого', *Труды отдела древнерусской литературы*, 59 (2008), 360–61.

116 A. Gorfunkel', '«Великая наука Раймунда Люллия» и ее читатели', in *XVIII век. Вып. 5* (Moscow: Nauka, 1962), pp. 337–8; V. Kul'matov, 'От «Ars magna» Р. Лулия к «Великой науке» Белобоцкого', *Verbum, Вып.5, Образы культуры и стили мышления: иберийский опыт* (Saint Petersburg: Izdatel'stvo SPBGU, 2001), p. 291.

to the text, heaven is inanimate, but has certain qualities such as wisdom, insight, love, intelligence, and even will, which drives it, while not being sensible or sensitive.[117] Heaven influences all the creatures that are below it in their hierarchy, but it does not affect the human will. In this context, Białobocki mentions the disputes between astrologers and theologians and supports the latter. However, he later provides purely astrological information, including the influence of celestial bodies and zodiac signs on human personality traits.[118] The model presented by *The Great Teaching* follows the geocentric Aristotelian-Ptolemaic system and mentions concentric spheres and epicycles. It could be even viewed as a certain novelty given that the concept of Ptolemaic corrections using deferents and epicycles was still largely unknown in Muscovy. However, since the text contains no mathematical tools to demonstrate how the corrections worked, this information can be seen as redundant. *The Great Teaching* also contains numerical information about the size of the Earth and other planets, as well as the distances between the Earth and other celestial bodies with references to Pythagoras and Cardano. Nevertheless, the data in the text are controversial and nonsensical, even for the work based on the medieval source.[119]

The cosmological part of *The Great Teaching* was based on the model preceding the scientific revolution and can be considered a peculiar sum of the foreign cosmological knowledge available to educated Muscovites at the end of the seventeenth century. The cosmology of *The Great Teaching* is densely mixed with astrology, lacks coherence, and is essentially based entirely on the writings of Western medieval authors. Due to the compilatory nature of the cosmological part of the text, it can be compared with Polotsky's *Crown of Faith*. The latter author, however, was much more careful in astrological matters, but he also provided conflicting numerical data. In this eclecticism, in serving the reader with diverse and often loosely connected information about the world, we see an iconic feature of Baroque literature, which created a motley mosaic universe, a world as a book in which opposites, paradoxes, and contradictions can coexist.

The Aristotelian image of the universe merged with scholasticism came to Russia belatedly. In some parts of Russian society, it was quickly replaced by the formal modernised education received in the educational institutions emerging in the early eighteenth century, including the Moscow School of Mathematics and Navigation (founded in 1701), the Artillery and Engineering School (1701), the Naval Academy, the Academy of Sciences (1724) and, finally, at the Moscow University (1755). The controversies of the Copernican system in relation to Christianity so widely disputed in the West in the seventeenth century were yet to be brought to Russia in the eighteenth century. And it was for Ukrainian intellectuals again, Stefan Iavorsky and Theophan Prokopovich, who were the

---

117 Moscow, Russian State Library, MS Buslaev (396) 22, fol. 158^{v}.

118 Moscow, Russian State Library, MS Buslaev (396) 22, fols 159^{r}–163^{r}.

119 Raikov, *Очерки по истории гелиоцентрического мировоззрения*, pp. 62–63.

first to occasionally express certain anti-heliocentric views in Petrine Russia.[120] Apparently, they borrowed their views from neo-scholastic sources that circulated in the Catholic West. However, the issue became topical later, when it was actively discussed in Russia as late as in the 1750s.[121]

Conclusion

Despite the significant lagging behind in scientific attitude, Early Modern Russia demonstrated a certain interest in cosmological matters. It might indeed have been less evident than the practical interest in technological novelties, however, it was still present. The absorption of the Western worldview was complicated in Muscovy by two aspects — the difficult attitude to the problem of cognoscibility of the world in Eastern Orthodoxy aligned with the lack of a unified accepted model of the universe. This problem was accompanied by the absence of institutionalised education. However, the seventeenth century prepared the ground for the cultural changes that Russia was to undergo under the rule of Peter the Great, and Ukrainian and Belarusian scholars had an important share in this process. We can see that the discussion over the structure of the world was demanded by various groups of society, be it the ones who adhered to Westernisation or looked up to *Pax Latina*, or those who felt threatened by the changes and tried to find comfort and solace turning to the medieval simplistic images of the world. The latter should not be neglected and looked down upon as 'retrogrades' or uneducated primitives. For in both parties, we see one thing in common — admiration for the beauty of the world that is above historical turmoil and perishable conflicts. And it was an Old Believer protopope Avvakum, who expressed the wisest idea regarding the world and the place of a human in it: '…a day is equally bright for everyone, and the Sun shines equally, so that [everybody] could live in love like in a common home.'[122]

120 R. Collis, *The Petrine Instauration. Religion, Esotericism and Science at the Court of Peter the Great, 1689–1725* (Leiden: Brill, 2012), pp. 239, 307.

121 Raikov, *Очерки по истории гелиоцентрического мировоззрения*, pp. 258–360.

122 Avvakum, *Снискание и собрание*, p. 107.

SANDY SAKORRAFOU

Introducing Western Natural Theology in the Greek Orthodox Christian World (Late Eighteenth to Mid-Nineteenth Century)

The Cases of Pamblekis, Lesvios, and Vamvas

▼ **ABSTRACT** The discourse surrounding Western natural theology found limited acceptance within the Greek Orthodox Christian theological discussions. However, a period of interest emerged during the late eighteenth to the late nineteenth century. This study focuses on three key figures — Christodoulos Pamblekis (1733–1793), Veniamin Lesvios (1759–1824), and Neophytos Vamvas (1776–1855) — to examine their religious and intellectual framework, motivations, and sources in engaging with Western natural theology. This initiative emerged during the so-called 'neo-Hellenic Enlightenment' movement preceding the Greek War of Independence in 1821 and continued beyond the establishment of the independent Greek state in 1830. Situated amidst the theological aspirations of Protestant expansionism in Greek-speaking regions and the influential centers of Orthodox Christian culture — the Ecumenical Patriarchate of Constantinople and the monastic Orthodox 'philokalic' movement — these authors were driven by a shared commitment. Their goal was to introduce the latest scientific and philosophical ideas to the Greek-speaking population. Within this context, natural theology could allegedly serve a multifaceted role: combating atheism, demonstrating the religious utility of the new philosophical and scientific ideas, and enriching the Orthodox Christian theological discourse.

Sandy Sakorrafou • Postdoctoral Researcher, National Hellenic Research Foundation

Orthodox Christianity and the Study of Nature, ed. by Kostas Tampakis and Ronald Numbers, Science and Orthodox Christianity, 4 (Turnhout: Brepols, 2024), pp. 105–132
BREPOLS PUBLISHERS 10.1484/M.SOC-EB.5.141382

Introduction

The inquiry of natural theology refers to the existence and attributes of God that can be known through observation of nature and the use of human reason, unaided by divine revelation. Typically contrasted with revealed theology of the Scripture, natural theology should be neither confused with religion itself nor with doctrinal theology. It is a philosophical inquiry dating back to ancient Greek thought: pre-Socratic philosophers frequently identified the term 'archē' (first principle) of the cosmos with the divine who assures the order of the universe and Plato later was the one to articulate the first cosmological and teleological arguments in elaborate detail to prove the divine's existence and providence.[1]

Natural theology flourished within the Christian cultural context, especially in the West, from Augustine to the Middle Ages, when natural theology became prominent. It was reconsidered in the late seventeenth century and throughout the eighteenth century as the natural sciences advanced and rationalism, deism, empiricism, and the critical reading of the Scripture gradually emerged. Practised independently from any church authority, natural theology relied on human rational and perceptual faculties to answer religious questions on the existence of God. By contemplating God's creatures, it could provide elements of knowledge of the divine: ontological arguments, from a priori and necessary premises derived from reason alone, demonstrated that no being greater than God — a supremely perfect or necessary being — can be conceived.[2] It could also explain the first cause of this world, a being that is absolutely necessary and whose own explanation is contained within itself: cosmological arguments were motivated by the existence of the cosmos as a whole and attempted to explain the cause of its existence. Finally, natural theology was perceived as a project of recovering the divine design of the creatures: teleological arguments displayed the natural world as having some sort of purposive design leading to the conclusion that there was a potent and intelligent designer.[3]

In 1781 Immanuel Kant talked about the unattainability of rational proofs of a supreme being in his *Critique of Pure Reason*, only to argue over morality and how it requires the assumption that God exists in the *Critique of Practical Reason* seven years later. While it was clear that faith did not necessarily draw legitimation from philosophical argumentation, at the dawn of the nineteenth century, natural

1 G. Naddaf, 'Plato: The Creator of Natural Theology', *International Studies in Philosophy*, 36. 1 (2004), 103–27.

2 A. Chignell, D. Pereboom, 'Natural Theology and Natural Religion', in *The Stanford Encyclopedia of Philosophy* (Fall 2020 Edition), ed. by Ed. N. Zalta <https://plato.stanford.edu/archives/fall2020/entries/natural-theology/>; G. Oppy, 'Ontological Arguments', in *The Stanford Encyclopedia of Philosophy* (Winter 2021 Edition), ed. by Ed. N. Zalta <https://plato.stanford.edu/archives/win2021/entries/ontological-arguments/>. [accessed 3 November 2022].

3 S. Peterfreund, *Turning Points in Natural Theology from Bacon to Darwin; the Way for the Argument from Design* (New York: Palgrave Macmillan, 2012), pp. 100–07.

theology was still perceived as an acceptable means of rational foundation for faith in Christian God, especially in Britain; a point well-proved by the popularity of Anglican clergy apologist William Paley's book *Natural Theology or Evidence of the Existence and Attributes of the Deity, collected from the appearances of nature* (1802).

Western natural theology argumentation was never truly adopted in the Greek Orthodox Christian theological discourse;[4] yet, for a while, from the late eighteenth to the end of the nineteenth century, there was an interest. This was not an outdated practice on the fringe of the periphery: a lot of European authors were still fascinated[5] by natural theology; thus, some Greek Orthodox Christian authors also felt captivated and justified enough to establish elements of Western natural theology in the Greek-speaking space. Though there was no great originality in their endeavour, they were participating somehow in the contemporary natural theological debate.

Restricted to the three representative cases of Christodoulos Pamblekis (1733–1793), Veniamin Lesvios (1759–1824), and Neophytos Vamvas (1776–1855), the present preliminary study explores the religious and intellectual framework, the motivation, and the sources employed in their engagement with Western natural theology. Their initiative emerged during the so-called 'neo-Hellenic Enlightenment' movement before the Greek War of Independence in 1821 against Ottoman rule and expanded even after the foundation of the independent Greek state in 1830. Caught between the theological aspirations of Protestant expansionism in the Greek-speaking regions and the two centres of the Orthodox Christian cultural setting, the Ecumenical Patriarchate of Constantinople and the monastic Orthodox 'philokalic' movement, and inspired by a common intellectual inheritance, the ancient Greek philosophy, it is argued that those authors shared a general commitment not only to introduce the latest scientific and philosophical ideas in the Greek-speaking population but also to get familiarise with European natural theology. The latter had, for them, a manifold role to play: to battle against atheism, reveal the religious utility of the new philosophical and scientific ideas, and fertilise the Orthodox Christian theological discourse.

4 D. Skliris, 'Reactions of Modern Greek Theologians to Natural Theology', in *Natural Theology in the Eastern Orthodox Tradition*, ed. by D. Bradshaw and R. Swinburne (St Paul, Minnesota: IOTA Publications, 2021), pp. 125–48; T. Dumsday, 'Experiential Objections Against Natural Theology in Some Recent Orthodox Thinkers', in *Natural Theology in the Eastern Orthodox Tradition*, ed. by Bradshaw and Swinburne (St Paul, Minnesota: IOTA Publications, 2021), pp. 149–74.

5 J. H. Brooke, *Science and Religion: Some Historical Perspectives* (Cambridge: Cambridge University Press, 2014), pp. 269–76.

Talking About 'Neo-Hellenic Enlightenment' in Modern Greek Historiography: A Short Review

Before reviewing the three representative cases it is essential to address a particular historiographical concern: the interpretative schema. According to the dominant historiographical account introduced by the historian Konstantinos Dimaras in the 1950s, the struggle for Greek independence culminated in the late eighteenth century with a cultural rebirth, the so-called 'neo-Hellenic Enlightenment'. During that ideological, philological, linguistic, and philosophical movement, the popularization and dissemination of new scientific and philosophical ideas were said to contribute to the intellectual recovery of the Greeks from illiteracy and superstition; a so-called prerequisite for the consolidation of the Greek national consciousness that led to political independence from the Ottomans. Neo-Hellenic Enlightenment was considered by Dimaras to be part of the European Enlightenment and highly influenced by the French ideals of Voltaire, the *Encyclopédie*, and the movement of *Idéologie*.[6] According to this view, the work of so-called 'progressive' Greek-speaking scholars was restricted to an eclectic, mere 'transfer' of philosophical and scientific knowledge from the 'scientific centre' (Western Europe) to the 'periphery' (Greek-speaking space).[7] The concept of transfer emulated the term 'metakenosis' by which, the prolific scholar and iconic figure of this movement, Adamantios Korais (1748–1833) had described the project of transfusion of modern European thought into the Greek-speaking world.[8] Indeed, most of the Greek scholars studied abroad and then returned to teach in an underdeveloped educational system. There they promoted a popular version of the latest advances in Europe through the translation of various philosophical and scientific textbooks into the Greek language. Regarding the Orthodox Christian thought, represented by the Ecumenical Patriarchate, the local bishops or priests, and the monks of Mount Athos, the 'transfer' schema presented them as having confronted the latest advances in philosophy and science

6 K. Th. Dimaras, *Νεοελληνικός Διαφωτισμός* [Neo-Hellenic Enlightenment] (Athens: Ermis, 1993 [1977]); Tr. Sklavenitis, ed., *Επιστημονική συνάντηση στη μνήμη του Κ. Θ. Δημαρά* (Athens: Center of Neo-Hellenic Research / National Hellenic Research Foundation, 1994), pp. 1, 10–12; M. Patiniotis, 'Origins of the Historiography of Modern Greek Science', *Nuncius*, 23 (2008), 265–89.

7 For the notions of 'scientific centre' and 'periphery', see e.g., M. Patiniotis, ed., ''Εθνος, επιστήμη, ταυτότητες. Η ιστοριογραφία της επιστήμης στην περιφέρεια της Ευρώπης', *Neusis*, 15 (2006), 3–16; M. Patiniotis, K. Gavroglu, 'The Sciences in Europe: Transmitting Centers and the Appropriating Peripheries', in *The Globalization of Knowledge in History* (Max Planck Research Library for the History and Development of Knowledge), ed. by J. Renn (Berlin: Edition Open Access, 2012), pp. 321–43.

8 A. K. Paraderos, *Μετακένωσις: Ελλάδα-Ορθοδοξία-Διαφωτισμός κατά τον Κοράη και τον Οικονόμο* (Athens: Akritas, 2010, pp. 70–72.

with hostility; appraised as more conservative, they allegedly conceived reason always in battle with religion and fought over deism and atheism.[9]

In the late 1990s, the historiographical account of 'transfer' was accused of a rather deprived and unprocessed view of the diffusion of the introduced knowledge within the Greek-speaking space and a narrow perception of the science-religion relationship in terms of a conflict. Instead, the historiographical account of 'appropriation' was developed to describe the flexibility of the intellectual atmosphere within the Greek-speaking world during that period and the dynamic procedure of confrontation or integration underway between traditional and new ideas. According to this interpretive schema, the neo-Hellenic Enlightenment was not simply a matter of mere transfer of new scientific and philosophical ideas differentiated by their degree of acceptability, but an ongoing procedure of modifying and adapting the new material to local cultural, social, religious, and political ends. Under these circumstances, the new ideas were received not unconditionally. In particular, the discourse used by scholars for discussing scientific issues had a strong philosophical character that was not associated with the experimental discourse developed in Western Europe. It was perceived as being a conceptual extension of the ancient Greek philosophy they inherited. Such appropriation, however, served their primary educational purpose: to make the new knowledge familiar to their Greek-speaking audience.[10]

What was further challenged by the 'appropriation' schema was how strong and wide the impact of the ideals of the French Enlightenment on the neo-Hellenic Enlightenment was. Following the historiographical turn in Enlightenment studies earlier in the 1980s, the neo-Hellenic was now regarded as one

---

9 R. Demos, 'The Neo-Hellenic Enlightenment (1750–1821)', *Journal of the History of Ideas*, 19.4 (1958), 523–41; G. Robertson Cragg, *The Church and the Age of Reason, 1648–1789* (London: Penguin, 1967); R. Clogg, 'Anti-clericalism in Pre-Independence Greece 1750–1821', *Studies in Church History*, 13 (1976), 257–76; S. Runciman, *The Great Church in Captivity: a Study of the Patriarchate of Constantinople from the Eve of the Turkish Conquest to the Greek War of Independence* (Cambridge: Cambridge University Press, 1968).

10 K. Gavroglu, M. Patiniotis, 'Η ρήξη που δεν έγινε: Επιστήμες και αρχαία ελληνική σκέψη στον ελλαδικό χώρο κατά τον 18ο αιώνα', *Synchrona Themata*, 64 (1997), 88–92; D. Dialetis, K. Gavroglu, and M. Patiniotis, 'The Sciences in the Greek Speaking Regions during the 17th and 18th Centuries: The Process of Appropriation and the Dynamics of Reception and Resistance', *Archimedes*, 2 (1999), 41–71; K. Gavroglu, M. Patiniotis,'Patterns of Appropriation in the Greek Intellectual Life of the 18th Century: A Case Study on The Notion of Time', in *Revisiting the Foundations of Relativistic Physics: Festschrift in Honor of John Stachel*, ed. by Ab. Ashtekar, R. Cohen, D. Howard, J. Renn, S. Sarkar, and Ab. Shimony (Dordrecht: Kluwer Academic Publishers, 2003), pp. 569–91; M. Patiniotis, 'Scientific Travels of the Greek Scholars in the 18th Century', in *Travels of Learning. A Geography of Science in Europe*, ed. by A. Simões, A. Carneiro, and M. P. Diogo (Dordrecht: Kluwer Academic Publishers, 2003), pp. 49–77; M. Patiniotis, 'Neo-Hellenic Enlightenment: In Search of a European Identity', in *Relocating the History of Science; Essays in Honor of Kostas Gavroglu* (Boston Studies in the Philosophy and History of Science), ed. by Th. Arabatzis, J. Renn, and A. Simões (New York: Springer International Publishing, 2015), pp. 117–30.

among several enlightenments not uniquely determined by the French. The schema of 'appropriation' exposed the geographical, social, and political character of the cultural movement of the neo-Hellenic Enlightenment, away from a mere understanding of the European Enlightenment either as a unique intellectual and philosophical movement unbound to time or place, a system of socially disembodied ideas, or in terms of the stereotypes derived from the French Enlightenment. Finally, this interpretative schema did not conceive the science-religion relationship as fully shaped by a clear conflict between reason and faith, or between progressive and conservative agents.[11] It agreed with the latest historical accounts on religiosity in Enlightened Europe: contrary to the mainstream narration most scholars retained a belief in God, even if they were hostile to the Church. Deism was not such a robust movement and very often the adoption of the universalizing philosophical discourse on God was the outcome of the unresolved doctrinal controversies which emerged earlier in Europe.[12]

The Orthodox Christian Cultural Setting

The introduction of Western natural theology into the Greek Orthodox theological discourse at the time had to encounter the two centres of Orthodox culture and their response to modernity[13]: the Ecumenical Patriarchate and the monastic community of Mount Athos.

As the political authority of the Byzantine Emperor was dismantled in 1453, the Ecumenical Patriarchate, whose hierarchy was re-established by the Ottomans in 1454, played a diverse social role in the life of the Christian populations in the Balkans. Within the context of the institution of 'ethnarchy,' the Patriarchate controlled education, exercised juridical practice, and collected taxes for the Ottoman state. For the Patriarchate, there was no contradiction in the transformation

11 This is an account already supported by other historians, see e.g., Makrides, Vasilios N., 'Science and the Orthodox Church in 18th and Early 19th-Century Greece: Sociological Considerations', *Balkan Studies*, 29 (1988), 265–82; P. M. Kitromilides, ed., *Enlightenment and Religion in the Orthodox World* (Oxford: Voltaire Foundation, 2016); P. M. Kitromilides, 'The Enlightenment and Religion in Europe', in *The Oxford Handbook of Religion and Europe*, ed. by G. Davie and L. N. Leustean (Oxford: Oxford University Press, 2021), pp. 118–36.

12 Sh. Gilley, 'Christianity and enlightenment: An historical survey', *History of European Ideas*, 1.2 (1981), 103–21; D. Outram, *The Enlightenment* (Cambridge: Cambridge University Press, 1995); S. J. Barnett, *The Enlightenment and Religion. The Myths of Modernity* (Manchester: Manchester University Press, 2003); D. Edelstein, *The Enlightenment. A Genealogy* (Chicago: University of Chicago Press, 2010); J. M. Sandberg, 'Religion and the Enlightenment(s)', *History Compass*, 8.11 (2010), 1291–98; J. Waligore, 'Christian deism in eighteenth-century England', *International Journal of Philosophy and Theology*, 75.3 (2014), 205–22.

13 Ef. Nicolaidis, *Science and Eastern Orthodoxy. From the Greek Fathers to the Age of Globalization*, trans. by S. Emanuel (Baltimore: The John Hopkins University Press, 2011), pp. 151–68.

of the Church, a primarily eucharistic community, into a political institution;[14] the political appeasement of the occupier and the forced collaboration with the Sublime Court was part of the Church's tradition that had been applied before and was based on certain theological grounds.[15] What further shaped this coalition was a common political interest: the Ottomans had a suspicious attitude towards the European powers and there was a growing anti-Latin feeling in the Orthodox Church since the Great Schism in 1504. This alliance ensured that the Church would have been protected both from Catholic and Protestant religious propaganda.[16]

Accordingly, the Ecumenical Patriarchate's stance towards the European Enlightenment had been dictated by the same fear. Initially, the Patriarchate was not hostile to the new ideas and no policies were undertaken through any synodical resolution: it was longing for Western-trained scholars to reform ecclesiastical education and fight against Roman Catholic propaganda. But the general policy altered after the French Revolution in 1789, when it felt that the system of ethnarchy was imperilled, a system which, allegedly, assured the cooperation and spiritual unity among the Christian populations, according to the Gospel's message of ecumenism. The case of 'The Athonite Academy' is a good example of how the Patriarchate changed its attitude. In 1749 the Academy was established by the brotherhood of Vatopedi monastery at Mount Athos to satisfy the needs of the Church for higher education. Soon it was placed under the Patriarchate's aegis. Evgenios Voulgaris (1716–1806), who was invited to direct the Academy in 1753,[17] taught an Enlightenment-oriented curriculum (a mixture of classics with modern European philosophy). The program did not survive the attacks by 'traditional' monks who felt that this was an intrusion against Orthodoxy. Voulgaris quit in 1758, although the modernist endeavour at Athos continued for a little while. Yet, through the years more patriarchates became unsympathetic or reluctant towards Enlightenment ideas and condemned those scholars who introduced them fearing the political effects. In 1798 patriarch Gregory V decided

---

14 T. Papademetriou, *Render unto the Sultan. Power, Authority, and the Greek Orthodox Church in the early Ottoman Centuries* (Oxford: Oxford University Press, 2015). It has been argued that the Orthodox Church was already a component part of the state structure in the late Byzantine period.

15 Romans 13.1: 'Let every soul be subject to the governing authorities. For there is no authority except from God, and the authorities that exist are appointed by God' and Acts 5. 29: 'But Peter and the other apostles answered and said: "We ought to obey God rather than men"'.

16 P. M. Kitromilides, *Enlightenment and Revolution. The Making of Modern Greece* (Cambridge, Mass.: Harvard University Press, 2013), p. 25.

17 G. Speake, '"The Ark of Hellenism": Mount Athos and the Greeks under Turkish Rule', in *Mount Athos. Microcosm of the Christian East*, ed. by G. Speake and K. Ware (Oxford: Peter Lang, 2011), pp. 137–60, on pp. 149, 151.

to create a censorship committee to evaluate the content of books. In 1819 mathematics and natural sciences were criticized as harmful to church life.[18]

The monastic community of Mount Athos, which is under the spiritual guidance of the Ecumenical Patriarchate, played a significant role in the formulation of Orthodox theological discourse from the late eighteenth century onwards. In particular, the ideas of a traditional group of monks at the Lavra Monastery, the Kollyvades[19], would later become part of the standard Greek Orthodox theological literature. In 1754[20] the Kollyvades had opened a debate questioning the theological validity of the commemoration of dead people held on Sunday (when Christ resurrected from the dead), a recent practice at Athos, instead of Saturday (when the dead Christ rested in the tomb) as it used to be according to the Orthodox 'typicon' (rite).[21] Their attempt to withdraw the new rite failed; they were formally condemned and excommunicated by the Patriarchate. Only a few young members, such as Nikodemos the Hagiorite (1749–1809), were able to stay in Athos. Years later, the Kollyvades would focus on another issue, that of the frequent (continual) holy communion.[22] By then they were no longer an organized monastic movement, only a weak network of ecclesiastic scholars who recognized Makarios Notaras (1731–1805) as their leading figure.

It has been argued that what led to this controversy in liturgical practices was the Kollyvades' desire to present themselves as protectors of the Orthodox Christian dogma against Western Christianity and the European Enlightenment. They had also a strong desire to achieve their independence from the Patriarchate. Although they never overpowered monks at Athos, they gradually succeeded in restoring their status in the ecclesiastic hierarchy introducing the exemplar of the elder in Orthodox Christian life and gaining access to laypeople who recognized them as the masters of monastic and patristic tradition. Their ideas became highly popular and influential through the publication of several theological books. The

---

18 D. Moschos, 'The Churches of the East and the Enlightenment', in *The Oxford Handbook of Early Modern Theology, 1600–1800*, ed. by U. L. Lehner, R. A. Muller, and A. G. Roeber (Oxford: Oxford University Press, 2016), pp. 500–13, on p. 506.

19 The name derives from the word 'kollyva', the boiled wheat used at memorial services for the dead.

20 e.g., N. Skrettas, *Η περί Θείας Ευχαριστίας και των προνομίων της Κυριακής διδασκαλία των Κολλυβάδων* (unpublished doctoral thesis, Aristotle University of Thessaloniki, 2001) <http://invenio.lib.auth.gr/record/2320/files/gri-2003-113.pdf>; L. Bargeliotes, 'Neophytos the Kaphsokalyvites (1713–1784) and the Movement of Kollyvadon', *ΣΧΟΛΗ. Философское антиковедение и классическая традиция*, 11.2 (2008), 253–64. <https://www.ceeol.com/search/journal-detail?id=277>. [accessed 15 October 2022].

21 Ch. Tzogas, *Η περί μνημοσύνων έρις εν Αγίω Όρει κατά τον ΙΗ' αιώνα* (unpublished doctoral thesis, Aristotle University of Thessaloniki, 1969) <https://thesis.ekt.gr/thesisBookReader/id/23473#page/1/mode/2up>. [accessed 20 December 2022].

22 e.g., H. Patapios, (Archbishop) Chrysostomos, *Manna from Mount Athos. The Issue of Frequent Communion on the Holy Mountain in the Late Eighteenth and Early Nineteenth Centuries* (Byzantine and Neohellenic Studies 2) (Oxford: Peter Lang, 2006).

most significant work was *Φιλοκαλία των Ιερών Νηπτικών* [Philokalia of the Holy Neptic Fathers],[23] a collection of ascetic and mystic texts written between the fourth and fifteenth centuries, which was published in Venice by the two editors, Notaras and Nikodemos, in 1782. The content of the book was not explicitly doctrinal, but the Orthodox doctrine and ecclesiology were prerequisites for all the texts.[24]

According to Kallistos Ware, the Kollyvades promoted the 'patristic ressourcement', a return to the authentic sources of Orthodox Christian tradition and followed the teachings of the mystical hesychast tradition of St Symeon the New Theologian (eleventh century) and St Gregory Palamas (fourteenth century).[25] The book *Philokalia* aimed to spread the practice of 'unceasing prayer' to a wider audience. Prayer was perceived as an ascetic and mystical exercise that should not be restricted to monks alone; it was a means by which, in synergy with the grace of the holy spirit, the intellect was purified and brought to perfection. For Dimitrios Moschos,[26] the reintroduction of neptic theology was about exploring a specific theological question: how it was possible to know God and what could be said about God's existence. God's essence was incomprehensible and inexpressible, according to apophatic theology. Human reason could not conceive of the divine essence and so human language could not properly describe God. Yet, as cataphatic theology stated, it was possible to speak for God, not through the divine presence in nature, since this was considered restricted, but through God's incarnation, the events of the life of Christ.[27] For the hesychastic tradition, communication with God was possible through the repetition and the subsequent

---

23 A. Radovic, *Η φιλοκαλική αναγέννησι του XVIII και XIX αι. και οι πνευματικοί καρποί της* (Athens, Goulandris Foundation, 1984); (Archimandrite and Abbott of the Holy Monastery of St Gregorios at Agion Oros) Georgios, 'The Neptic and Hesychastic Character of Orthodox Athonite Monasticism'<http://www.greekorthodoxchurch.org/neptic_monasticism.html>. [accessed 3 November 2022].

24 St Nikodimos of the Holy Mountain and St Markarios of Corinth, eds, *The Philokalia: The Complete Text Volume 1*, [(ed. and transl. by G. E. H. Palmer, Ph. Sherrard, and K. Ware) (London: faber and faber, 1979), p. 15.

25 K. Ware, 'St Nikodimos and the Philokalia', in *The Philokalia: Exploring the Classic Text of Orthodox Spirituality*, ed. by B. Bingaman and B. Nassif (Oxford: Oxford University Press, 2012), pp. 9–35; A. Louth, 'The Influence of Philokalia in the Orthodox World', in *The Philokalia: Exploring the Classic Text of Orthodox Spirituality*, ed. by B. Bingaman and Bradley Nassif (Oxford: Oxford University Press, 2012), pp. 50–60, on p. 58; N. A. Chrissidis, 'The World of Eastern Orthodoxy', in *The Oxford Handbook of Early Modern European History, 1350–1750: Volume I: Peoples & Place*, ed. by Hamish Scott (Oxford: Oxford University Press, 2015), pp. 626–51.

26 D. Moschos, 'An alternative "enlightenment": The Philokalia', in *Enlightenment and Religion in the Orthodox World*, ed. by P. M. Kitromilides (Oxford: Voltaire Foundation, 2016), pp. 63–72; Moschos, 'The Churches of the East and the Enlightenment', pp. 500–13.

27 e.g., A. Louth, 'Apophatic and Cataphatic Theology', in *The Cambridge Companion to Christian Mysticism* (Cambridge Companions to Religion), ed. by A. Hollywood and P. Z. Beckman (Cambridge: Cambridge University Press, 2012), pp. 137–46.

internalization of the Jesus prayer ('Lord Jesus Christ, Son of God, have mercy on me'); the praying person could further participate in God through divine 'energies' (actions) which were distinct from the divine 'essence'.

The 'philokalic' revival is usually perceived as a conservative turn endorsing 'theosis' (deification) through prayer and asceticism over the use of human reason, but as Moschos has argued, the restoration of hesychastic theology did not result in the rejection of secular Western knowledge. Hence, for instance, Nikodemos adapted Roman Catholic works to the Orthodox tradition, such as *The Unseen Warfare*, his version of the *Combattimento Spirituale* (1589) by Lorenzo Scupoli.[28] Paul Collins also points out that Nikodemos was influenced by Roman Catholic spirituality, canon law, and theology. He even succumbed to Western pietistic moralism when he wrote his *Chrestoethia of Christians*. But Collins insists that the *Philokalia*, which was written for the 'benefit of all the Orthodox' as it is stated in the introduction,[29] was not a simple transmission from the Counter-Reformation; the purpose of the *Philokalia* collection was to reply to the new learning of the Enlightenment and the different theological perspectives of Roman Catholicism and Protestantism.[30] The point is further elaborated by Paschalis Kitromilides who argues that any loans from Western religious thought were minor and supplementary to the Orthodox spiritual revival. According to him, by appealing to the two fundamental practices, 'nepsis' (the virtues of sobriety, lucidity, and vigilance) and 'hesychia' (the inner stillness of the heart), the editors of the *Philokalia* summoned the Greek-speaking people to recall their distinctive theological heritage and stay strong against the advances of the missionaries and the secular philosophy of the European Enlightenment. For Kitromilides, the *Philokalia* was still a product of the European Enlightenment, but it represented an example of 'Counterenlightenment': the Kollyvades' mystical worldview was promising not only the intellectual renewal of the Greek-speaking people but most importantly the salvation of their soul.[31]

The return to the patristic tradition and the re-emergence of the hesychastic spirituality was the Kollyvades' answer to the modernity which allegedly infiltrated into the Greek-speaking world threatening the Greek Orthodox way of life. What probably pushed them further to a program that, according to them, secured the Orthodox Christian theological discourse were their concerns about the Catholic and Protestant missionary advances in the Greek-speaking regions.[32]

---

28 Moschos, 'The Churches of the East and the Enlightenment', pp. 507–09.

29 St Nikodemos of the Holy Mountain and St Markarios of Corinth, eds, *The Philokalia of the Neptic Fathers* (Venice, 1872), title page and p. 8.

30 P. M. Collins, *Partaking in Divine Nature. Deification and communion* (New York: T&T Clark International, 2010), pp. 88–91.

31 Kitromilides, *Enlightenment and Revolution*, pp. 299–300.

32 G. Podskalsky, *Η Ελληνική θεολογία επί Τουρκοκρατίας 1453–1821. Η Ορθοδοξία στη σφαίρα επιρροής των δυτικών δογμάτων μετά τη Μεταρρύθμιση* (trans. in Greek by G. Metallinos) (Athens: National Bank of Greece Culture Foundation, 2005).

The movement felt obliged to differentiate the Orthodox Christians from the rest of Christianity, persisting in the significance of traditional Orthodox Christian values as well as spiritual over secular knowledge. Their fear of Western Christianity and the European Enlightenment that encouraged religious tolerance and anti-clericalism had led them even to wrong decisions, like adopting a polemic against the Greek revolutionary movement and embracing the Ottoman empire as a God-appointed protector of Orthodoxy.[33]

The Intellectual Framework

The Revival of the Classical Philosophical Heritage and the example of Korydallism

The Orthodox Christian cultural setting described above coexisted with a general intellectual framework which seemed to vindicate the engagement with Western natural theology: the revival of the classical philosophical heritage and the reminiscence of Korydallism's example.

Greek Orthodox Christian scholars must have felt confident in introducing Western natural theology in the Greek-speaking space since the practice of natural theology was already part of their past local philosophical discourse: from its pre-Socratic origins, when the notion of divinity was found in the concept of 'physis'(nature), to Plato who identified the divine with 'nous' (intellect) immanent to the nature and then to Aristotle who argued for the prime mover of all other movers which was not itself moved. Plato was the one to launch the cosmological-teleological argument of the soul which was identified with nature; the soul was neutral and what established its goodness was the 'nous'. The latter was the one which sustained harmony and motion in the natural world. Thus, the cosmos was a product of rational design, a point demonstrated by mathematical rather than observational astronomy. Aristotle, in what has been interpreted as a deistic view of God, conceived the divine as separated from matter and identified as 'noeses' (intellect) that was neither the creator nor its preserver but the final cause of the world.[34] As previously mentioned, the revival of the native philosophical output was a requisite for the new philosophical and scientific ideas to be appropriated: by perceiving the European contemporary scientific discourse as a conceptual extension of the ancient philosophical heritage it was easier for the Greek-speaking audience to receive them. In the case of the Western natural

---

33 Speake, 'The Ark of Hellenism', pp. 148–59; S. D. Petmezas, 'On the Formation of an Ideological Faction in the Greek Orthodox Church in the Second Half of the Eighteenth Century: the Kollyvades', *Bulletin de correspondance hellénique moderne et contemporain*, 2 (2020) <https://journals.openedition.org/bchmc/416>. [accessed 19 November 2022].

34 e.g., Naddaf, 'Plato'; A. C. Thiselton, *Systematic Theology* (Cambridge: Wm. B. Eerdmans Publishing, 2015), p. 96.

theological discourse, the scholars must have felt justified in talking about it since natural theology was already a cultural component of the Greek-speaking population.

Primarily perceived within the system of metaphysical knowledge, the study of natural theology could also be defensible by the fact that Western metaphysics and Orthodox Christian theology had come together before. In 1624 Ecumenical Patriarch Cyril Lucaris (1572–1638) appointed Theophilos Korydalleas (1570–1645) as director to reorganize the Patriarchal Academy of Constantinople. For Lucaris, higher education for the clergy and the whole Orthodox population would prevent philosophical ventures and Roman Catholic propaganda from threatening the Orthodox dogma. Korydalleas, a student[35] of Cesare Cremonini, introduced neo-Aristotelianism and renewed the interest in natural philosophy and cosmology in Greek-speaking intellectual life. For the Orthodox Church, Korydalleas' philosophical program was useful because it blended elements of ancient Greek thought with Christian Orthodox tradition and conflicted with the Roman Catholic scholastic account. Nevertheless, it could never fully integrate with Orthodox theology: neo-Aristotelians were interested in a naturalistic interpretation of Aristotelian philosophy. The natural world was explained outside the context of a religious doctrine in terms of causality, though they never proceeded to the development of the experimental discourse.[36] As Manolis Patiniotis suggests, the introduction of neo-Aristotelianism in the Greek-speaking world was a marriage of compromise and a political movement made by Lucaris who sought to create a strong intellectual identity—that would differ from the Catholic one—for all the Orthodox Christian populations under the Ottoman rule.[37]

Until the late eighteenth century there were other Greek Orthodox Christian authors who wanted to renew their interest in natural philosophy, but as Patiniotis suggests they knew that the philosophical programme of Korydalleas has been surpassed by the latest advances in European philosophy; any attempt to bring together faith and reason had to come through the channels of modernity.[38] This is right, but at the same time, I argue that Korydallism could play a heuristic role: Korydalleas' inheritance, which lasted throughout the eighteenth century,

35 N. Agiotis, 'Greek Aristotelianism in the Seventeenth Century: Uncovering Cesare Cremonini in the Works of Theophilos Korydalleus', *Byzantine and Modern Greek Studies*, 43.1 (2019), 105–16.

36 M. Patiniotis, 'Ευσεβής επιστήμη: Ο Νεύτωνας στην Ορθόδοξη Ανατολή του 18ου αιώνα', in *The Proceedings of the Conference Orthodoxy, Nation, and Ideology* (Athens: Moraitis School, 2007), pp. 63–81, on. pp. 70–71; M. Patiniotis, 'Οι *Pestifarae Questiones* του Κυρίλλου Λουκάρεως και η ανάδυση του κοριδαλλικού προγράμματος' [The *Pestifarae Questiones* of Cyril Lucaris and the Rise of the Program of Korydaleus], in *Byzantium — Venice — Modern Hellenism: A Wandering in the World of Greek Scientific Thought* (conference proceedings), ed. by G. Vlahakis and Ef. Nicolaidis (Athens: National Hellenic Research Foundation, 2004), pp. 211–44; Kitromilides, *Enlightenment and Revolution*, pp. 26–28.

37 Patiniotis, 'Ευσεβής επιστήμη', pp. 63–81.

38 Patiniotis, 'Ευσεβής επιστήμη', pp. 74–75.

suggested that cooperation between metaphysics and Christian theology was possible (even if it was incomplete), making the metaphysical study of natural theology look promising as a tool of innovation of Orthodox theological discourse.

The Patristic Legacy

As it will be shown later, the Greek Orthodox Christian scholars did not really draw upon the Scripture[39] or the Greek Church Fathers' work for natural theological argumentation, though they were certainly familiar with these, as they were educated as priests or monks.

Greek Church Fathers moderately relied upon arguments of natural theology in their apologetic works, either to defend the Christian belief or for the benefit of a Christian audience. In any case, their cosmological and teleological arguments were shaped by empirical observation and the study of the laws of nature, making them conclude the existence of God as the cause of all things. For instance, Apologist Athenagoras (133–190) proceeded to an analogy between the Creation and an artificial object, a guitar, to prove that the world is an excellent work of the creator; Basil the Great (329–379) articulated arguments based on the observation of the harmony in the universe in his work *Hexaemeron*; Gregory of Nazianzus (329–390) demonstrated that it was more reasonable to explain the perfect harmony of the universe and its regular motions by an intelligent creator than by chance, and lastly, John of Damascus (675–749) pointed out the need for an efficient cause in his talk about the creation of the world.[40]

Hence, in patristic thought, natural theology was acceptable as long as it did not contradict a certain presupposition: that God remains incomprehensible and unapproachable to human contemplation. Theologizing was possible not by contemplating God's essence but by being in communion with God within the Church, detached from the natural world. For Apologists, the knowledge of God was a product of divine revelation to the human mind. Philosophy was still important as a tool to demarcate and illuminate the content of the Christian faith against the attack of pagan philosophers. When, later, the Cappadocian Fathers

---

39 There are a few references that look like natural theology within the Scripture. For instance, in the Old Testament, God is presented not only as the world's creator but also as its preserver: Psalm 19 ('The heavens declare the glory of God and the firmament proclaims his handiwork'), or Psalm 104 ('O Lord my God, You are very great: You are clothed with honour and majesty, Who cover Yourself with light as with a garment, Who stretch out the heavens like a curtain [...]' See, Barr, James, *Biblical Faith and Natural Theology: The Gifford Lectures for 1991: Delivered in the University of Edinburgh* (Oxford: Clarendon Press, 1995), pp. 81–101.

40 D. Bradshaw, 'Introduction', in *Natural Theology in the Eastern Orthodox Tradition*, ed. by D. Bradshaw and R. Swinburne (St Paul, Minn.: IOTA Publications, 2021), pp. 1–22; A. Fokin, 'Natural Theology in Patristic Thought: Arguments for the Existence of God', in *Natural Theology in the Eastern Orthodox Tradition*, ed. by D. Bradshaw and R. Swinburne (St Paul, Minn.: IOTA Publications, 2021), pp. 23–64.

determined the object of theology to be the doctrine of the Trinity and God's revelation as Three Persons within human history, it was important for theology to maintain its independence from philosophy. Hence, the latter had a specific mission to accomplish: to improve the human perception of the world and specify the articulation of the Christian doctrine. At the same time, a philosophical analysis of God's existence was still legitimate since human rationality was said to reflect 'the image of God'. For Apologist Justin (first century), all people possess a portion of the seminal Logos that enable them to grasp the divine Logos. Gregory of Nazianzus and earlier Irenaeus (first century) regarded the existence of God as a truth that can be comprehended by human reason which was implanted by God. But, for as Maximus the Confessor (580–622) underlined, humans cannot know God's essence, only God's existence; the distinction between the supernatural and natural revelation of God is acceptable, though the latter is fully grasped only in the light of the former.[41]

Within this framework, natural theology, an activity based on experience and human reason, could not determine what was known for God; due to the ontological difference between creation and God, one might infer God's existence by observing the created natural world but could never expose God's uncreated essence.[42] Nevertheless, since a philosophical analysis of God's existence could coexist with an experiential understanding of God's essence through faith and church life, as implied by Church Fathers, then natural theological argumentation was still a legitimate way to speak only for God's existence or attributes.

The hesychastic tradition was even more hesitant to introduce natural theological arguments in the Orthodox theological discourse, as the Kollyvades movement indicated. According to hesychasm, secular philosophy does not refer to absolute truth. It requires not only a perpetual struggle to enter the mystery of true being but also divine illumination to advance the obstacles set by the cognitive limitations of the human mind. Secular knowledge is not posed by the holy spirit and therefore, it is never complete and there is always the danger of logical fallacies. Thus, it is not quantitatively or qualitatively appropriate for reaching divine transcendence. In this sense, the hesychastic tradition does not perceive human knowledge as an intermediate path to the discovery of the inner self or God, but as an irrelevant element to the process of the deification of the human person. Accordingly, natural philosophy refers to the study of natural things alone, not of the inner truth of beings. It is not excluded, but it is considered a tool that reveals the secrets of natural laws and hence, leads to the knowledge of God who created the world. Does that leave an open door to natural theology? Given

---

41 J. Pelikan, *Christianity and Classical Culture: The Metamorphosis of Natural Theology in the Christian Encounter with Hellenism* (Gifford Lectures Series) (New Haven: Yale University Press, 1993).

42 S. Agouridis, *Ο Χριστιανισμός έναντι Ιουδαϊσμού και Ελληνισμού κατά το β' αιώνα μ. χ. - Αποστολικοί Πατέρες και Απολογητές* (Athens: Ellimika Grammata, 1997); Triantari-Mara, Sotiria, *Ιστορία της Φιλοσοφίας (πρώτος τόμος) από την αρχαιότητα στον Μεσαίωνα* (Thessaloniki: Stamoulis, 2005).

the superiority of the biblical worldview, according to hesychastic discourse, the natural world cannot be fully comprehended and appreciated outside the context of divine revelation. Still, a window is left slightly open for natural theological practice, as a useful but restricted activity, since natural philosophy and theology are not regarded as contradictory or mutually exclusive. In Orthodox tradition, the human person exercises his/her intellectual faculties as a psychosomatic unity. The latter reflects the basic dogmatic assumption: there is no sharp distinction between the human and divine plane and for that reason, the human person may come to deification.[43] Gregory Palamas himself, the acknowledged intellectual leader of the movement of hesychasm, was very careful to not explicitly develop natural theological arguments: one may infer God's Goodness by observing how things are good or how opposing elements are balanced in nature; that was another way to demonstrate God's existence, though not a very trustworthy one. Such knowledge is said to be incomplete and insufficient. It is only through prayer and faith that one might be led to illumination and then to spiritual knowledge; only those purified in their heart could achieve a kind of demonstration of God's existence.[44]

The Motivation and the Sources

Perceived as a cultural element inherited by local philosophical discourse and patristic theology, I argue that the Greek Orthodox Christian scholars' plan to introduce Western natural theology was reminiscent of Korydalleas' attempt to bring Western metaphysics and Orthodox Christian theology together. Their objective, the introduction of Western natural theology, was part of the ongoing secularization aiming for the intellectual revival of the Greek-speaking population. Contrary to the way the 'philocalic' movement and the Orthodox Church perceived this renewal, in terms of the salvation of the soul, those authors were anxious to exhibit the importance of secular knowledge. Contrary to the usual Western practice[45] in which scientific knowledge was used to legitimate natural theology, in their case natural theology could be used to establish the status and legitimacy of the newly introduced scientific and philosophical knowledge; the new ideas were supposed to have some religious utility.

Their effort never really led to a concrete atheistic movement or a straightforward anticlericalism that put the Orthodox clergy into disdain; the introduction

---

43 A. Antonopoulos, Ch. Térézis, 'Aspects on the relation between faith and knowledge according to Gregory Palamas', *Byzantinische Zeitschrift*, 101.1 (2008), 1–20; D. Costache, 'Theology and Natural Sciences in St Gregory Palamas', in *God, Freedom, and Nature*, ed. by R. S. Laura, R. A. Buchanan, and Am. K. Chapman (Sydney: Body and Soul Dynamics, 2012), pp. 132–38.

44 D. Bradshaw, 'Natural Theology in St Gregory Palamas', in *Natural Theology in the Eastern Orthodox Tradition*, ed. by D. Bradshaw and R. Swinburne, pp. 51–63.

45 Brooke, *Science and Religion*, p. 32.

of Western natural theology was confined to an unfulfilled desire to enhance (even to redefine) the content of the Orthodox theological discourse with elements of philosophical theology. As it will be displayed later, primarily perceived within the system of metaphysical knowledge, years after Kant investigated its limits and scope, they were probably convinced that natural theology could provide them with the proper premises for the Orthodox theological discourse to continue, this time within the context of rational theology. They were interested in showing how faith and reason are allied through the practice of natural theology and how the latter was a reliable apologetic tool against atheism.

There were several sources from which Greek Orthodox Christian scholars abstracted those elements of natural theology to include them in their books (which were usually creative compilations of a variety of philosophical works). Their choices were relevant to the period they were active. Just before the beginning of the Greek War of Independence in 1821, when anticlericalism was still intensive and the Enlightenment enterprise was flourishing in Europe, the deistic views found in the French *Encyclopédie*, the German Christian Wolff's and Leibniz's, or English Samuel Clarke's metaphysical accounts of natural theology were very popular. Years after the foundation of the independent Greek state in 1830, different needs emerged and new conditions were imposed: the intellectual movement of the neo-Hellenic Enlightenment fell into oblivion, the University of Athens, and the faculty of Theology were established in 1837, and the autocephalous of the Orthodox Church of Greece was declared by the Ecumenical Patriarchate in 1850. The Wolffian account of natural theology was still feasible and the most influential British book on natural theology, Paley's *Natural Theology*, was translated; this was a work that had already been used as a rhetorical tool to persuade the religious audience or soothe the fear of the rising scientific materialism, but it also underlined the significance of the latest technological and scientific advances which could enforce the Christian worldview.[46]

The Three Representative Cases

Christodoulos Efstathiou from Acarnania or Pamblekis (1733–1793)

Pamblekis lived and acted before the Greek War of Independence. After becoming a monk Pamblekis studied at Athonite Academy when Voulgaris was the director. During the Kollyvades' debate, he did not interfere in favour of one side or the

46 N. O' Flaherty, *Utilitarianism in the Age of Enlightenment: The Moral and Political Thought* of William Paley (Cambridge: Cambridge University Press, 2019), p. 132.

other. He spent his life writing, translating, and working as a private tutor for wealthy families of the Greek communities at Liebig and Vienna.[47]

In his first book *Αληθής πολιτική* [True Politics] (Venice, 1781), a translation of the French book *La veritable politique des personnes de qualité* (Paris, 1692) written probably by Nicolas Remond des Cours, Pamblekis presented himself as a pious Orthodox Christian. Years later, in 1793, in his unfinished work *Απάντησις*[48] *ανωνύμου προς τους άφρονας αυτού κατηγόρους επ' ονομασθείσα περί θεοκρατίας* [*The Anonymous Reply to the ignorant accusers of the so-called theocracy*] published by his students in Leibig, he was presented as a radical[49] deist (a follower of Spinoza): Pamblekis' ferocious criticism resulted in his post-mortem ex-communication by the Ecumenical Patriarchate.[50]

Between these two works his portrayal, a combination of theistic and deistic views, emerges within the pages of his book *Περί Φιλοσόφου, Φιλοσοφίας, Μεταφυσικών, Πνευματικών και Θείων Αρχών* [On Philosopher, Philosophy, Natural and Supernatural, Intellectual, and Divine Principles], published in Vienna in 1786. The book was condemned by his former teacher Voulgaris who in his old age was especially concerned about the social and political consequences of the new ideas introduced in the Greek-speaking space, especially after the French revolution.

It is within this context that Pamblekis discusses natural theology in the first chapter 'On Philosophy' of his book *On Philosopher*, where he introduces Wolff's account of natural theology found in *Theologia Naturalis Pertractata* (Frankfurt, 1736–1737). According to Wolff, natural theology is a part of theoretical philosophy (the other two being cosmology and psychology). Theoretical philosophy is conceived as a discipline that examines the actuality (which was learned from the experience) but also the possibility of things (all actuals were possible). It is related to the human capacity for knowledge and its purpose is to define these abstract possibilities of things. Natural theology can establish God's existence, attributes, and operations from the contemplation of natural things, through certain cosmological, moral, and ontological arguments. Following Wolff's description, Pamblekis names natural theology as the 'meta-physikotheological' (μεταφυσικοθεολογικόν) part of philosophy (metaphysics) and defines it as 'the science of all that is possible through God, which exist in him and which are

47 For a novel narration of Pamblekis' life, see El. Priovolou, *Βαθύ το σκοτάδι πριν την αυγή* (Athens: Kastaniotis, 2024).

48 This book was written as a reply to an abusive and extreme text (Ακολουθία'[Sequence]), a diatribe against Pamblekis and primarily against the European Enlightenment.

49 e.g., P. M. Kitromilides, *The Enlightenment as Social Criticism: Iosipos Moisiodax and Greek Culture in the Eighteenth Century* (Princeton: Princeton University Press, 1992), p. 101; Kitromilides, 'Spinozist Ideas in the Greek Enlightenment', *Balcanica*, L (2019), 105–11; N. Siniossoglou, *Αλλόκοτος Ελληνισμός* (Athens: Kichli 2016), pp. 161–90.

50 For a thoroughly account of Pamblekis' philosophical ideas, see G. Barbarousis, *Οι Φιλοσοφικές απόψεις του Χριστόδουλου Παμπλέκη στο πλαίσιο του Νεοελληνικού Διαφωτισμού* (PhD thesis, National and Kapodistrian University of Athens, 2011).

understood to be able to occur through him'[51] or 'that part of philosophy which refers to God and the beginning of those [creations] that they are known by him, is called natural philosophy and it is named as the teaching on God'.[52] Pamblekis must have felt comfortable with Wolff's philosophical system, a revised version of Aristotelian/Leibnizian metaphysics;[53] Wolff's attempt to modernize the Aristotelian tradition in metaphysics (for example he agrees with what Aristotle was saying on the necessity of simple[54] beings but in the light of the Leibnizian account) could be easily embedded within the local philosophical discourse, where the new philosophical ideas were regarded in continuity with ancient Greek thought, but at the same time, they were introducing something new.

In the same chapter, Pamblekis denies atheism and accepts that there is supernatural involvement in the world, promoting a rather theistic view of God: he refers to the attributes of God and insists that God's existence is supported by the undoubted proofs that physics delivers.[55] Physics or philosophy of nature is defined as a science that studies the features of the physical bodies, revealing the laws of nature and fighting superstition.

As it has been argued, the content of the book is heavily based on the translation of various entries in the *Encyclopédie* for the very first time in the Greek language.[56] In the sub-chapter 'On Theology'[57] Pamblekis adapts the deistic entry 'Théologie' (1745) where one reads the following definition of natural theology: 'Natural theology is the knowledge that we have of God and his attributes, by reason and nature, by considering the works that can only come from his hands'.[58] He also adopts the scholastic distinction made by the author into two kinds of

51 Pamblekis, *On Philosopher*, p. 26.

52 Pamblekis, *On Philosopher*, pp. 27–28.

53 e.g., Ch. Wolff, *Preliminary Discourse on Philosophy in General*, (transl. by R. J. Blackwell) (Indianapolis, 1963); Wolf, *Theologia Naturalis Pertractata*, 2 vols (Frankfurt, 1736–7); Ch. A. Charles A., 'The Existence of God, Natural Theology and Christian Wolf', *International Journal for Philosophy of Religion*, 4 (1973), 105–18; J. B. Schneewind, ed., *Moral Philosophy from Montaigne to Kant* (Cambridge: Cambridge University Press, 2003), pp. 331–52; S. Gaukroger, *The Natural and the Human: Science and the Shaping of Modernity, 1739–1841* (Oxford: Oxford University Press, 2016).

54 G. Robert, 'Simples, Representational Activity, and the Communication among Substances: Leibniz and Wolff on pre-established Harmony', in *From Leibniz to Kant* (Logical Analysis and History of Philosophy, 21) ed. by. K. L. Dunlop and S. Levey (Paderborn: Brill/mentis, 2018), pp. 92–128.

55 Pamblekis, *On Philosopher*, pp. 34–36.

56 P. Ch. Noutsos, 'Χριστόδουλος ο εξ Ακαρνανίας και *Encyclopédie*', *The Eranistis*, 17 (1981), 13–24.

57 Pamblekis, *On Philosopher*, pp. 51–61.

58 Unknown, 'Theologie', in *Encyclopédie, ou dictionnaire raisonné des sciences, des arts et des métiers, etc.*, vol. 16, ed. by Denis Diderot and Jean le Rond d'Alembert (Paris, 1754), pp. 249–251 (University of Chicago: ARTFL Encyclopédie Project (Autumn 2022 Edition (ed. by R. Morrissey and G. Roe) <http://encyclopedie.uchicago.edu/.Encyclopédie, p. 249>. [accessed 18 October 2022].

theology: natural theology and supernatural theology. These are argued to be two equal means through which humans know about God: one is knowing through natural evidence and the other is believing in the truths of faith. Later, in the final chapter 'On God, metaphysically and theologically'[59] Pamblekis delivers a partial translation of the entry 'Dieu'[60] (1754). The entry contains Samuel Clarke's (1675–1729) metaphysical arguments found in his book *A Demonstration of the Being and Attributes of God* (London, 1709), where he demonstrated the certainty of the existence of God and deduced the essential attributes of God's nature. Among Clarke's metaphysical arguments, there is an 'argument from design'[61] where the beauty, variety, order, and symmetry in the universe, and its purpose in it, prove the intelligence of a prime being. Pamblekis at the end of this translation underlines what Clarkes has noticed, that natural and experimental philosophy can give great help to religion, and this seems possible through the discourse of natural theology.[62]

Natural theological arguments complied with Pamblekis' deistic understanding of human reason: for him, divine testimony is a sufficient but not a necessary condition to demonstrate the teachings of the Scripture, only reason and natural theology may prove Scripture's truths.[63] Natural theology, a philosophical activity, was important, for him, not only as a justifiable source of knowledge of God and nature but also as a supportive device in his rational reappraisal of the Orthodox dogma, away from any hermeneutics made by the clergy. Pamblekis is resentful of the clerical interpretation of Christianity; he thinks that clerics and political leaders have been personally profiting by teaching a false kind of religion.[64]

Gradually Pamblekis renounced all the central Christian doctrines in his later book *On Theocracy*. But already in his book *On Philosopher* he was undermining the relationship between philosophy and Christian thought by endorsing the existence of a personal God, conceiving the divine in terms of philosophy, and speaking of a rational secular kind of religion. Pamblekis was an exceptional case among a few; within the European framework during the eighteenth century

59 Pamblekis, *On Philosopher*, pp. 367–412.

60 J. H. S. Formey, 'Dieu', in *Encyclopédie ou Dictionnaire raisonné des sciences, des arts et des métiers*, vol. 4, ed. by D. Diderot and J. le Rond d'Alembert (Paris, 1754), pp. 976–83 [trans. by S. Emanuel (*The Encyclopedia of Diderot & d'Alembert Collaborative Translation Project*) (Ann Arbor: Michigan Publishing, University of Michigan Library, 2013) <http://hdl.handle.net/2027/spo.did2222.0003.041>. [accessed 12 October 2022].

61 For the 'argument from the design', e.g., Peterfreund, *Turning Points in Natural Theology*; R. Collins, 'The teleological argument: an exploration of the fine tuning of the universe', in *The Blackwell Companion to Natural Theology*, ed. by W. L. Craig and J. P. Moreland, (Chichester, West Sussex: Wiley-Blackwell, 2009), pp. 202–81; K. E. Himma, *Design Arguments for the Existence of God* <https://iep.utm.edu/design/>. [accessed 2 December 2022].

62 Pamblekis, *On Philosopher*, pp. 390–99.

63 Pamblekis, *On Philosopher*, pp. 58–60.

64 Pamblekis, *On Philosopher*, pp. 374–75.

which was almost exclusively Christian in origin and orientation, many radicals and deists continued to subscribe to some sort of Christianity, or religion exemplified by elements found in the Bible.[65]

Veniamin of Lesvos or Lesvios (1759/62–1824)

Lesvios was a monk on Mount Athos. He studied at the University of Pisa, at École Polytechnique in Paris, and then in England. He worked as a director in several academies. His aversion to biblical cosmology, reinforced by personal antagonisms, led to charges of atheism against him. He was condemned in absentia by the Holy Synod of the Ecumenical Patriarchate in 1803. Thanks to the intervention of influential friends, the patriarchal condemnation was never carried out. He became later a member of *Φιλική Εταιρεία* (Society of the Friends), the secret organization for the liberation of Greeks from the Ottomans.

In his book *Στοιχεία της Μεταφυσικής* [*Elements of Metaphysics*][66] (Vienna, 1820) Lesvios was among the first to emphasize the primacy of epistemology over ontology in Greek-speaking space. Having abandoned Aristotelian metaphysics, he introduces John Locke's empirical account: metaphysics is now defined as that kind of science that studies the principles and the causes of sciences and arts.[67] For Locke, humans communicate with nature through the senses, and it is through their cognitive abilities such as reason, fantasy, memory, and consciousness that they can create a relationship with nature and a reference to it. Thus, it is their cognitive powers that help humans to comprehend the principles and the causes of natural phenomena, in other words, to make science.[68] Lesvios makes this point clear when he argues that cosmological and teleological arguments on the existence of God are interrelated, 'since the hypothesis of final cause is a hypothesis of logical cause'; in other words, experience does not suffice for any physico-theological proof to stand, it has to be justified as well as by human reason.[69]

Lesvios defines natural theology as 'metaphysics of metaphysics',[70] recalling Wolff's definition of natural theology as 'the science of all that is possible through God': natural theology provides a bottom-up justification for metaphysics which

---

65 J. Waligore, 'Christian deism in eighteenth-century England', *International Journal of Philosophy and Theology*, 75.3 (2014), 205–22.

66 Veniamin of Lesvos or Lesvios, *Elements of Metaphysics* (Vienna, 1820). For more on Lesvios' thought see R. D. Argyropoulou, 'Βενιαμίν Λέσβιος και η ευρωπαϊκή σκέψη του 19ου αιώνα' (Athens: Center of Neohellenic Research/ National Hellenic Research Foundation, 2003).

67 P. M. Kitromilides, 'John Locke and the Greek Intellectual Tradition: An Episode in Locke' s Reception in South-East Europe', in *Locke's Philosophy. Content and Context*, ed. by G. A. J. Rogers (Oxford: Clarendon Press, 1994), pp. 217–35.

68 Lesvios, *Elements of Metaphysics*, p. ια΄ [11].

69 Lesvios, *Elements of Metaphysics*, p. 440.

70 Lesvios, *Elements of Metaphysics*, pp. θ΄ [9].

is concerned with all actual beings of a created (possible) reality. In that case, Lesvios regards natural theology as being the last part of his epistemology: natural theology can lead to the knowledge of God, who is the beginning and the cause of all things visible and invisible, and hence to the knowledge of the nature of the human soul. He also accepts the distinction between natural and supernatural theology found in the deistic entry of 'Théologie', as Pambekis did before. Lesvios even argues that natural theology could replace the revelation which he does not consider to be a justifiable source of knowledge of God. What, however, justifies the knowledge obtained by natural theology, are the senses and reason; both validate humans to think that the creator exists as a rational and wise being. For him, the senses have been placed by nature as guards of human beings; it is from the senses that the existence of external things is announced to the soul and from these perceptions, ideas are created, and therefore, from this world, one may infer the existence of the creator of all beings.[71]

Lesvios elaborates his views on the following cosmological and teleological arguments. In the chapter 'On Soul', he argues that humans do not see God or God's properties, but they do see the wondrous sphere of the earth. Hence, it is God's providence and God's existence that are attested by the accommodation and order of the visible world. Recalling Leibniz's[72] 'principle of sufficient reason' his argument goes like this: the Earth was created by God, and so it has a beginning. Given God's rationality, we may infer that the Earth was created for a purpose, that being the glory of God. Since the Earth lacks reason and sensibility, we may infer further that the Earth was not created for itself but for someone else. Hence, for Lesvios, the Earth was created for humans who share the faculty of reason (being the image of God); they have consciousness of their existence, as materialists would also accept, and therefore, they know about the creator of nature and the existence of beings.[73]

Later in the chapter 'On the Existence of God', Lesvios claims that it is from the observation of the heavens, of the motion of the planets around the sun, that we may infer that the cause of all these is God. It is also the observation of the sequence of the seasons and the significance of the sun for the sustenance of life that makes us confess that God is the lawmaker of nature. For Lesvios, works of nature have a final cause for their existence, like the human artifacts which are attributed to the existence of an artisan. Therefore, the idea of the final cause is part of the very idea of God.[74]

71 Lesvios, *Elements of Metaphysics*, pp. 440.

72 e.g., A. R. Pruss, 'The Leibnizian cosmological argument', in *The Blackwell Companion to Natural Theology*, ed. by William Lane Craig and James Porter Moreland (Chichester: Wiley-Blackwell, 2009), pp. 24–100; M. V. Griffin, *Leibniz, God and Necessity* (Cambridge: Cambridge University Press, 2013); L. Strickland, ed., *Leibniz on God and Religion. A Reader* (London: Bloomsbury, 2016).

73 Lesvios, *Elements of Metaphysics*, pp. 429–31.

74 Lesvios, *Elements of Metaphysics*, pp. 435–36.

An 'argument from design' closes the chapter: it is from the observation of the human body and the way the heart and the nerves work that we may infer that all have been created by a rational creator. Lesvios gives a fine description of the organ of the human eye (in analogy to the human artifacts of spectacles and telescopes), reminding the description found in Paley's book. For Lesvios, the wondrous structure of the human eye can only be explained by the existence of a wise artisan; the hand of the creator is revealed in the complexity of the human body's design.[75]

Once accused of being an atheist by the Orthodox Church, Lesvios uses natural theology to fight against atheism, especially materialistic views. For him, religion is a universal phenomenon, found in every nation, though it is not an implanted idea. Recalling Clarke's arguments on the existence of God, Lesvios argues about the divine qualities (infinity, eternity, independency, necessity, simplicity, immutability, goodness, etc.).[76] He regards atheistic accounts as insanity since from the observation and the study of the order of the universe the existence of God is evident; by watching the infinite diversity of beings, the wondrous connection and arrangement in the world anyone can see the wisdom and the power to invent the light, the air, and the force of gravity. Therefore, for Lesvios, the endurance of the atheistic views is due primarily to obstinacy and inadequate knowledge of natural phenomena. For example, he notes that an atheist might consider motion as substantial while studying the motion of planets. But this account is false since the velocity of planets differs — the motion of each planet is pertinent to its centre of rotation and its distance from the sun — therefore, motion is an accidental feature of matter. Being closer to Locke's and Newton's empirical accounts than to French philosophes' materialistic positions, Lesvios rejects the concepts of a mechanism of a 'blind automaton' or Descartes's notion of the creation of the world because of a vortex of matter. Even when he does translate the deistic entry of "Dieu" from the *Encyclopédie*, flirting with materialism, he insists that the matter is indifferent to motion or rest unless acted upon by an external force, that is God. He seems even to combine Clarke's Newtonianism with Aristotle's notion of the prime mover: not only does he consider the laws of motion of matter as effects of God continually acting upon matter, but he also thinks that the distinction between the moving part and the mover explains the beginning of a world that does not possess a self-motion power.[77]

In the end, according to Lesvios, atheistic views are due to the false conception that the idea of God has been forcibly and deceptively placed on people by priests and state leaders. For him, this is a moral argument that ignores the fact that the idea of God came first before the priesthood, or that natural religion existed before any organized political system. But then, for him, a moral argument is not relevant to the question of God's existence; only arguments

75 Lesvios, *Elements of Metaphysics*, pp. 437–40.
76 Lesvios, *Elements of Metaphysics*, pp. 443–53.
77 Lesvios, *Elements of Metaphysics*, pp. 441–44.

that provide a sufficient or legitimate explanation of how nature works may be. Lesvios concludes that neither ignorance nor fear leads someone to believe in the existence of God, only philosophical (therefore, natural theology) and scientific knowledge is a prerequisite for interpreting the mysteries of Creation.[78]

Neophytos Vamvas (1776–1855)

Vamvas, a well-known scholar and an influential figure of modern Greek education, was ordained a deacon at an early age and then he went to study in France where he met Korais. Later he taught in several important schools. He became director of the School of Philosophy and then dean at the newly founded University of Athens in 1837. He was ordinated as an archimandrite.

Just before the beginning of the Greek War of Independence, Greece was swarmed by Protestant missionaries who were after proselytism.[79] Vamvas was the principal of the School of Chios in 1820 when two American Protestant missionaries gave him Paley's *Natural Theology* as a gift for teaching them Greek.[80] Vamvas also started to learn English and soon he was helping them with the translation of their religious leaflets.[81] Years later he was able to translate Dugald Stewart's *Outlines of Moral Philosophy* during his year of teaching philosophy (1829–30) at the Ionian Academy at Corfu.[82]

In 1893, thirty-eight years after Vamvas's death, a manuscript, which includes the translation of the first eight chapters of Paley's book, was published under the title *Φυσική Θεολογία και Χριστιανική Ηθική* [Natural Theology and Christian Ethics] in Alexandria where the Greek community was the largest around the Patriarchate. The publication was sponsored by the eminent Greek-Alexandrian Choremis family. The book was edited by Konstantinos Papazis, Doctor of Philosophy, under the supervision of Nektarios Kephalas, Metropolitan of Pentapolis of Libya at the time, who had bought the manuscript. It is unknown when Vamvas

78 Lesvios, *Elements of Metaphysics*, pp. 456–68.

79 Paraderos, *Metakenosis*, pp. 223–61.

80 Th. Saloutos, 'American Missionaries in Greece: 1820–1869', *Church History* 24.2 (1955): 152–74; Giannuli, Dimitra, '"Errand of Mercy": American Women Missionaries and Philanthropists in the Near East, 1820–1930', *Balkan Studies*, 39.2 (1998), 223–62; S. Yemenedzi-Malathouni, 'American Philhellenes, Protestant Missionaries and the "Orphans" of the 1821 Hellenic War of Independence: The Case of Christodoulos Evangelides', *European Journal of American studies*, 17.1 (2022), 1–15. By the year 1843, most Protestant missionaries had left Greece because of the growing opposition, see Fitzgerald, Thomas E., *The Ecumenical Movement: An Introductory History* (Westport, Conn.: Praeger, 2004), pp. 71–72.

81 R. Clogg, 'O Parsons και ο Fisk στη Χίο το 1820' [Parsons and Fisk at Chios in 1820], *The Eranistis* 5 (1967), 177–93.

82 A. Glykofridy-Leontsini, 'The Impact of Scottish Philosophy on Modern Greek Philosophy via French Eclecticism. A Study of Intercultural Impacts and Exchanges in the History of Philosophy', in *Comparative and Intercultural Philosophy. Proceedings of the IIP Conference Seoul 2008* (Philosophy in International Context, Vol. 5), ed. by H. Lenk (Berlin 2009), pp. 93–118.

decided to translate Paley's book. Papazis assumes that since the work was left unfinished, Vamvas had to edit the final copy, but then he died in 1855. So it is only assumed that he translated it around the late 1840s or the early 1850s. Vamvas's manuscript on natural theology was not the first to be printed in the Greek-speaking space. The translation of Joseph Buttler's book *The Analogy of Religion Natural and Revealed to the Constitution and Course of Nature* by an unknown author, was already published in Athens in 1844.[83]

Vamvas' engagement with natural theology might start with Paley's book translation but he was probably already familiar with elements of it found in the Scripture. Assigned by the British and Foreign Biblical Society[84] to translate the Bible into the vernacular Greek language, he had the chance to thoroughly study the Bible. His translation, published during 1833–1844, was accused of introducing Protestantism in the Church of Greece. It was disapproved in 1834, and its use and its distribution were forbidden. Two years later the Ecumenical Patriarchate issued an encyclical condemning any attempt to translate the Bible into the vernacular Greek language.[85] Vamvas, being also a person in the clergy, was also possibly familiar with the fragments of natural theological argumentation found in Patristic thought. Finally, having translated Stewart's *Outlines of Moral Philosophy*, Vamvas was already acquainted with philosophical arguments on the existence and the moral attributes of God.

In his introduction to the book *Natural Theology and Christian Ethics*, Vamvas seems also to recall a Wolffian definition and describes theology as the 'science of God and the divine things, or science of beings in God and those are, or possible be through God'.[86] He also accepts the scholastic distinction of supernatural and natural theology as independent ways of how God's attributes are revealed, as Pamblekis and Lesvios did before. For Vamvas, 'supernatural', the very kind of apocalyptic theology, refers to the essence and the attributes of God, and 'natural' theology which is attained through human reason refers to God as the first cause of things in the visible world, an account that, as he points out, recalls Aristotle's notion of first cause.[87]

Vamvas feels confident in the capacity of the human cognitive faculty of reason to philosophically justify the divine existence through God's operations in nature; as he indicates, several wise minds from Plato to Newton have been persuaded about God's existence. He also thinks that it is absurd and irrational to question

83 Patiniotis, 'Ευσεβής επιστήμη', p. 80.

84 British and Foreign Bible Society members came to Greece as early as 1827. See Ch. A. Frazee, *The Orthodox Church and Independent Greece 1821–1852* (Cambridge University Press, 1969), pp. 136–39.

85 See J. Krasovec, ed., *The Interpretation of the Bible: The International Symposium in Slovenia* (Sheffield: Sheffield Academic Press, 1998), pp. 313–14; William A. Ross, W. Edward Glenny, *T&T Clark Handbook of Septuagint Research* (Bloomsbury Publishing, 2021).

86 Vamvas, *Natural Theology*, p. 1.

87 Vamvas, *Natural Theology*, pp. 1, 95, 97–98.

and fight the revealed knowledge of the Scripture since all beings exist through divine free will and it is God's knowledge that directs; God's will that decides; and God's omnipotence that executes. For Vamvas, human reason is unable to discern the relation of things within the world and as one may see by reading Romans 11.33, the divine decisions remain unexplored.[88]

Yet, for him, human reason is vital for the fortification of Christian belief; he refers to Romans 1.20 to emphasise his point: 'For since the creation of the world His invisible attributes are clearly seen, being understood by the things that are made, even His eternal power and [a]Godhead,[...]'.[89] Therefore, he encourages theologians to borrow the principles of various disciplines, such as those of natural sciences, and employ them in the explanation of theological issues.[90]

As Vamvas notices, the subject matter of natural theology refers to God's 'ποιήματα' (creatures)[91], which are observable in the visible world. Nature is 'the mirror through which deity's existence and attributes reflect'.[92] The translation of Paley's book is useful for Vamvas since it explains how exactly this mirror works. In it, there is an early formulation of the 'argument from design'.[93] Paley is talking about the similarities of the mechanical properties between human machinery and mechanisms in nature. In his explanation, for example, of the movement of the neck, this is not done simply by analogy with the telescope mechanism, but by considering the neck itself as a mechanism. In this way, Paley is trying to preserve the superiority in the works of the creator that the argument of a simple analogy could not serve. For him, it is important to break away from the old-fashioned celestial mechanics and Newtonian natural theology. By emphasizing the mechanical principles of human anatomy and nature, Paley is assured that his natural theologian argument has been updated after the complex mechanical technology of the industrial revolution of his time.[94]

As Vamvas points out, natural theology could help clerics comprehend supernatural theology and fight against the resisting atheists. It could help those 'people

---

88 'Oh, the depth of the riches both of the wisdom and knowledge of God! How unsearchable are His judgments and His ways past finding out!'

89 Vamvas, *Natural Theology*, p. 96.

90 Vamvas, *Natural Theology*, pp. 2, 91, 100–02.

91 Vamvas, *Natural Theology*, p. 96.

92 Vamvas, *Natural Theology*, pp. 9, 96.

93 G. T. Ryan, 'The Argument from Design: A Guided Tour of William Paley's *Natural Theology* (1802)', *Evolution: Education and Outreach*, 2 (2009): 602–11, on p. 602. Paley's famous design (teleological) argument explores how from the complex organization of a watch, which was constructed to perform a specific function, one derives that the watch must have had a maker, who designed it for a specific purpose. Accordingly, eyes provide a particularly illuminating exemplar of organic design like something artificial, like a telescope, 'there is precisely the same proof that the eye was made for vision, as there is that the telescope was made for assisting it' (Paley, *Natural Theology*, p. 20).

94 Gillespie, Neal C., 'Divine Design and the Industrial Revolution: William Paley's Abortive Reform of Natural Theology', *ISIS*, 81 (1990), 214–29.

who might admire natural phenomena and respect the laws and order that prevail in the cosmos, yet at the same time they doubt that the cause of them is God'.[95] In this case, 'Physics always gives the evidence in need for the existence of God and his attributes by observing the order and the motion of all things in the world', a point also made in Psalm 19.1, as Vamvas notices.[96] Therefore, for him, it is mania and insanity for anyone to claim that the wondrous harmony of the world is due to random events of some eternal matter, without will, logic, and knowledge.[97]

Vamvas' motivation in translating Paley's book must have been his conviction that natural theology can deliver a satisfactory proof of God's existence, as well as be a resourceful rhetorical apologetical tool which appeals to emotions, imagination, and even reason, as it had done before in Britain.[98] It could be also a useful weapon against the denial of the revelation's supernatural character and the rising scientific materialism in Europe in the late 1830s.[99] Once again natural theology could become a weapon against atheism and a shield to protect the autocephalous Orthodox Church of Greece recognized in the Patriarchal Tomos in 1850.[100] Nektarios' motivation in publishing Vamvas' manuscript later in 1893 must have been similar since the Greek Christian theologians at the time had to confront scientific materialism during the introduction of Darwin's theory of evolution in Greece.

Yet, inspired by Kant's introduction of the ethical argumentation for religious belief, Vamvas also delivers a pious analysis of Christian ethics in the same book. This was more than an attempt to accord with the latest trends in apologetics; Vamvas argues that since metaphysical proofs result from the observation of the human soul and the innate idea of the first cause of beings (God), an idea inscribed in human nature, it is important to study the latter. His doctrine on Christian ethics re-introduces the 'inscription-in-the-heart' argument by focusing on the moral ability of humans; for what 'testifies now the power, the wisdom, and the will of God, that be the human being, the most divine element of the visible world, created in the image of God and after his likeness'.[101] Along with the motive that led him to translate Paley's work, to embellish Orthodox tradition with

95 Vamvas, *Natural Theology*, p. 2.

96 Vamvas, *Natural Theology*, p. 9 ('The heavens declare the glory of God; and the firmament shows His handiwork').

97 Vamvas, *Natural Theology*, p. 12.

98 J. Topham, 'Science, Natural Theology, and the Practice of Christian Piety in Early-Nineteenth-Century Religious Magazines' in *Science Serialized: Representations of the Sciences in Nineteenth-Century Periodicals*, ed. by G. Cantor and S. Shuttleworth (Cambridge, Mass.: MIT, 2004), pp. 37–66.

99 F. Gregory, *Scientific Materialism in Nineteenth Century Germany* (Studies in the History of Modern Science, 1) (Berlin: Springer: 1977).

100 Topham, 'Science, Natural Theology, and the Practice of Christian Piety in Early-Nineteenth-Century Religious Magazines', pp. 37–66.

101 Vamvas, *Natural Theology*, p. 96; N. Eliopoulos, 'Neophytos Vamvas and Religious Toleration', *The Historical Review*, 8 (2011), 61–83.

Western natural theological arguments, it seems that he was also eager to revive the spiritual life in the Orthodox Church by introducing elements of Western pietism.[102]

Conclusion

The introduction of Western natural theology in the Orthodox Greek-speaking space from the late eighteenth century up to the middle of the nineteenth century emerged within a certain intellectual setting: there was the tacit acknowledgement that the practice of natural theological argumentation was already part of the ancient Greek philosophical heritage sustained later through the Greek Church Fathers. Moreover, Korydallism's inheritance was promising a potential coexistence between a metaphysical account and the Orthodox Christian dogma.

The project was driven by specific motives and objectives. As was shown above, in the late eighteenth century Pamblekis felt comfortable raising his critique against the clergy, championing a combination of theistic and deistic views, fighting against atheism, and advocating a kind of rational religion developed at the time in Europe. Natural theology was part of his enterprise. In the 1820s, for Lesvios, natural theological argumentation was vital not only in his fight against atheistic (materialistic) views but probably a useful means to show the way the empirical features of the world (and how they have been scientifically explained) constituted an explanation of the existence of God. Moreover, he probably thought that the cosmological and teleological arguments provided by natural theology could innovate the Orthodox theological discourse by being in dialogue with the latest philosophical theological discussion. In the middle of the nineteenth century, Vamvas contributed further to the introduction of natural theology with his decision to translate Paley's popular book on natural theology. For Vamvas, natural theology, assisted by the latest scientific advances, was not only a means for interpreting divine revelation and embellishing the Orthodox Christian tradition but also a useful apologetic tool against atheism (materialism). However, when Vamvas' edited manuscript was published in 1893 the tradition of natural theology in Europe had already receded in the face of the systematic development of natural history, the teleological model of explanation had been set aside and the weaknesses of analogical reasoning had become apparent.[103]

Within this context, I argue that introducing Western natural theology in the Greek Orthodox Christian space was part of the dynamic procedure of integrating the latest philosophical and scientific theories into the local culture during the neo-Hellenic Enlightenment. Western natural theology was perceived

102 J. Pinnington, *Anglicans and Orthodox: Unity and Subversion 1559–1725* (Herefordshire: Gracewing, 2003), p. 216.

103 A. Fyfe, 'Publishing and the Classics: Paley's Natural theology and the Nineteenth-Century Scientific Canon', *Studies in History and Philosophy of Science Part A*, 33.4 (2002), 729–51.

more in terms of metaphysics and could allegedly play not only the apologetic role fighting over atheism, as the early Christian Apologists and Fathers did before by appropriating their contemporary philosophical views but also the role of a useful device that could further establish the status and legitimacy of the newly introduced philosophical and scientific knowledge, attributing to them a religious utility. Natural theological arguments were allegedly built by observing nature and contemplating it, promoting a dialogue between natural sciences, philosophy, and theology. In this light natural theology could probably help to achieve a potential revitalization of the Orthodox theological discourse by introducing the energy of Western rationality.

Greek Orthodox Christian enthusiasts of natural theology and rational religion responded to what the time called for, a change. They were devoted to what was considered at least a moral imperative: bringing intellectual revitalization and secularization through the introduction of the latest European philosophical and scientific advances to a population that struggled under superstition and analphabetism for the formation of its national identity; a procedure undertaken before its war of independence and wishfully later after a state was founded. The same moral imperative was to be found, however, among the Orthodox Church and the enthusiasts of the 'philokalic' movement. But this time the road to the renewal of the Greek-speaking population went through what was considered for them the fundamental feature of its national character, Orthodoxy, perceived primarily within the suffocating and distorted notion of ethnophyletism.

It would be wrong, however, to ignore how nervous the Ecumenical Patriarchate and later the autocephalous Orthodox Church of Greece were about the Protestant missionary involvement in the eastern Mediterranean which was strong throughout the nineteenth century. It would be also unfair to not notice that the intrusion of Protestant or theistic and deistic elements within the Orthodox tradition was leading to an intrinsic contradiction. According to the Orthodox tradition, the ecclesiastical experience was more important than any intellectual access to theological truth. It was through faith, not knowledge, that the question of the existence of God could be answered and no metaphysical venture was reliable enough to do that. Thus, natural theology, and human knowledge, in general, were based on fallen reason leading to a false religion. Considered a genuine continuation and renewal of the patristic tradition the 'philokalic' movement felt obliged to restore the Orthodox principles, which they considered as the distinctive theological heritage of the Greek-speaking people. Within this context, any attempt keen to address contemporary issues had to be confined to the Orthodox tradition, any other option was bound to fail; as was evident in the case of the Western natural theology argumentation which was never truly adopted in the Greek Orthodox Christian theological discourse.

DANIEL P. TODES AND
ELEONORA FILIPPOVA

Prayer as Paradigm

*Aleksei Ukhtomsky, the Dominanta, and the Psycho-Physiology of Salvation**

▼ **ABSTRACT** We explore the fundamental role of Eastern Orthodoxy in the origin, content, and development of Aleksei Ukhtomsky's psycho-physiological principle of the dominanta, which brought him renown and election to the USSR Academy of Sciences, became an important current in Soviet and post-Soviet physiology and psychology, and, since the fall of communism, the subject of an expanding literature in Russian philosophy, literary studies, and theology as well. Ukhtomsky dedicated his intellectual life to uniting the 'two paths' of science and religion to a single, ultimate Truth. Using his diaries, correspondence, laboratory notebooks, and marginalia in books, we explore his changing conceptualization of this goal as his religious and scientific lives

* We are profoundly indebted to N. I. Nikolaev and Lawrence Principe for years of patient, illuminating, expert consultation about Eastern Orthodoxy and Western Christianity, and to them and N. F. Matveev, Alex Pavuk, Jay Schulkin, Irina Sirotkina, Roger Smith, Nikolai Krementsov, and Yakov Gall for stimulating discussions of the themes and drafts of this article. Kostas Tampakis and Ronald Numbers kindly included us in the conference on 'Orthodoxy and Science' in Athens, June 2022, where they and other participants provided us newcomers a broad, rich and engaging view of that subject. Irina Tunkina, Director of the Archive of the Russian Academy of Sciences, St Petersburg branch, has provided invaluable support for our research. We are very grateful to the Fulbright Scholar Program (2015–2016) and the archivists and librarians of the Archive of the Russian Academy of Sciences, St Petersburg branch; the St Petersburg State University Library, Section of Rare Books and Manuscripts; the Library of the Academy of Sciences, Section of Rare Books; the Literary Museum of the Institute of Russian Literature (Pushkin House), the Memorial House–Museum of academician A. A. Ukhtomsky, and the State Archive of the Yaroslav Region.

Daniel P. Todes • Johns Hopkins University

Eleonora Filippova • Independent Scholar

Orthodox Christianity and the Study of Nature, ed. by Kostas Tampakis and Ronald Numbers, Science and Orthodox Christianity, 4 (Turnhout: Brepols, 2024), pp. 133–183
BREPOLS PUBLISHERS 10.1484/M.SOC-EB.5.141383

developed and intertwined during four periods: 1. (1894–1900). At Moscow Theological Academy, and then after engaging R. Avenarius's philosophy, he proposed unification of physiological method and 'new thinking' in theology to study the 'organ of religion' and religious experience; 2. (1910–1911). Ukhtomsky's 'two paths' intertwined during simultaneous work on articles about church singing and a dissertation on the physiology of coordinated nervous processes. The metaphor 'attention of the nervous system', which originated from reading the Holy Fathers to understand church singing, resolved the central conceptual problem in his dissertation; 3. (1911–1921). As he overcame spiritual crisis through prayer and studies of the Holy Fathers, Ukhtomsky's understanding of 'attention' and mind/body dynamics became embedded in Eastern Orthodox doctrine about The Fall, Salvation, and the 'governing mind'; 4. (1921–1927). Integrating his religious development with experimental studies, he conceived the scientific-religious principle of the dominanta – a holist psychophysiological replacement for the mechanistic reflex concept, an epistemology of human experience and guide to humane relations, and, most important for Ukhtomsky, a scientific corollary to the Holy Fathers' teachings about prayer, The Fall, and Man's struggle for Salvation. Having replaced 'the will' with 'attention', and 'attention' with 'dominanta' – he had finally integrated his 'two paths' to Truth.

Introduction

In this essay we explore the fundamental role of Eastern Orthodoxy in the origin, content, and development of Aleksei Ukhtomsky's psycho-physiological principle of the dominanta, which brought him renown and election to the USSR Academy of Sciences, became an important current in Soviet and post-Soviet physiology and psychology, and, since the fall of communism, the subject of an expanding literature in Russian philosophy, literary studies, and theology as well.[1]

---

1 Among the many publications about Ukhtomsky (the great majority since the collapse of the Soviet Union in 1991) are: V. L. Merkulov, *A. A. Ukhtomskii: Ocherk zhizni i nauchnoi deiatel'nosti* (Moscow-Leningrad: Akademiia Nauk, 1960); N. I. Nikolaev, 'Pechatny Prolog s zapisiami A. A. Ukhtomskogo iz sobraniia M. S. Lesmana,' *Trudy Otdela drevnerusskoi literatury*, 50 (1996), 817–24; V. P. Zinchenko, 'Aleksei Alekeevich Ukhtomskii i psikhologiia,' *Voprosy psikhologii* 2000, 4, pp. 79–97; S. E. Reznik, *Protiv Techeniia: Akademik Ukhtomskii i ego*

Ukhtomsky (1875–1942) dedicated his mature intellectual life to uniting the 'two paths' of science and religion to the ultimate single Truth. His conceptualization and pursuit of that goal evolved as his religious and scientific lives developed and intertwined.

We concentrate upon four junctures in that process: First, we briefly address the origins of this goal in his early life, and its initial intellectual framing at the Moscow Theological Academy and, shortly thereafter, under the influence of Richard Avenarius. Ukhtomsky envisioned a combination of 'new thinking' in theology and physiological method to address the 'organ of religion' and religious experience. We then turn to the intertwining of Ukhtomsky's 'two paths' during his simultaneous work, as a religiously devout graduate student in St Petersburg, on two essays about church singing and a Master's thesis on the physiology of coordinated actions (1910–1911). To resolve a central problem in that thesis, he relied upon the 'attention of the nervous system' — a metaphor drawn from his study of church singing and his related reading of the Holy Fathers of Eastern Orthodoxy. During our third juncture (1911–1921) Ukhtomsky responded to a spiritual crisis by devoting himself to intensive ascetic prayer and study of the Holy Fathers. His approach to 'attention' and mind/body dynamics became deeply embedded in the Eastern Orthodox narrative of The Fall and Salvation, leading him to a new emphasis upon the 'governing mind' that played a central role in Man's recovery of communion with God. During our final juncture (1921–1927), he returned to scientific research at the end of Russia's civil war and

---

biograf (St Petersburg: Aleteiia, 2015); V. E. Khalizev, 'Intuitsiia sovesti: teoriia dominanty A. A. Ukhtomskogo v kontekste filosofii i kul'turologii XX veka,' *Problemy istoricheskoi poetiki*, 6 (2001), 21–42; L. V. Sokolova, *A. A. Ukhtomskii I Kompleksnaia Nauka o Cheloveke* (St Petersburg: St Petersburg University, 2010); N. I. Nikolaev 'Akademik kniaz' A. A. Ukhtomskii kak deiatel' i myslitel' staroobriadchestva,' *Trudy Otdela drevnerusskoi literatury*, 64 (2016), 576–86; K. B. Zuev, 'Rol' religioznogo mirovozzreniia A. A. Ukhtomskogo v sozdanii printsipa dominanty,' *Sovremennye problemy nauki i obrazovaniia*, 6 (2014), at https://science-education.ru/ru/article/view?id=169963 [viewed 2 November 2022]; and six volumes of published documents: F. P. Nekrylov, comp., A. A. Ukhtomskii v vospominaniiakh i pis'makh (St Petersburg: St Petersburg University, 1992); A. A. Ukhtomskii, *Intuitsiia sovesti*, L. V. Sokolova, G. M. Tsurikova, I. S. Kuz'michev, comps., with introduction by G. M. Tsurikova and I. S. Kuzmichev (St Petersburg: Peterburgskii Pisatel', 1996); A. A. Ukhtomskii, *Zasluzhennyi sobesednik*, with introduction by L. V. Sokolova and G. M. Tsurikova (Rybinsk: Rybinskskoe Podvor'e, 1997); A. A. Ukhtomskii, *Dominanta dushi: Iz gumanitarnogo naslediia*, L. V. Sokolova, G. M. Tsurikova, I. S. Kuz'michev, comps. (Rybinsk: Rybinskoe Podvor'e, 2000); A. A. Ukhtomskii, *Litso drugogo cheloveka*, I. S. Kuzmichev, comp. (St Petersburg: Izd. Ivana Limbakha, 2008); and A. A. Ukhtomskii, *Stat'i i vystupleniia raznykh let: Zametki na poliakh*, L. V. Sokolova, ed. (St Petersburg: St Petersburg University, 2015). For Ukhtomsky's continuing importance for psycho-physiology, see Mihai Nadin, ed., *Anticipation: Learning From the Past: The Russian-Soviet Contributions to the Science of Anticipation* (New York: Springer, 2015). Ukhtomsky is widely recognized as both devoutly religious and an important scientist for his concept of the *dominanta*, but the present article is the first systematically to explore the relationship between the two.

combined experimental studies with the framework developed through his religious practice and meditations to conceptualize the principle of the dominanta.

Ukhtomsky's dominanta offered an integrated, holist psycho-physiological principle to replace the mechanistic reflex concept, an epistemology of human experience, a guide to humane relations, and, most important for its author, a scientific corollary to the Holy Fathers' teachings about prayer, The Fall, and Man's struggle for Salvation. Over the course of his long quest he had replaced 'the will' with 'attention' and then with the 'dominanta' — finally integrating, in his own way, the 'two paths' to Truth.

Early Life: Two Paths to Truth

Prince Aleksei Alekseevich Ukhtomsky was born in 1875 in the central Russian village of Vosloma to an aristocratic family that traced its origins to Rurik, the legendary ninth century founder of Kievan Rus. Aleksei's parents were not wealthy, but remained comfortable despite the challenges to the aristocracy after the 1861 emancipation of the serfs. When he was one year old, they gave him for upbringing to his aunt Anna Nikolaevna, and he lived with her in 'the old home of my grandfather' in the nearby provincial town of Rybinsk, on the Volga River.

Anna Nikolaevna, toward whom Aleksei always related with great love and reverence, was deeply religious. He felt through her, as he put it, 'an ancestral, blood connection with the world of God's eternity'.[2] Their modest home apparently became a place of refuge for peasant Old Believers from the area, and Ukhtomsky later recalled the profound influence upon him of their 'strict spirit': 'I was in large measure brought up with the traditional wisdom of this closed, and, at the same time, primordially Russian peasantry [...] From early childhood I felt myself at one with them, and not with the police, priests, professors and ministers — though I myself became a professor'.[3] He absorbed from his aunt the Old Believers' skepticism about humanitarian theories expressed in words rather than deeds, and their general distrust of abstract intellectualism[4]. He would return

2 Arkhiv Rossiiskoi Akademii Nauk, Sankt-Peterburgskii filial, f. 749 op. 1 d. 92 (3), list 26; f. 749 op. 147(1), list 128rev., f. 749 op. 2 d. 447, list 100-100rev. [Hereafter: St. Petersburg, ARAN 749.1.92 (3): 26].

3 Letter from Ukhtomsky to F. G. Ginzburg, 17–18 November 1927, in Ukhtomskii, *Dominanta Dushi*, pp. 372–74.

4 Old Belief originated during the 1650s and 1660s as a protest against Patriarch Nikon's changes to various liturgical texts and practices, by which he intended to achieve uniformity between the Russian and Greek Orthodox churches and to claim Russia's status as the 'Third Rome'. The many Russian clergy and parishioners who rejected these changes were condemned as schismatics and subjected to mass persecution by the official Orthodox Church. Over time, the cultures of the official Church and Old Believers grew increasingly different, as the former became part of the state apparatus, increasingly bureaucratic and modernized, while the latter evolved in conditions of isolation and repression. The Old Believers came for many

to this beloved home summers and winters until 1920, and, in deeply resonant memories, for the rest of his life.

Ukhtomsky's parents hoped he would follow the family tradition of military service and so enrolled him at age thirteen in the prestigious Arakcheev Military School. There Ivan Dolbnia, a widely-read mathematician, took him under wing, impressing upon him the importance of scholarly inquiry and supervising his study of science and philosophy. Already religiously devout, Ukhtomsky learned from Dolbnia that 'aside from the truth that I knew only by *feeling* (religious truth) there is the truth of the *mind* (scientific truth) and these correspond by their mood and by that which they should provide to Man. That means that Christ and the apostles and such people as Socrates, Plato, Spinoza, Newton, and Kant are all going in the same direction. Truth is one.'[5] Under Dolbnia's tutelage he studied the works of Slavophile philosopher Aleksei Khomiakov, reinforcing his conviction that, in Ukhtomsky's words, 'Religious truth is in the church, [...] the truth of the mind is in science; so the church and science should coincide.'[6] This thought, as applied both to him personally and to knowledge in general, precipitated (in Ukhtomsky's words) 'a great internal break' and came to define his unwavering goal.

Upon graduation from the Military School, he was pronounced physically unfit for military service (he detected here the 'finger of God') and at Dolbnia's suggestion entered the Moscow Theological Academy.[7] Unlike his older brother Aleksandr, who completed the Academy and became a famous activist priest, Aleksei hoped to receive there an education in the history of ideas about true Christianity' to further his search for the 'single Truth' uniting feeling and reason, religion and science.[8]

He would recall his years at the Academy (1894–1898) as 'most happy and beneficial for his spiritual development'.[9] There he encountered faculty members and students who shared the widespread perception that 'science' — in the broad sense of the German *Wissenschaft* and the Russian *nauka* — based on the intellectual and spiritual traditions of the Holy Fathers was central to revitalizing the Eastern Orthodox Church. Particularly important was escaping the Western bias toward philosophizing and fully embracing their faith's traditional emphasis on experience.

---

to embody longstanding traditions and values of the 'true Russian folk' — including communal democracy and fundamentalist traditions, morality, and spirituality. At the beginning of the twentieth century Old Believers numbered about 15 million compared to 80 million in the official Church.

5 St. Petersburg, ARAN 749.1.93: 30.

6 St. Petersburg, ARAN 749.1.93: 30.

7 Letter from Ukhtomsky's father A. N. Ukhtomsky to A. L. Polovtsovoi, n. d., ARAN 749.2.447: 163-164rev.

8 St. Petersburg, ARAN 749.1.138: 33rev–34.

9 A. A. Zolotarev memoirs, St Petersburg, ARAN 749.3.90: 7.

Ukhtomsky composed his graduation essay *Cosmological Proof of Divine Existence* very much in that spirit. The submitted version of that essay, written hurriedly amid his aunt's failing health, incorporated only a fraction of his notes and thoughts toward his original plan, which set out the basic directions that he would pursue in subsequent years.

Fundamental to his 'new thinking' in theology was experience and experiential knowledge. Developing the ideas of Immanuel Kant, J. S. Mill, and William James, Ukhtomsky considered the subjective-psychological qualities of Man, including will and faith, as the basic source of 'all of our knowledge', both scientific and religious. In both, Man and reality, subject and object, were constantly joined in dynamic interaction.[10]

From this perspective, the very idea of a 'cosmological proof' of God's existence was an 'untenable and futile' creation of the 'Hellenistic spirit' with its rationalist bias. Nor, from a 'religious-ethical' standpoint, was such a proof desirable, as it rendered God 'an absolute of sky-high metaphysics' and separated Divine Truth from its 'subjective-psychological' foundations and sources of development.[11] The real task, rather, was to investigate the history of the personal religious experience, of the religious sense itself: 'what we have seen, what we have heard, what our hands have grasped'.[12] 'The question of the existence of God is not a question resolved once and for all. It is a practical question, a question of life, of religious consciousness, and as such is constantly resolved by us anew in every new life circumstance'.[13]

Scientific knowledge, too, had its specific goal and features. Science was valuable for, but limited to, the investigation and description of empirical, material reality. Ignoring the important, subjective factor in knowledge, however — viewing nature as independent of 'the world of human thought' — it provided only 'temporary-conditional', 'partial, descriptive truth'. Long constrained by a mechanistic view of nature, the 'true scientific spirit' was now finding expression in the science of energetics, which promised to free it from materialism through a view of nature as dynamically connected processes.[14]

This understanding of the particularities and goals of each approach to nature dissolved the false antagonism between them, revealing them as allies in the search for Truth. For Ukhtomsky, this conclusion was both intellectually exciting and personally gratifying, since, as he wrote in his diary, he 'loved' both prayer and science, and experienced the same spiritual uplift from each.[15]

The common path of science and religion resided in the inclusion of religious experience as a scientific subject, as a form of human knowledge of reality. 'One

10 St. Petersburg, ARAN 749.1.147 (1): 83rev–85.
11 St Petersburg, ARAN 749.1.93: 24; 749.1.87(2): 2rev.
12 Ukhtomskii, *Kosmologicheskoe dokazatel'stvo Bytiia Bozhego*, in ARAN 749.1.92 (6): 104.
13 St Petersburg, ARAN 749.1.147 (1): 128.
14 Ukhtomskii, *Kosmologicheskii*, ARAN 749.1.92 (6): 91rev.
15 St. Petersburg, ARAN 749.1.92 (8): 9rev–10.

can think of religious subjects completely without touching metaphysical questions — one can and must think scientifically because these [religious subjects] urgently demand from us an account of ourselves, a calm, truly scientific report.'[16] The scientific investigation of religious experience was properly the province of Psychology, which should approach it as an independent phenomenon in dialectical interaction with reality. That project would produce 'not a proof of God's existence,' but a 'psychology of religion'.[17]

The issues that Ukhtomsky engaged in *Cosmological Proof* set the terms for his intellectual quest in subsequent years. He confided in a diary entry of the early 1900s that the essay 'still constitutes [...] my comfort and hope that I will succeed in achieving something. I was groping for promising scientific ground upon which human knowledge can finally attain a single, all-encompassing truth.'[18] Consulting it frequently, he jotted notes in its margin into the mid-1920s.

His advisor at the Theological Academy observed that his approach to knowledge resembled that of Richard Avenarius, and urged him to read the work of this German-Swiss empiriocritical philosopher. Ukhtomsky found in Avenarius a method for studying experience as the product of the dynamic physiological relationship between individual and external world.

> We are accustomed to think that physiology — as a specialized science necessary for physicians — is not necessary for "development of a worldview". But this is as untrue as the proposition that it is not the task of a physician, but rather specifically of a priest or metaphysician to develop a worldview. One must understand that the division of "soul" and "body" is only a historically based psychological product; that the task of the "soul" [...] requires the laws of the "body".[19]

Physiology, then, offered a method for investigating the religious experience described by the Holy Fathers, who viewed soul and body as a single whole.

Critically evaluating that method, Ukhtomsky adapted it to encompass the particularities of religious experience. For him, Avenarius' most important error was ignoring the active psyche in interactions with reality, reducing those responses to adaptations for the mere 'preservation of life'. For Avenarius, psychological states were determined by physiological processes, and so not fundamentally distinct from them. Ukhtomsky, however, insisted that the active psyche determined the content of experience. A fact was not 'a concrete phenomenon of the world', 'a completely determined, complete, clear reality', but rather 'a phenomenon of extraordinary complexity', 'a psychological threshold of thought'.[20]

16 St. Petersburg, ARAN 749.1.147 (1): 31.
17 St. Petersburg, ARAN 749.147.1: 159–159rev; 749.1.92 (6): 103rev–104.
18 St. Petersburg, ARAN 749.1.93: 30–30rev.
19 St. Petersburg, ARAN 749.1.92 (4):14.
20 St. Petersburg, ARAN 749.1. 92 (4):23–23rev.

The process and content of adaptation in religious experience also differed fundamentally from Avenarius' conception. For the believer, adaptation consisted, not in 'preservation of life' or attaining a 'neutralizing calm', but rather in remaking one's life and person through 'new experience', which was attained by the aspirational movement toward God and higher Truth.[21] This point, of course, related to Ukhtomsky's religious convictions and struggles, and he would develop it further — making it qualitatively more central to his thinking — as those struggles developed.

For Ukhtomsky, 'the absolutizing active principle' in religious experience and the aspiration toward God — the principle that negated 'any passivity' and motivated 'actions not justified by reason' — was the will. This was the basis and distinctive feature of religious consciousness, and also 'an integral part of psychological experience'. As he constantly noted in his diaries and marginalia to books, the will was 'the first and main proposition of the forthcoming Psychology of religious feeling'.[22]

After graduation from the Theological Academy, he was enthusiastic about these ideas, grief-stricken by the death of his aunt, and conflicted about his future. Whatever the ultimate prospects for the unity of science and religion, those two passions now pulled him in different directions. His brother Andrei and closest advisors at the Academy urged him to enter a monastery or another form of Church service, but Ukhtomsky hungered to continue his studies. In 1898 he spent almost four months at a monastery outside Moscow, reading science and philosophy, and reflecting on his future. There he came to the conclusion that even as a mere worshipper, let alone a monk with regular duties, he might easily succumb to the 'depressing,' 'lazy,' intellectually inert monastery tradition.[23] He was not an activist, but a 'meditator', so 'My true place is the monastery. But I cannot imagine living without mathematics, without science. So, I must create my own cell — with mathematics, with spiritual freedom and peace. I think that this is my true place'.[24]

He decided to create that 'cell' in St Petersburg, where he could spend his time most productively by studying 'the new biological sciences' while working toward a Master's thesis in Theology about the subject that so animated him:

> Two paths, two treasure-houses of thought known to me and contemporary mankind — in which they can find an answer to the questions of life: the first, bequeathed to me by my upbringing and the best times of youth — the path of Christian and patristic philosophy; the second, in science, which is primarily a method. Why and how did there occur this fatal division of

21 St. Petersburg, ARAN 749.1.92 (4): 61rev–62.
22 St. Petersburg, ARAN 749.1.92 (4): 12rev.
23 St. Petersburg, ARAN 749.3.90:7; 749.1.92 (8): 39rev–40; 749.1.92 (4): 22–22rev.
24 St. Petersburg, ARAN 749.1.147 (3): 65.

> paths that have before them a single goal? Do not these two paths essentially constitute one? This is the question [...] that preoccupies me most of all.[25]

Avenarius, as Ukhtomsky put it, 'blessed my going to the university' and reinforced his intention to concentrate upon physiology there.[26] He moved to the capitol in 1899, and, after some difficulty negotiating entry to the University's physico-mathematical division, began full-time studies in 1900.

The Paths Converge: Church Singing and Master's Essay in Physiology

Ukhtomsky's 'two paths' to Truth came together in 1910–1911, when he labored simultaneously on two subjects that emerged from his two preoccupations — participation in the Nikol'skaia Church and physiological research at St Petersburg University.

Arriving in St Petersburg, he had been horrified by the secular culture of the city and university. Monastery life had been depressingly inert, but here 'I encounter another, stronger and more ancient, more terrifying and powerful, more profound and depressing tradition of the meaningless, dark, deathly-inert, carelessly-physiological `secular' passing of time.' On the streets, in the classrooms and labs, he was surrounded by people for whom 'there is almost no Gospel'.[27]

'Entirely alone with my books,' he sought a religious community. Deeply critical of the official Church for departing from Eastern Orthodoxy's spiritual traditions, he gravitated toward the Old Believers. In 1904 he joined Nikol'skaia Church of Co-Religionists (*edinovertsy*), who retained Old Believer rituals but recognized the authority of the official Church. This choice probably reflected both Ukhtomsky's desire to avoid the practical consequences of being labeled a schismatic and his commitment to the mission (which he shared with his brother Andrei and advisors from the Theological Academy) to use the Co-Religionists as intermediaries to bring Old Believers back to the official Church in order to revitalize it.[28] Active at Nikol'skaia Church, where he became an elder (*starosta*) in 1912, he was deeply unsettled by attitudes and practices there that mirrored the 'pernicious orientations' in the official Church. Only in the nearby Old Believer

25 St. Petersburg, ARAN 749.1.92 (8): 53–53rev.
26 St. Petersburg, ARAN 749.1.92 (8): 3.
27 St. Petersburg, ARAN 749.1.92 (8):39rev; 749.1.92(4): 42rev.
28 As his contact with Old Believers deepened, Ukhtomsky rejected this missionary project.

prayer house did he feel truly 'at home' and 'able to rest spiritually'.[29] Here, too, he found 'living material' for his scholarly interest in authentic religious experience.[30]

While still an undergraduate, Ukhtomsky began studying with and assisting Professor of Physiology N. E. Vvedensky, and in 1906 began formal graduate work. Vvedensky adopted him as protégé and lab assistant, and Ukhtomsky internalized his mentor's functionalist theory of 'parabiosis' — viewing excitation and inhibition, not as ontologically distinct, but rather as two phases of the same nervous process under different conditions. In 1908 Vvedensky invited him to collaborate on an experimental study of the nervous dynamics of coordinated movements. That collaboration foundered upon irreconcilable differences in the interpretation of experiments. The basic disagreement between teacher and increasingly self-confident student concerned whether the results of their experimental trials — the response of the cats' flexors and extensors to stimulation of the sensory nerves — could be explained, in Vvedensky's view, as various phases of local nervous phenomena, or rather, as Ukhtomsky often insisted, as the effects also of central nervous processes within the integrated organism.[31]

These experiments and interpretive clashes led Ukhtomsky to engage deeply with the scientific literature on the coordination and reciprocal relations of nervous processes. Most important for him was British physiologist Charles Scott Sherrington's *The Integrative Action of the Nervous System* (1906). For Ukhtomsky, Sherrington's broad synthetic approach, like that of William James, contrasted with Vvedensky's narrowly local perspective and dogmatic mechanicism, and 'makes possible the holistic structure of knowledge that is the real goal of science'.[32] As he put it in a lab notebook 'One must borrow from physiology its fundamental idea and methods, by which it studies the functions of one or another life organ, and from there seek real knowledge about the organs by which the human soul lives, including the organ of religion.'[33]

In 1910–1911 his religious activities and scientific studies generated two projects that became conceptually intertwined: 1) two essays on church singing

29 St. Petersburg, ARAN, 749.3.25: 115rev; 749.3.19: 313–313rev. Ukhtomsky's copy of K. P. Pobedonostsev, *Moskovskii sbornik* (Moscow: Sinodal'naia Tip., 1897). Nauchnaia biblioteka im. M. Gor'kogo Sankt-Peterburgskogo gosudarstvennogo uiniversiteta. Otdel redkikh knig i rukopisei [hereafter: St Petersburg, ORK], U/249 (marginalia on p. 208).

30 St. Petersburg, ARAN 749.1.93: 31.

31 For these differences, see St Petersburg, ARAN 749.1.8 and 749.1.141; and the resultant article: N. E. Vvedenskii and A. A. Ukhtomskii, 'Refleksy Antagonisticheskikh Myshts Pri Elektricheskom Razdrazhenii Chuvstvuiushchego Nerva', *Raboty fiziologicheskoi laboratorii SPB universiteta za 1908*, 3, (1909), 145–84; in I. M. Sechenov, I. P. Pavlov, N. E. Vvedenskii, *Fiziologiia nervnoi sistemy. Izbrannye Trudy.* K. M. Bykov, editor. Vol. 3, 1 (Moscow: Gos. Izd Meditsinskoi Literatury, 1952), pp. 234–59.

32 St. Petersburg, ARAN 749.1.141: 163.

33 St. Petersburg, ARAN 749.1.87 (8): 23rev.

(1910, 1912); and 2) his Master's thesis *On the Dependence of Cortical Locomotive Effects upon Corollary Central Influences* (1911).[34]

The central concept in both projects was 'attention' (*vnimanie*). That concept had deep roots in the works of the Holy Fathers of Eastern Orthodoxy and also in secular Psychology (but *not* in Physiology). Ukhtomsky's metaphorical use of 'attention' in addressing both religious and physiological subjects brought his thinking about them into close, dialectical relationship. For Ukhtomsky-as-Believer, attention was 'the unity of will' — a central theme in the Holy Fathers' approach to the mind/body relationship, the very soul of ascetic prayer, and the purpose of authentic church singing. For Ukhtomsky-as-Physiologist, attention was 'a powerful and stable' center of excitation that actively influenced the changing functional connections within the living organism. The 'attention of the nervous system' resolved the central issue in his Master's thesis. For Ukhtomsky as Believer-Physiologist, the metaphorical interpenetration of these understandings of 'attention' enriched his thinking and inaugurated a new stage in his quest to unite science and religion.

In the spring of 1910, while conducting experiments for his Master's thesis, he was writing about church singing 'as if there were no tomorrow'.[35] He had been spurred to action by the new choir at Nikol'skaia Church, which appalled him as yet another reflection of the 'mournful decline' of Russian Orthodoxy, of the degradation of traditional liturgical practices in pursuit of 'progress' and greater attendance. For him, the 'over-musicalization' of services in Eastern Orthodox churches — the preoccupation with musicality and tonality, the choice of songs according to what was pleasant to the ear, the inclusion of operatic musical forms, and the exclusion of hallowed canons in order to reduce the length of the service — deprived the liturgy of its spiritual content, power, and purpose.[36]

In his first essay on that subject, Ukhtomsky explained that art and culture exercised a powerful, often hidden influence on people, eliciting 'feelings, moods of the will, ideas that simple logical speech cannot convey and instill with sufficient force and clarity'. Authentic church art had emerged 'historically, genetically' from Eastern Orthodox teachings and experience. Authentic liturgical singing (znamenny chant) integrated music with meaning and feeling, and, together with icons and the rhythms of the authentic service, cultivated 'the states of the will

34 A. Ukhtomskii, *O zavisimosti kortikal'nykh dvigatel'nykh effektov ot pobochnykh tsentral'nykh vliianii* [1911], facsimile edition: (Moskva: Kniga po Trebovaniiu, 2016); 'O Tserkovnom Penii', *Sankt-Peterburgskie vedomosti*, June 10, 1910 (#128), in St. Petersburg, ARAN 749.3.41: 12rev–17rev; and brochure *O Tserkovnom Penii* (St. Petersburg: Sankt Peterbrugskie Vedomosti, 1910). We cite below the brochure, with Ukhtomsky's marginalia, in St. Petersburg, ARAN 749.3.5.

35 Letter from Ukhtomsky to Father Grigorii Dribintsev, August 15, 1910, St Petersburg, ARAN 749.1.141: 104rev–106rev.

36 Ukhtomskii, *O Tserkovnom Penii*, pp. 16–18.

and ideas of Christ's Church,' thus facilitating 'the deepening of attention to the ideas of the Church.'[37]

Especially in view of human failings after The Fall, the role of Church art was 'enormous': it facilitated the 'gradual healing in Christ' of worshippers who had not 'entirely internalized with their entire being Christ's way and Spirit'. Church art 'would not be necessary for me if I had once and for all internalized the Spirit of Christ, not only with my mind but with my entire being; if, weighed down by the inertia of my nature, I did not renew the Image of Christ in my memory only when I hear Kassia's hymn and participate in the liturgy.'[38]

There was, he argued, a 'radical, primordial dissonance' between the spirit and content of the liturgy and the 'concert music' adopted by the choir at Nikol'skaia Church and many other places of worship to attract and please modern elites.[39] Set to religious themes and pleasing to the ear, this 'decadent art' communicated 'the sensation of beauty [...] independent of the content of ideas,' eliciting a mood conducive to the passive experiencing of content-less pleasurable sensations and 'wobbliness' of the will. Rather than 'preparing the worshipper for the Church idea of the liturgical moment,' such performances 'carry the attention and soul of people far away' from it.[40]

Many worshippers, he concluded sorrowfully, were now so accustomed to decadent fare that they had deteriorated beyond the reach of 'a strict, humble service without external sparkle'. Most important was to preserve at least some places of worship where 'the church service is observed in all its inviolate fullness' and a lover of true 'church beauty could find spiritual peace'.[41]

In early June 1910 he was waiting impatiently for his essay to appear in a leading newspaper and was planning to republish it as a brochure. His thinking had developed further, and he already found this hurried effort inadequate, but this intervention in a 'burning question' remained 'very dear to me'. Meanwhile, he was struggling with his Master's thesis in physiology. His recent experiments had failed, he had not yet tackled the 'sea of literature' related to his subject, nor had he resolved the conceptual problem at its center. As he confessed to his religiously devout friend and confidante Varvara Platonova, 'It is difficult to find the Ariadnean thread that would guide me in the labyrinth'.[42]

Both article and brochure on church singing appeared within two weeks, by which time Ukhtomsky was afire with ideas about deepening it — and he annotated his copy of the brochure accordingly. These notes referred to passages from the works of the Holy Fathers about the centrality of attention in prayer that

37 Ukhtomskii, *O Tserkovnom Penii*, pp. 8–9, 11.
38 Ukhtomskii, *O Tserkovnom Penii*, pp. 9–10.
39 Ukhtomskii, *O Tserkovnom Penii*, p. 7.
40 Ukhtomskii, *O Tserkovnom Penii*, pp. 11–15.
41 Ukhtomskii, *O Tserkovnom Penii*, pp. 29–32.
42 Letter from Ukhtomsky to V. A. Platonova, June 6–7 1910, St Petersburg, ARAN 749.3.69: 9–11rev.

would provide the 'Ariadnean thread' for both his second essay on church singing and his Master's thesis in physiology.

In these notes Ukhtomsky explored what he termed synonymously the 'psychological' or 'psycho-physiological' dynamics of art's influence upon people — the relationship between bodily, sensory experience, on the one hand; and human moods and ideas, on the other. He now emphasized that Church singing was part of a liturgy that was essentially a prayer, and that the importance of church singing (like other bodily dimensions of the authentic service) resided in its contribution to an essential quality in prayer: 'attention'. The many references in his first essay to 'the will' were largely replaced in the second by the more specific, psycho-physiological 'attention,' as were those to 'art,' 'service,' and 'singing' by 'prayer,' which he now conceptualized as their essence.[43]

Ukhtomsky's work on this second essay reflected a qualitatively different relationship to the Holy Fathers, whom he was now beginning to study and interpret, not just for personal solace and inspiration, but as an intellectual resource. This is evident, not only in his annotations to his first essay and the text of the second, but in the marginalia to his copy of the second volume of *Dobrotoliubie* (the Russian version of *Philokalia*) — especially to works by the ascetic monk-theologians Efrem Sirin (Ephrem the Syrian), Isaak Sirin (Isaac of Nineveh), and Ioann Lestvichnik (John Klimax, John of the Ladder), which he mined simultaneously for this second essay on church singing and his Master's thesis.

Ukhtomsky's second essay began with the observation that the essential difference between services at an Old Believer and mainline Eastern Orthodox church resided, not in the ceremonial differences between them, but rather in their 'spirit and mood'. The 'unhurried regularity' of the Old Believer service, the careful observation of its every detail, created a 'completely unique combination of artistic depth with profound virtue and, so to speak, musical humility in znamenny chant — all this creates the special church-artistic elemental force [*stikhiia*] of the Old Believer temple'.[44]

Drawing upon citations from the Holy Fathers, he explained that the powerful Old Believer service embodied historical experience and wisdom in managing the mind/body relationship for attaining concentrated attention in prayer. 'The great psychological significance' of the 'general regularity' of the service — with its powerful rhythm, chanting, 'timely placed bows' and 'timely silencing of every external movement' — was that it compelled the worshipper's attention: 'In order to

43 In the 1910 essay he used variants of the Russian words for 'will' or 'volition' 7 times, 'attention' 3 times, and 'prayer' not once; in the 1912 essay he used variants of the words 'will' or 'volition' only once; 'attention' (which replaced 'will') 29 times, and 'prayer' (which displaced 'art', 'singing', and 'service') 26 times.

44 'Doklad kn. Ukhtomskogo o tserkovnom penii,' *Trudy Pervogo Vserossiiskogo S'ezda Pravoslavnykh Staroobriadtsev (Edinovertsov)* [1912] (Moscow: Pravoslavnyi Sviato-Tikhonovskii Gumanitarnyi Universitet, 2012), pp. 99–116; pp. 100–03.

be reliably and firmly permeated by the church service, and to subordinate oneself to its guidance, what is needed is not impulses and outbursts of inspiration, but that `even wind', restraint, constant self-discipline, patience, unrelenting attention and `regularity' to which the ascetics and fathers of the church return repeatedly and insistently in their works.'[45] While services in the great majority of mainline churches featured 'a patchwork of impressions', Old Believer worshippers experienced and internalized 'that evenness of flow and unity of attention about which the fathers taught'. This facilitated, as Diadokh (Diadochus) expressed it, 'that `unwavering, undreamy movement of the soul' toward Divine love, which will entice the very body into the depths of this unspoken love'.[46]

> Prayer, as presented by the fathers of the church, is the heightened activity of the mind; church art and singing promotes the task of prayer when, by means of the feelings, it constantly strengthens the flagging attention of the mind; sensation is only a means for this, to strengthen attention, the center of which stands upon the activity of the mind, that is, upon the act of prayer.[47]

Authentic religious singing thus prepared the sensory ground for the internalization of religious truths. The great majority of worshippers — who possess 'an ordinary psycho-physiological organization' — could accomplish this only with great effort and rigorous discipline, gradually internalizing the faith and transforming themselves 'in both spirit and action'.[48] Christianity was not algebra or metaphysics — it could not be understood 'abstractly'. 'To incorporate it requires elevating the entire human nature'.[49]

This analysis was grounded in Ukhtomsky's engagement with the Holy Fathers' experience-based counsel regarding the ascetic's struggle against the fragmentary thought forms (*pomysel, pomysly; logismos, logismoi*) that interfered with prayerful attention in the quest for communion with God.[50] The notes on his copy of the 1910 brochure reference relevant densely-annotated and cross-indexed passages in his copy of *Dobrotoliubie* by these 'great psychologists' (as he termed them).

This initial dialogue with the Holy Fathers is exemplified by his annotations to Ephrem the Syrian's 'On the Struggle with the Eight Great Passions' and 'General Lessons about Ascetic Life'.[51] For example, he underlined Ephrem's description

---

45 'Doklad,' pp. 103–04.

46 'Doklad,' pp. 104–05. He is citing here the fifth century ascetic Diadochus from works included in *Dobrotoliubie*, vol. V.

47 'Doklad,' p. 105.

48 From Ukhtomsky's marginalia to his 1910 essay, St Petersburg, ARAN 749.3.25: 129.

49 From Ukhtomsky's marginalia, St Petersburg, ARAN 749.3.25: 129–129rev.

50 The struggle against these fragmentary thoughts and images was an important theme in the works of the Holy Fathers. We are deeply grateful to Alex Pavuk for identifying the Greek equivalent of the Russian term and suggesting the English translation 'thought form'.

51 Sviatoi Efrem Sirin, 'O bor'be s vosem'iu glavnymi strastiami, voobshche' and 'Obshchie uroki o podvizhnicheskoi zhizni' in *Dobrotoliubie*, vol. II, second edition (Moscow: Tip-Litografiia I.

of the Devil's efforts to disrupt prayer by exploiting individual inclinations: 'The unclean spirit of arrogance is devious and inventive, making every effort to prevail over all; it catches the wise with wisdom, the strong with strength, the wealthy with wealth, the beautiful with beauty, the artist with art.' Ukhtomsky wrote in the margin: 'Why religion and science can decline, why art declined'; then jotted that same note on the cover of his 1910 brochure, along with the observation that 'the singers [of decadent music] only *interfere* with the worshippers in church being worshippers'.[52]

For the Holy Fathers, church services were but one part of constant prayerful silence, so the issues related to singing and prayer applied also to life in general. Ephrem the Syrian wrote that 'The man who leads his life negligently, deceives himself' — forgetting God's rewards for righteousness and punishment for sin because 'carnality has settled in his mind, darkened his eyes'. Ukhtomsky reflected: 'This way of life is also the absence of attention, of unity of the will, it is the dispersion by the needle of sensations, [what the Biblical prophet Habbakuk termed] `corruption by the murk''.[53] (Note the equation of attention with 'unity of the will'.)

In another passage, Ephrem discussed the constant challenge during prayer of combatting inattention. 'If during prayer something [...] disperses your mind, know that this is the act of the enemy.' Satan 'directs the thought form like an arrow, shot at random'; to prevent the worshipper from communion with God, 'the enemy leads him to dispersed thinking, so his thought flew like an arrow aimed at hail'. For Ukhtomsky, here 'The energy of attention is squandered, [directed] elsewhere or everywhere.'[54]

Studying Ephrem's writings in *Dobrotoliubie* while rethinking his essay on church singing, Ukhtomsky filled the margins of the religious text with such observations. Underlining Ephrem's injunction to 'Be attentive to yourself, in order not to lose your time in negligence and dispersion', he commented that 'this is accomplished by attention', and cross-referenced this passage with Ephrem's repeated injunction to 'Clean up your thought forms (be attentive)'.[55] He annotated similarly Ephrem's many passages on the struggle against thought forms: 'absence of attention', 'again about attention', 'singing in mental attention', and 'significance of the church service for attention'.[56] Discussing 'our terrible traitor

Efemova, 1895), pp. 377–429, 430–78. We cite Ukhtomsky's annotated copy, in the Memorial'nyi Dom-Muzei Akademika A. A. Ukhtomskogo in Rybinsk [Hereafter: Rybinsk, MDM], RBM-45380.K-5588.

52 Ukhtomsky's note on his copy of *O Tserkovnom Penii* (1910), ARAN 749.3.25: 114, 121; and on his copy of *Dobrotoliubie*, vol. II, p. 423.

53 Ukhtomsky's copy of *Dobrotoliubie*, vol. II, p. 460. Ukhtomsky also emphasized Sirin's passage about the 'murk' of inattention on pp. 356 and 457; and jotted a reminder on his brochure (St Petersburg, ARAN 749.3.25: 122).

54 Ukhtomsky's copy of *Dobrotoliubie*, vol. II, p. 474.

55 Ukhtomsky's copy of *Dobrotoliubie*, vol. II, pp. 465, 455.

56 Ukhtomsky's copy of *Dobrotoliubie*, vol. II, pp. 473 (Sirin's passage), 483, 485.

— the dispersed eye [*oko*]', Ephrem enjoined monks to 'make of all yourself an eye [*okom*], unceasingly observing and on guard against the plunders of the evil one.' Ukhtomsky noted that 'eye [*oko*] is a symbol of attention'.[57]

Given Ukhtomsky's synthetic cast of mind, his determination to unite the 'two paths' of religion and science, the metaphorical stretch afforded by the concept of 'attention', and the simultaneity of his work on church singing and the physiology of coordinated movements, it is hardly surprising that he extended this analysis of church singing, and his interpretation of Ephrem and other Holy Fathers, to the pressing physiological subject on his mind.

The goal of his thesis was to develop his ongoing disagreement with Vvedensky by demonstrating that excitatory and inhibitory processes were constantly dependent upon the dynamic nervous relations that united the organism in a single, integrated whole. To this end, he conducted experiments on cats' responses to various combinations of stimuli: on the interaction between the locomotive reflexes elicited by stimulation of the cortical points and sensory nerves controlling the extensor and flexor muscles of the paws (that is, of 'antagonistic muscles'), on the results of both separate and simultaneous cortical and sensory stimulation of these muscles, and of the relationship between the results of these local stimulations and other, ongoing reflexive reactions in the broader organism. In Chapter Two, he demonstrated that electrical cortical excitation of the antagonistic muscles of the paw elicited very different responses depending upon the strength and duration of the current, which influenced the extent to which these excitations remained localized or spread to other nervous centers.[58] He then argued in Chapter Three that localization doctrine, while containing an important kernel of truth, was 'too mechanical': exciting the same cortical point did *not* always produce the same effect, since excitation of a cortical-spinal path often spread to other cortical centers, creating a 'certain state of excitation' that affected reflexive reactions. The functional state of the nervous centers was constantly changing, and these centers themselves interacted continually.[59] In Chapter Four, he reported experiments during which, in various combinations, he simultaneously stimulated both the cortical points and sensory nerves controlling the flexor and extensor muscles of the paw. He had expected to find that, for example, doing so to both cortical and sensory points for the flexors would prove 'synergetic', while stimulating one for the flexor and another for the extensor would not — but the results, while supporting his general argument about the complex fluidity of nervous responses, failed to reveal any 'continuous and stable' relations.[60]

57 Ukhtomsky's copy of *Dobrotoliubie*, vol. II, pp. 396–97.

58 Ukhtomskii, *O zavisimosti*, pp. 38–42. See the experiments of May 1910 in Ukhtomsky's lab notebook, St Petersburg, ARAN 749.1.142: 7–23rev.

59 Ukhtomskii, *O zavisimosti*, pp. 51–52, 74–75, 82.

60 Ukhtomskii, *O zavisimosti*, pp. 87–107 (discussion of experiments), 126–28 (summary of conclusions).

In late June and July 1910 he conducted the experiments for what proved the centerpiece of his thesis, which followed up a chance observation that had long intrigued him: Preparing a dog for Vvedensky's lecture on localization in 1904, he had stimulated the cortical centers known to elicit movement of the fore-paws. The dog indeed initially moved its paws, but this reaction then faded and was replaced by movements of the tail and anus, and then, as he increased the current, grew stronger and culminated in a forceful act of defecation. In Ukhtomsky's later telling, Vvedensky dismissed this observation as insignificant, and Ukhtomsky himself, as a mere undergraduate lab assistant, lacked the training to investigate its 'physiological sense'. Yet it had seemed to him important and he had kept it constantly in mind. Now, seven years later, he viewed it hopefully as the centerpiece of his continuing argument with his mentor. The ongoing 'chain reflex' leading to defecation, he reasoned, had inhibited the cortical stimulation of the paws — and so could provide a compelling example of the dynamic relationship between local and central nervous processes.[61]

His attempts to replicate his earlier observation experimentally, however, proved disappointing. Reproducing the chain reflex leading to defecation was very difficult to accomplish in the lab, and impossible without extravagant procedures that introduced a mass of uncontrolled variables. Even his best trials failed to produce the 'complete inhibition' that he recalled from 1904.[62]

Seeking another chain-reflex that might inhibit cortical locomotive effects, he settled upon swallowing, which was more easily elicited in the lab. He found the experimental results, though inconsistent and sometimes fleeting, extremely encouraging. On a 'completely fresh specimen', during the first 15–20 seconds after swallowing began, the locomotive effects of cortical stimulation remained unchanged (that is, as in the locomotive act, the contralateral flexor continued to contract and its antagonist, the extensor, continued to relax), but after 30–40 seconds this 'changed noticeably': locomotive movements either completely disappeared or both muscles relaxed. This state continued for up to 25 seconds, even after the swallowing movements had ceased. During these trials, importantly, the

61 There is no contemporary record of this 1904 observation. Writing seven years later in his Master's thesis, Ukhtomsky claimed that he had spontaneously perceived this as 'elementary attention'. We believe that the religious-intellectual interests that brought him to the University indeed prepared him to focus upon this phenomenon, but we doubt that he thought of it at the time as 'attention'. That word occurred very rarely in his diaries and correspondence at the time, and appeared in his scientific lexicon only in June 1910 (see below). We suspect that in 1904 he was intrigued by the dog's response as a manifestation of elementary 'will,' a term he used quite frequently until it was displaced in 1910 by 'attention'. He would have had good reason to reframe his recollection in this way in his Master's thesis. See Ukhtomskii, *O zavisimosti*, p. 4; Ukhtomskii, 'Dominanta kak factor povedeniia', in A. A. Ukhtomskii, *Izbrannye Trudy*, E. M. Kreps, ed. (Leningrad: Nauka, 1978), p. 66.

62 Ukhtomskii, *O zavisimosti*, pp. 129 (failure of experiments), 175–84 (discussion of results). For his experimental trials of 17 July and 27 July 1910, see St Petersburg, ARAN 749.1.142: 28–31, 32rev–34.

cortical locomotive centers had *not* simply become refractory. Rather, stimulation of them produced an atypical reaction and (often) further excited the swallowing process.[63]

These results served as the point of departure for 70 pages of discussion in which Ukhtomsky synthesized his results to propose a functionalist interpretation of the dynamic interaction of nervous centers throughout the organism. Sometimes a weak stimulation excited a powerful response, and sometimes the opposite occurred; sometimes cortical stimulation excited the paws, sometimes it did not; under some conditions defecation and swallowing inhibited cortical or local excitation of the paws, and under others they did not; sometimes stimulating the rear extremities elicited heightened swallowing responses (the common term for this reciprocal excitation was *Bahnung*); sometimes it elicited instead 'reciprocal inhibition'. As he put it in one proposition for his Master's committee: 'Defined acts of excitation that take place in the organism and demand time for their fulfillment can, in a given moment of time, completely change the functional significance of a particular cortical focus of the so-called locomotive region of the cerebral cortex.'[64]

The central question, then, was: How to understand the dynamic interaction between excitation and inhibition in 'such apparently separate nervous apparatuses'? How could the nervous processes of swallowing or defecation, on the one hand, and the cortical or sensory locomotive apparatus of the paws, on the other, influence one another in the form of reciprocal excitation or inhibition?

Sherrington's concept of the 'common final path' provided an elegant, synthetic framework for thinking about this, and, in general, about the coordination of the complex, nervous processes that welded the organism into a functioning individual. Having studied *Integrative Action* closely, Ukhtomsky explained Sherrington's central concept: The great variety of possible excitatory acts cannot be implemented simultaneously since, at every level of the nervous system, there exists a relatively large number of afferent nerves carrying sensory impulses to the spine and a relatively small number of efferent nerves to carry the resulting locomotive impulses from the spine to the body's muscles and glands. At the higher levels in the central nervous system charged with coordinating nervous acts, these afferent nervous paths come together in groups — in 'common paths' — that travel toward one general, common path. This 'general principle in the architecture of the complex nervous system (principle of the final common path)' permitted only one reflexive act along any common path to be completed simultaneously. Impulses from other afferent paths, and especially impulses exciting the common path in an opposite direction, were thus inhibited. Reflexive

63 Ukhtomskii, *O zavisimosti*, pp. 135–36, 173.

64 Ukhtomskii, *O zavisimosti*, pp. 141–42; and propositions 7 and 9 for his formal thesis defense in St. Petersburg ARAN 749.1.142: 74 (draft), 749.1.138: 72–76 (draft), and the printed copy inserted in Avenarius, *Kritik der Reinen Erfahrung* (Leipzig: Fues, 1888), at St Petersburg, ORK U/264.

apparatuses in each path were structured in groups that were either 'allied' or 'antagonistic'. A reflex act irradiated excitation only to allied central apparatuses, and that excitation (by occupying the common path) inhibited antagonistic ones. So, under normal conditions, an efferent apparatus could not simultaneously be excited by two antagonistic innervations — for example, the innervation of the flexor is always accompanied by inhibition of the extensor. Sherrington referred to this interaction as the 'reciprocal innervation' of antagonistic apparatuses.[65]

Here, however, Ukhtomsky confronted a problem: How could Sherrington's 'architectural' concept of the common path explain the interaction between two nervous processes — swallowing and locomotion-by-paw — that were neither anatomically nor functionally related? How could two such 'extraordinarily separate' nervous processes compete for one of Sherrington's common paths?[66]

He would not respond to that central question for some 50 pages — and then only implicitly and indirectly — but he now prepared the reader for his response: Properly understood, 'the entire content of [Sherrington's] concept of reciprocal innervation is applicable'. The energetic excitation of the swallowing apparatus '*creates*' from this apparatus an 'antagonistic center' in relation to the cortical innervations of the apparatus of locomotion.[67]

The physiologist, he explained, must avoid an 'overly-schematized' understanding of Sherrington's 'common path'. The relations between 'mechanistic, anatomical antagonists' (such as flexor and extensor muscles) were not permanently fixed, but rather were determined by 'the dynamic conditions of the time; they can be synergetic in one act and antagonistic in another'. Ukhtomsky claimed that Sherrington had never proposed a strictly anatomical, 'once-and-for-all' conception of the reciprocal relations between nervous centers, but conceded that the British physiologist had, 'in order to emphasize his basic ideas, formulated propositions in a 'too-schematic fashion' — asserting, for example, that flexors and extensors seemed never to contract at the same time. If interpreted *unschematically*, however — that is, freed of its fixed anatomical architectural framework — the principle of the common path was broadly applicable even beyond the sphere of 'anatomical, mechanistic antagonists'. Here Ukhtomsky was preparing the reader for his own fundamental revision of Sherrington's concept.[68]

He now broke off this discussion to introduce a concept that he had not theretofore even mentioned in his thesis — one that he had decided only in the last stages of his writing to include, that he knew would elicit a hostile reaction from his advisor, but which provided the conceptual means to reinterpret Sherrington's 'common path' in a manner explaining his experimental findings.

That concept, which had come to play a central role in Ukhtomsky's analysis of church singing and prayer, was 'attention'. A psycho-physiological refinement

65 Ukhtomskii, *O zavisimosti*, pp. 157–61.
66 Ukhtomskii, *O zavisimosti*, p. 164.
67 Ukhtomskii, *O zavisimosti*, p. 164.
68 Ukhtomskii, *O zavisimosti*, pp. 162–63.

of the theological 'unity of the will,' this term first appeared in Ukhtomsky's lab notebook in June 15, 1910 — that is, one week after he complained to Platonova of his inability to find the 'Ariadnean thread' that would make sense of his experiments, a few days after he received a published copy of his first essay on church singing (and began intensively to annotate it for revision), and *before* he began his experiments on defecation and swallowing. Ruminating about an experiment that day in which stimulation of the cortical locomotive centers failed to produce the expected movements of the rear paws, but rather elicited movements of the neck and shoulder muscles, he considered several possible explanations and ended with this one: 'One could understand it in this way — that now `the attention' of the nervous system is concentrated on the front part of the body, and so there proceeds a slight reciprocal excitation there, with inhibitions in the other parts of the body.' (The quote marks around 'attention' indicate his recognition that he was using that term metaphorically.) Struggling to write his thesis at home in Rybinsk from late summer through winter, he decided to explain his experimental findings by replacing Sherrington's anatomical structures with the 'attention of the nervous system'.[69]

The same works by the Holy Fathers that were so important to his thinking about attention in church singing and prayer guided him in this reinterpretation of Sherrington's 'common path,' which Ukhtomsky reformulated as a psychophysiological *process* that was not limited by the physical architecture of the body, acted beyond the sphere of 'anatomical, mechanical antagonists,' and was 'created' by bodily attention to a purposive act.

This is clear, not only from the parallel reasoning uniting his work on church singing and reciprocal innervation, but also in his physiological commentary to those same works by the Holy Fathers. We can here only touch briefly upon a few illustrative examples. Reading 'The Struggle With the Eight Main Passions' by monk-theologian Ioann Kassian Rimlianin (John Cassian), Ukhtomsky emphasized this passage:

> The desires for real things cannot be suppressed or rejected unless other, redemptive [desires] are accepted instead. Therefore, if we wish to eradicate carnal lusts in our hearts, then in their place we should as quickly as possible plant spiritual desires, so that our spirit, always striving toward them, has neither the urge nor the time to pay attention to the charms of the transient joys of the earth.

He interpreted this in physiological language: 'one inhibits the other: *the action of attention*'.[70]

69 St Petersburg, ARAN 749.1.142: 26rev. '*Korraboratsii*' was Vvedensky's term for 'reciprocal excitation', '*Bahnung*'.

70 Ioann Kassian Rimlianin, 'Bor'ba s vosem'iu glavneishimi strastiami,' in Ukhtomsky's copy of Dobrotoliubie, vol. II, pp. 42–43 (marginal note on 43).

He developed this thought in commentary to several passages of Ephrem the Syrian's 'General Lessons about Ascetic Life'. For example, Ephrem wrote: 'One of the Saints said: `think of the good, so as not to think of the bad; for the mind does not suffer to be idle'. Why do we dedicate our mind to teachings in the words of God, to prayer, and to good deeds. Because occupation with vain thoughts gives birth to vain deeds; and occupation with good thoughts gives good fruit'. Ukhtomsky interpreted this in psycho-physiological terms: 'On the limitedness of consciousness, on antagonistic innervations, on attention as a consequence of the limitedness of consciousness'.[71]

Here he was using Ephrem to reformulate Sherrington. The 'limitedness of consciousness' and selective attention played the same role in Ephrem's instruction to the monk as did the limited carrying capacity of Sherrington's common path. Just as the occupation of Sherrington's common path by excitatory impulses to the extensor prevented the simultaneous transmission of excitatory impulses to the flexors, so, due to the 'limitedness of consciousness,' did the mind attentive to the good avoid the bad. As in his above-cited note to John Cassian's work: 'one inhibits the other: the action of attention'.

As Ephrem continued to counsel the monk about the struggle to protect his virtue, Ukhtomsky continued to ponder that advice with an eye to psycho-physiological processes. Ephrem wrote:

> Leaving your cell for service or for conversation with someone, protect your eye [*oko*] and excite your heart with a pious thought, saying: 'you are not a painter, and you did not come out in order to learn how to copy images from people.
> Be attentive to yourself. If your attention is occupied with the sensory, how can you be of pure mind, as in a mirror, to contemplate the heavenly, to delight and to please yourself with mindfulness of God?'

Ukhtomsky commented: 'psychophysiologically, the more one learns to inhibit one's excitations, the sharper is one's ability for attention, the more capable one is to coordinate one's actions'. For Ephrem, 'a pious thought' directs attention away from the merely 'sensory'. For Ukhtomsky, this is an example of attention's inhibitory role and its importance in coordinated actions. Here again, prayer served as paradigm.[72]

In his marginalia to Isaac of Nineveh's 'Ascetic Teachings' Ukhtomsky explicitly related the seventh-century ascetic-theologian's discussion to the physiological processes of reciprocal excitation (*Bahnung*) and reciprocal inhibition. For example, Isaac explained that a spiritually-awakened soul 'fervently attracts to itself the content of [virtuous] speech', while the heart 'preoccupied by the earthly' remains unaffected by it. Ukhtomsky commented: 'The thought of Scripture,

---

71 Efrem Sirin, `Obshchie uroki,' in Ukhtomsky's copy of Dobrotoliubie, vol. II, p. 444 (passage 48).

72 Efrem Sirin, 'Obshchie uroki,' in Ukhtomsky's copy of Dobrotoliubie, vol. II, pp. 449–50.

having become attached to the seed that is found in the soul, fans it like a coal and returns it (Bahnung)'.[73] Here he connects the ability of a 'spiritually awakened soul' to attract and be further awakened by virtuous speech with his experimental observation that, if a cat's swallowing apparatus has been excited, it will attract to itself and be further excited by the stimulation of the cat's paw. As for reciprocal inhibition: In one passage, Isaac explained that temptations arose constantly from 'the things of this world', but that 'a man who has received Divine grace, who has tasted and felt something higher' does not allow them 'to enter his heart, because there has prevailed in their place another, better desire'. These temptations therefore remain 'outside and inactive'. Ukhtomsky underlined the next sentence in emphatic black ink: 'They are forbidden entry into the heart, not by struggle, but as a consequence of the satiety of consciousness and the quest for spiritual knowledge, which fills the soul'. This, he noted, was a 'picture of the reciprocal inhibition of one spiritual state by another'.[74]

He was now able to close the circle between Sherrington's concept of the common path and his experimental data. The common path was created by 'the attention of the nervous system', or, in more acceptable physiological language, as he put it in his lab notebook, 'A very distant group of centers can *become, make of itself* inhibitors in relation to cortical locomotion in particular circumstances — for example, when this distant group of centers is in the process of preparing an intensive discharge in a particular act'.[75]

Before submitting his thesis, Ukhtomsky discussed his conclusions with Vvedensky, whose response left him in a cold fury. An angry entry in his lab notebook revealed clearly the origin and deeper meaning of the attention metaphor for him:

> The old physiology, which calls itself "mechanistic", was afraid to talk about "attention" and such higher functions, like an old spinster with faded charms that no longer excite anyone's desires [...] Just so, Vvedensky considers that, if I referred to attention and its physiological substrate, then I did so "in vain" [...] Yet what could be more natural than to see the physiological analogue of attention in those instances where the inhibition of some apparatuses is firmly connected with the Bahnung of others, and when this connection integrates a holistic purposive act! Were I not to speak of this analogy — of attention — somebody else would do so in 2 weeks, 2 months or two years [...]
> "Narrow your perspective as much as possible and then get to work" — that's what the precepts of the "exact method" come down to as edited by grey mediocrities. This is the same metamorphosis by which Christian thought among the Athonite monks of Gregory Palamas's time had been reduced to

73 Ukhtomsky's copy of *Dobrotoliubie*, vol. II, p. 647. He cross-referenced this with Sirin's two other illustrations of *Bahnung* on pp. 702 and 737–8. In a marginal note to this same passage about one decade later, Ukhtomsky reinterpreted it as an illustration of the dominanta.

74 Ukhtomsky's copy of *Dobrotoliubie*, vol. II, p. 670.

75 St Petersburg, ARAN 749.1.142: 111rev. Emphasis in original.

> the contemplation of heavenly light in one's own navel. Thus we commonly see that physiological investigation stumbles about within the confines of some little theory [...] and the eyes do not see, the ears do not hear [...]
> Clearly, one cannot avoid having some perspective — one needs points of departure [...] The only difference is that in one case people will be departing from the experience established by generations of great thinkers and psychologists, and in the other from some little theory that helps take data about the frog's limb and "pull it by the ears" toward the facts of cortical activity. That is the distinctive feature of these "strict minds".[76]

His 'generations of great thinkers and great psychologists' were of course the Holy Fathers. And it is telling that Ukhtomsky identified here, not with bold scientists such as Newton, Faraday, or Helmholtz, but with Gregory Palamas, the celebrated monk-theologian whose defense of the hesychasts of Mount Athos in the fourteenth century led him, in his *Triads*, to criticize Hellenistic distortions of the faith and to systematize Eastern Orthodox epistemology and doctrine.

After his confrontation with Vvedensky, Ukhtomsky hurriedly revised his thesis in preparation for his defense. He repeatedly rewrote key phrases as he attempted to craft a version that would both make his essential point and mollify his mentor. In his concluding section on 'attention,' Ukhtomsky's always complex, elliptical narrative style became defensive, convoluted and difficult to follow. His line of argument was buried in discussions of theories of inhibition, of the few passing references he could find among physiologists to justify his key 'figure of speech,' and in an extended critique of psychologist William McDougall's overly mechanistic 'drainage theory' of attention. He entirely omitted 'attention' from the propositions submitted to his committee for his formal defense.[77]

Yet the conclusions were there — for posterity and his own future development: At the most general level, he advanced a dynamic, functionalist view of physiological processes based, not upon 'once-for-always determined and constant functional statics of various `foci', as the bearers of separate functions,' but rather 'on the unceasing intra-central dynamics of excitations in the cortical, subcortical, medular and spinal centers, determined by the changing functional states of all these apparatuses.' As for the central phenomenon he sought to explain: When the organism is engaged in the act of swallowing and defecation, and this inhibits the nervous apparatus in the paws, 'this connection of reactions corresponds to that functional moment which one must propose as the physiological foundation of so-called reflexive attention.'[78]

Ukhtomsky's use of this metaphorical 'attention' brought into creative dialectical relationship his thinking about unconscious, spinal nervous processes and the Holy Fathers' conceptualization of the conscious struggle during prayer for

76 St Petersburg, ARAN 749.1.142: 38–39.
77 See p. 138 n. 64.
78 Ukhtomskii, *O zavisimosti*, pp. 204, 211.

communion with God. For him, there was indeed a fundamental connection between these two realms of phenomena, a connection that he did not fully understand but planned to explore (that, in a nutshell, is the nature of metaphor in science). He alluded to this connection in his thesis:

> The chain reflexes arising in the digestive tract, like the act of swallowing or the act of defecation, are in all probability only a weak analogy of those connected intra-central currents of excitation that occur in the cortex during the purposive activity of consciousness. But from this analogy we can guess how powerful these higher cortical acts can be in their influence on the rest of the nervous apparatus of the animal in both its excitatory and inhibitory moments.[79]

He put it more forthrightly in his copy of *Integrative Action of the Nervous System*, where he translated into religious language Sherrington's physiological description of the integration of the organism by adding an epigraph from fourth-century ascetic Ioann Zlatoust (John Chrysostum): 'The soul occupies everything and imparts to the diversity a certain unity'.[80]

Religious Recentering and Communion With the Holy Fathers, 1911–1921

Ukhtomsky successfully defended his thesis in May 1911, though the response to it was tepid at best. Vvedensky recognized the work as a continuation of their longstanding disagreement, but praised its virtues and continued to view its author as his talented protegé.[81]

The strains of Ukhtomsky's religious mission in an alienating secular culture reached the breaking point one month later. Setting aside his experiments, he confessed in his lab notebook to a 'cracking' and 'splitting' in his inner life, and plunged into spiritual crisis.[82] A friendly reunion with mentors from the Moscow Theological Academy had elicited memories of his former classmates there that cast a disturbing light on his own life. Those classmates, now all in service to the church, radiated fulfillment and joy, while he was mired in an 'altogether oppressive and dark spiritual state,' suffering 'extreme dissatisfaction with my affairs, with my moral task [and…] all this life in a foreign milieu.'[83] Returning to Rybinsk, he sought spiritual guidance in prayer, meditative reading of religious philosopher

79 Ukhtomskii, *O zavisimosti*, p. 189.

80 Ukhtomsky's copy of Charles Scott Sherrington, *The Integrative Action of the Nervous System* (London: Constable and Co., 1911), St Petersburg, ORK У/603. His earlier edition of this volume has been lost.

81 Ukhtomskii, 'Parabioz i Dominanta,' *Izbrannye Trudy*, p. 93.

82 St Petersburg, ARAN 749.1.152: 1.

83 Letter from Ukhtomsky to Platonova, 25 July 1911, St. Petersburg, ARAN 749.3.69: 19–34rev.

Sergei Bulgakov's *Two Cities*, and the writings of Holy Fathers Gregorii Sinait (Gregory Sinaites), Simon Blagogoveinyi (Symeon Eulabes, Symeon the Devout), and Maksim Ispovednik (Maximos The Confessor).

In a soul-searching letter to Platonova, he shared his conclusions. He was guilty of 'selfness' (*samost'*), of devaluing his relationship to God in favor of worldly, egotistical preoccupations. 'There is too much 'self,' too much 'my way''; an 'internal subtle pride' manifest in his desire to contribute a 'new word, as if salvation resided in novelty'. The world was awash in such self-serving novelties — masses of papers, books, letters, and, referring to his own scientific efforts, 'So many [...] perished frogs — unfortunate, stupid frogs, having been bloated to bursting for the new word'.

> I would be completely peaceful if I were certain that the path upon which I embarked and the path that [his former classmates] walk is essentially one and the same — the path of Christ — and differ only by external form. But I was horrified by the thought that I, as if subordinated to the prevailing spirit, departed from the real path of Christ, and, without noticing, gradually distanced myself from it.[84]

Absorbing Bulgakov's discussion of Christian asceticism (*podvizhnichestvo*), Ukhtomsky identified the essence of his problem, not in his scientific work, but rather in the egotistical spirit in which he pursued it. Christian asceticism demanded renunciation of self and the quiet, determined fulfillment of one's duty to God as captured by the concept of 'obedience'. That concept, Ukhtomsky observed, was 'applicable beyond the monastery to any work whatsoever'; one could be 'obedient' in science as well.

Bulgakov's description of Christian asceticism defined the way back: 'To turn your life into invisible self-denial, obedience, to perform your work with one's greatest effort, self-discipline, and self-control, and to see in it and in yourself only an instrument of Providence'; to transform one's personal will, experience and life 'by a continuous and unflagging feat to the fullest possible penetration by the will of God.' This was a struggle for transformation of 'the human person itself'.

Ukhtomsky described in this letter his effort to combat selfness through prayer, guided now by intensive study of the Holy Fathers, to whom he referred again as 'great psychologists'. Bulgakov's passages on the three virtues in prayer (restraint, silence and humility) made a deep impression. He now dedicated himself to their constant practice. Commenting on Gregory Sinaites's 'Instructions to the Prayerfully Silent,' he reminded Platonova that these virtues 'must be faithfully observed in prayerful silence, and every hour we must test whether we are always in them, so that we are not caught in oblivion and somehow venture outside them'.[85]

---

84 Letter Ukhtomsky to Platonova, 25 August 1911, St. Petersburg, ARAN 749.3.69: 35–46rev.

85 Letter from Ukhtomsky to Platonova, 25 August 1911, St. Petersburg, ARAN 749.3.69: 35–46rev.

This self-conscious turn from 'selfness' toward 'obedience' constituted an important turning point in his intellectual and spiritual life. As he put it retrospectively ten years later, he now consciously rejected his previous way of being — of leading his life and seeking truth through a self-centered 'spontaneity' originating 'from some intimate depths of the spirit' — and struggled to replace it with a constant 'vigilant effort of attention and the `governing mind', 'prayerful listening to and permeation by the Single and Exclusive Truth'. 'I returned to the search for and acquisition of Truth *in prayer*.' People had noticed, he confided in his diary, how profoundly he had changed.[86]

As earlier with Church singing, Ukhtomsky's participation in Old Believers' services played a central spiritual, experiential and epistemological role. As he put it in an autobiographical note: 'The most invaluable opportunity to delve deeply into the spirit of Russian Orthodoxy is through acquaintance with the Old Believers. The fullness of the specific psychology of Orthodoxy is revealed here, whereas in the `ruling Church' one senses only the consumptive remnants of something that once lived a full and good life, but now has withered in the rough hands of rude people'.[87]

This religious recentering defined Ukhtomsky's religious and scientific life during the years 1911–1921 — that is, between his Master's thesis and his conception of the dominanta. From a biographical perspective, those were tumultuous and harrowing years: he lived through world war, revolution and civil war; witnessed the sacking of churches and the arrest, exile, and execution of priests and other members of his circle; and was himself arrested by the Bolsheviks and imprisoned for more than a month.

Yet for Ukhtomsky — who viewed all this as but the latest chapter in an Apocalypse that had begun with the corruption of the Church in the seventeenth century — this decade was, above all, characterized by the experiential and intellectual synergy between constant prayer and a deep intellectual engagement with the Holy Fathers. Prayer matured as paradigm, and his thinking about 'attention' became fully embedded in the Eastern Orthodox narrative of The Fall and Salvation — a narrative in which he assigned the dynamics of attention a central role.

One central theme of this narrative was Christian dialectics applied to subject and object, mind and body, Man and God. Ukhtomsky's basic view is captured in an aphorism that he attributed to the fourth-century theologian and Archbishop of Constantinople Grigorii Bogoslov (Gregory of Nazianzos), and cited repeatedly: 'Matter is given us for the exercise of freedom'.[88] In marginal notes to his copy of *Prolog* (the Eastern Orthodox Book of Days), he expressed the two sides of this view: 'Both the body of Man and the soul are divine angels put at his service,' yet '"All the senses in Man resist faith" [...] "The earthly tendencies of

86 St Petersburg, ARAN 749.1.148: 5rev–8rev (diary entry of 3 May 1921).

87 St Petersburg, ARAN 749.1.93: 31.

88 For example, in his copy of Dobrotoliubie, II, p. 12.

the flesh pull away from the path [...] and the feat of attention [to God]" and toward purely sensual pleasures'.[89] The body — and matter in general — was not evil, but rather part of Divine existence (*bytie*) and a contradictory resource and field of action for human will.

For Ukhtomsky, the dynamics of this mysterious unity resided at the heart of the 'entirely new, exclusively Christian doctrine of the body'. He was commenting here on a sentence in *Prolog* about Mother Mary: 'The embodiment of God's Word, in her resides the entire bodily habitat of Divinity'.[90] The body was capable of aspiring toward God and even of resurrection. In humans, conscious will (or, in his lexicon of these years, attention) played an essential role in gathering up and directing this bodily resource.

Ruminating about one passage in the work of John Chrysostom, whom he read with particular intensity during the civil war years (1918–1921), Ukhtomsky wrote in his diary:

> For us Christians, matter, the body, the concrete reality of history — in its opposition to the idea — is not `evil'. Matter resists the idea not because it is fundamentally inert and bad while the idea is fundamentally marvelous and spiritual [...] Matter can also resist the idea because the idea itself can be the deed of a perverse will, and [matter's] resistance inculcates a new and better one! Matter resists. Human will also resists — resists the evil idea it has constructed.[91]

That play between matter and consciousness related to a general issue of direct concern to him as a Christian and scientist: 'On the struggle of Saint Ioann [Chrysostom] with the idea of `Fate'. For human life, mechanistic determinism becomes `fatalism'. And if they want to avoid fatalism, then for human life there is required determinism of another order.'[92]

This 'determinism of another order' rested upon the interplay of natural law and God-given free will, with the latter itself subject to the development of individual spiritual qualities and Man's struggle to overcome the legacy of The Fall. Here he recognized both the influence of outside stimuli upon human moods and actions, *and* the important, active role of human agency (in the form of attention). Decadent music interfered with prayer and corrupted worshippers, but those inspired by znamenny chant *actively* attended to the spiritual content of the

---

89 Ukhtomsky's copy of *Prolog*, ch. 1, at St Petersburg, Literaturnyi Muzei Instituta russkoi literatury (Pushkinskii Dom) Rossiiskoi Akademii Nauk, rukopisnyi otdel, pp. 55, 203. (For this and other volumes paginated in Old Slavonic, we provide Arabic numbers.)

90 Ukhtomsky's copy of *Prolog*, p. 13. Our thanks to Lawrence Principe for discussion of Mother Mary and the theology of body and spirit.

91 St Petersburg, ARAN 749.1.148: 99. Diary entry dated 5 November 1921.

92 Ukhtomsky's copy of *Polnoe sobranie tvorenii sviatogo ottsa nashego Ioanna Zlatousta., arkhiepiskopa Konstantinopol'skogo, v russkom perevode*. Tom 6, book 1 (Sankt-Petersburg: Sankt-Peterburgskaia Dukhovnaia Akademiia, 1900), p. 85. St Petersburg, ORK U/1951.

service; some individuals responded carnally to the sight of a beautiful woman, but others with appreciation of God's creation. Preoccupied now with prayer and the quest for salvation, Ukhtomsky put increasing emphasis upon the active moment in this dialectic.

His epistemological thinking took firm form around this same perspective. He read extensively in philosophy, theology, and history during these years, peppering his books with references to the Holy Fathers.[93] For him, existence (*bytie*) and God had an immanent reality, but the human relationship to it depended upon experience and faith. Idealist philosophy and mind/body dualism were expressions of the arrogant philistinism of the intelligentsia. 'When I understood that it is not my idea that dictates to the world its laws, I immediately ceased to be an `idealist'. Since an idealist is the dictator of Existence.'[94] Yet reality did not simply imprint itself upon human consciousness in the form of sense impressions, ideas, or science to produce trustworthy and unlimited knowledge. The positivist accumulation of facts failed to penetrate to 'the meaning of existence,' and scientists were mistaken to identify that 'meaning' with atoms, the ether, or the vibration of electrons. 'The meaning of existence is provided by [...] forces beyond positivism!'[95] Human knowledge was tendentious and shaped by sense organs, ideology, and, especially, the presence or absence of faith.

Ukhtomsky put the dialectics of mind/body and attention at the very heart of the Eastern Orthodox narrative of 'the natural Adam', The Fall, and Salvation. This new synthesis reintegrated his religious faith and science in a manner fully consonant with his religious recentering, with the active, God-centeredness of the Christian ascetic. We can here only illustrate this briefly through Ukhtomsky's commentaries on that narrative.

Having found solace and direction in the writings of Maximos The Confessor during his spiritual crisis of summer 1911, he engaged deeply with the landmark study of the seventh-century ascetic theologian by S. L. Epifanovich, who portrayed Maximos in 1915 as a central figure in Eastern Orthodoxy, comparable to St Augustine in the West, and used him to provide a synthetic view of the faith's mystical tradition.[96]

Many of Ukhtomsky's detailed notes concerned Epifanovich's account of Maximos The Confessor's exegesis of the history of Man's relationship to God: As a 'microcosm', the 'natural Adam' (pre-lapsarian Man) possessed everything

93 See, for example, Ukhtomsky's copy of P. D. Uspenskii, *Tertium Organum: kliuch k zagadkam mira*, 2nd ed. (Petrograd: N. P. Taberio, 1916), ORK U/466; of N. F. Kapterev, *Patriarkh Nikon i ego protivniki v dele ispravleniia tserkovnykh obriadov*, 2nd ed. (Sergiev Posad: Tip. Sv.-Troits. Sergievoi lavry, 1913), St Petersburg, ORK U/560.

94 Ukhtomsky's copy of Pobedonostsev, *Moskovskii sbornik*, marginal note on inside cover.

95 Ukhtomsky's copy of Uspenskii, *Tertium Organum*, p. 125 (positivism and sense); 9, 126 (scientific elements).

96 Ukhtomsky's copy of S. L. Epifanovich, *Prepodobnyi Maksim Ispovednik i vizantiiskoe bogoslovie* (Kiev: Izd. M. S. Elova, 1915). St Petersburg, ORK U/474.

necessary for his task of uniting created and uncreated nature, and achieving his own deification (*obozhenie*, theosis). Ukhtomsky focused upon this passage:

> [Man] was supposed to subordinate all these elements one to another, the lower to the higher, and in this way to unite them into a single aspirational movement toward God [...]
> If Man had fulfilled this assignment — if, subordinating in himself the lower desires to the higher, the body to the mind, and the mind to God, he would have [...] established the corresponding harmony of the parts of the world [...] in one large, well-structured organism. United in the end with God, he would have been flooded by the currents of deification, and through Him all of worldly existence.[97]

Alongside the first paragraph, he wrote 'The unification of everything in the unity of Attention'; in the margin of the second: 'Sacred Attention and Unification as the means of the gravitation of existence toward God'; and 'Man — master of all the created and its attention to God and deification in Him!' The word 'attention,' which did not appear in this passage of Epifanovich's text, had special significance for Ukhtomsky as the central psycho-physiological theme in his essays on Church singing and his Master's thesis, and now in his developing reflections about Eastern Orthodoxy.

In passages that Ukhtomsky underlined heavily in emphatic red, Epifanovich explained the monk-theologian's vision of the disordered, unnatural state that resulted from The Fall:

> Under the influence of sensory knowledge of nature as a source of pleasure and suffering [...] carnal attractions prevailed. Reason lost the power of moral guidance over the unreasonable (passionate) forces of the soul — lust and irritability — and was subordinated to their erratic movements, which caused Man only to seek pleasure and avoid suffering, and thus he fell under the power of carnal self — love [...] caused by the struggle for worldly goods, for the sources of pleasure [...] Thus developed these basic unnatural passions, and others from them.[98]

His interpretation of this passage again rested upon that same keyword: 'Decay, disintegration of the self as a consequence of sin. The falling away from attention, from the real unity of spirit and being, into an indefinite multiplicity of self-assertions in everything'. (Note his definition of attention as 'the real unity of spirit and being'.)

He studied carefully Epifanovich's account of Maximos The Confessor's exposition on the struggle against the passions unleashed by the unnatural Adam's disordered and fragmented state, a struggle that defined the 'path toward salvation'

---

97 Ukhtomsky's copy of Epifanovich, *Prepodobnyi Maksim*, p. 59.
98 Ukhtomsky's copy of Epifanovich, *Prepodobnyi Maksim*, pp. 62–63.

(salvation itself was an act of Divine grace) and mystical communion with God. Christian asceticism was directed toward this end through 'the gradual recovery of the forces of the soul,' which relied upon the contemplative and mystical powers of 'the governing mind' and its disciplining of bodily habit.[99]

The locus of this mind/body interaction was prayer, in which, as Ukhtomsky put it in a marginal note to *The Lenten Triodian*, 'The method of salvation begins with the discipline of attention'.[100] Prayer became the constant, pervasive center of his daily life. Describing himself as a 'monk in the world', he strove to make prayer as central to his existence as it was to monastery ascetics.[101] Prayer practice thus also became paradigmatic for his thinking about mind/body dynamics. The preliminary, undeveloped insights of his essays on church singing and his Master's thesis matured during the decade within that context, and so within what he termed the Holy Fathers' 'culture of attention,' 'empiricism' (analysis of experience), and 'physiologism'.[102] Developing the 'entirely new, exclusively Christian doctrine of the body', he discovered that these 'great psychologists' were 'much more physiologists than is generally thought'.[103]

Ukhtomsky's personal papers, letters, and marginalia; and the memoirs of his occasional guests, offer fragments of information about his prayer practices. He strove to be *constantly* at prayer. One must, he wrote to Platonova, practice 'the constant unrelenting discipline of prayer in our daily life'. The Holy Fathers taught that 'prayer is not, of course, connected with either place or circumstances [...] It is for always and everywhere [...] an internal thought in the heart'.[104] His letters to her also included texts of the prayers he composed himself. Scattered notes attest to his effort to maintain an inner prayerful silence when compelled to participate in secular society. For example, reading Ephrem the Syrian's injunction to monks against interrupting their prayer when traveling lest 'their mind coarsen', Ukhtomsky added the marginal note: 'Poor Professors!' — perhaps having in mind the challenges to prayerful silence during his days at the university.[105] In his copy of *Prolog*, he jotted another marginal reference 'To Professors' — this time alongside a text that warned of the perils of empty talk and counseled that, when interrupting prayer for a necessary conversation or meeting, one must ask God's

99 Ukhtomsky's copy of Epifanovich, *Prepodobnyi Maksim*, pp. 84–86, 91–93.

100 Ukhtomsky's marginal note to *Triod' Postnaia*, p. 442, at St. Petersburg, ORK U/967.

101 A. V. Kazanskaia (Koperina), in *A. A. Ukhtomskii v vospominaniiakh i pis'makh* (St Petersburg: St Petersburg University, 1992), p. 65; A. A. Zolotarev, St Petersburg, ARAN 749.3.90: 2, 5, 7, 10, 19.

102 'Culture of attention,' in Ukhtomsky's marginal note to A. S. Orlov, *Iisusova molitva na Rusi v XVI v.* (St Petersburg: Tip. M. S. Aleksandrova, 1914), p. 1. St Petersburg, ORK U/858; 'empiricism' and 'physiologism,' in ARAN 749.1.92 (6): 104.

103 Letter from Ukhtomsky to A. I. Miakutin, 1913 (fragment), St Petersburg, ARAN 749.4.26: 184rev–185.

104 Letter from Ukhtomsky to Platonova, August 25 1911, in St Petersburg, ARAN 749.3.69: 44–45rev.

105 Ukhtomsky's copy of *Dobrotoliubie*, vol. II, p. 460.

blessing to do so and cleanse one's heart in preparation. He apparently attempted to maintain a prayerful state (perhaps through the internalized rhythms of the Jesus Prayer) during those activities.[106] His personal papers contain frequent comments about liturgies he attended at Nikol'skaia Church, and also at Old Believer and mainline Eastern Orthodox churches (some of these, particularly Easter services, inspired his thinking). Until November 1923 he made confession to his spiritual father, hegumen Irodion of St Petersburg's Kievo-Pecherskoe Courtyard.

He was generally silent about his experiences during prayer, as 'an inner voice, a secret instinct, impels one [...] to leave this for the most secret corners of his conversations with the Almighty'.[107] He did, however, leave occasional references to bodily experiences during prayer, and to his struggles to achieve the disciplined attention that prayer demanded. He wrote at length to one fellow Old Believer intellectual, for example, of the need to read holy texts 'while standing'. 'In order to experience the writings of the ancient Fathers, one must [engage] the entire person, both heart and will; and this requires a certain tension, a certain `tonus' of the spirit — as physiologists and neurologists would put it'.[108]

One room of his small apartment was devoted to prayer, and here only a few closest friends could enter. Their memoirs mention the traditional rhythmic singing of his domestic liturgies, with prayers and psalms — many of which he knew by heart, and some which he composed himself. He painted icons for himself, and iconic images that used to illustrate the specifications for those he ordered from leading Old Believer iconographers for the room's iconostasis. That room contained also his sizeable collection of religious texts. Reading and rereading them (sometimes during prayer), he filled his diaries and correspondence with long commentaries on John Chrysostom, Gregory of Nazianzos, John Klimax, Maximos The Confessor, Isaac of Nineveh, and Simeon Novyi Bogoslov (Symeon The Theologian, Symeon The New Theologian).

Ukhtomsky's diaries, letters and marginalia of 1911–1920 are permeated by references to the dynamics and over-riding significance of attention in prayer. In his copy of volume 2 of *Dobrotoliubie* alone, that word appears in 44 separate marginal notes.[109] For him, 'attention' was defined by its nature and role in the Eastern Orthodox narrative: it was the remnant of the natural Adam's communion with God and the instrument (or 'organ') for recovery of that state. He underlined this passage from Isikhii (Hesychios of Jerusalem): 'Attention is the incessant silence, free of any thought form, of the heart, by which it breathes always, continuously and unceasingly, of Jesus Christ, the Son of God, and Him

106 Ukhtomsky's copy of *Prolog*, ch. I, p. 266.

107 Letter from Ukhtomsky to Platonova, July 7, 1917, St Petersburg, ARAN 749.3.70: 1–4rev.

108 Letter from Ukhtomsky to A. I. Miakutin, 1913 (fragment), St Petersburg, ARAN 749.4.26: 184rev–185.

109 Even in his marginalia to religious texts, the word 'attention' largely replaces his earlier references to 'will' and 'spirit' (as it had earlier in his second essay on church singing).

Alone.' This prayerful state, Ukhtomsky commented here, 'reveals the path toward new experience' — that is, toward restored communion with the Divine.[110] 'It is specifically in prayer', he wrote in his copy of *Prolog*, 'that Man attains his [pre-Fall] norm, like a tsar, commanding his entire being with his governing mind and enslaved to nobody'.[111]

Studying 'Chapters on Prayer' by Nil Sinaiskii (Nilus of Ankyra), Ukhtomsky underlined the sentence 'Attention seeking prayer will acquire prayer'. In standard theological language, this referred to '*umnoe delanie*' — literally, 'mental acting', that is, the unified contemplative action of mind/body under the rule of the mind. In a marginal note, Ukhtomsky termed it '*umnoe vnimanie*' — that is, 'mental attention'. There follow several underlined passages to which he added the marginal note '*delanie vnimaniia*' — that is, the 'action [result] of attention', of that specific psycho-physiological force in 'mental acting' essential to prayer.[112] Ukhtomsky internalized the Holy Fathers' emphasis on the unceasing effort, discipline, and self-training required to attain and sustain attention, the mind/body unity of attention in action, and the active role of attention in the struggle against thought forms and other obstacles to prayerful communion with God.

He fastened upon a passage of John Cassian's 'The Struggle with the Eight Main Passions' in which the monk-theologian explained that to attain a virtuous state one must combine fasting and 'bodily labor and handicraft' (which 'keep the heart from wandering') with 'repentant contrition of the spirit and relentless prayer against the [Devil]' and the reading of Scripture. Only then can 'mental acting' triumph over passion. Ukhtomsky commented in the margin: 'bodily discipline', 'intense labor toward the discipline of attention'.[113]

He cross-referenced that last note with a page in Nilus of Ankyra's 'Chapters on Prayer': Emphasizing that prayer requires a mind free of thoughts and images, the monk-theologian explained the importance of bodily activity:

> Inaction nourishes passions and gives freedom to wishes that aspire toward subjects related to them — which is especially manifest during prayer, since then the attention of the mind is completely consumed by whatever occupies the heart, and [...] among its thought forms it picks out those suggested by the passion that has come into motion, rather than conferring with God. Those who reject handicraft under the pretext that one must constantly pray actually do not pray either. Thinking that by idleness they free the soul from preoccupations, they entangle it in a labyrinth of inescapable thought forms and so render it incapable of prayer. The body laboring upon a task holds to itself the thought, which no less than the eyes should observe what is being done and assist the body for infallibility in action.

110 Ukhtomsky's copy of *Dobrotoliubie*, II, p. 158.

111 Marginal note to *Prolog*, ch. 2, p. 45.

112 Ukhtomsky's copy of *Dobrotoliubie*, II, p. 225.

113 Ukhtomsky's copy of *Dobrotoliubie*, II, p. 33.

Ukhtomsky responded with the marginal note: 'Toward the psychology of prayerful acting. During prayer, attention, having descended within [the heart], sorts out the states and contents, as in a mirror'. And he added: 'From this the need for physical labor during prayer is again understandable — the significance of ritual, bowing, their measuredness, etc'.[114]

Reading closely John Cassian's 'The Struggle with Thought Forms and through them with Evil Spirits', Ukhtomsky emphasized the monk's point that God left fallen Man with the ability to control these fragmentary thoughts and feelings, to attain a prayerful state through 'constant exercise and constant training'. He added: 'Somebody said 'Genius is attention.' But there it is a gift, an inborn quality. Here it is an achievement — the norm of life — the establishment of the governing mind over all the rest of the being'.[115]

In two letters to Platonova, he briefly summarized his understanding of the centrality of attention in prayerfulness: 'The truth is not a matter of naked mind as such nor of the passive experience of the heart, but of active, ascetic, intense attention to one's mind and heart, to the `purification of thoughts,' as the Fathers said'. Writing from his own experience, he put it this way: 'It takes great attention to enter your internal world and sort out what there is reliable stone that can be used in construction, and what is straw, barely glued together with manure, which will burn as soon as it is touched by fire!'[116]

Attention was also 'the angelic guardian' of the united Church in the struggle against Satan. He jotted that phrase in reference to the Feast Day of Archistratege Michael and Other Bodiless Heavenly Forces, which celebrates the Angelic Synaxis by which Archangel Michael united the angels in holy war against Satan and other fallen spirits. In his copy of *Prolog* Ukhtomsky underlined the account of St Michael's gathering of the angels and his instruction that they not even ponder those who had joined Satan, but rather 'Let us be attentive and scrupulous in our tasks before the Creator'. Alongside the declaration that 'Such a gathering was called the angelic assembly, which is attention and unification,' Ukhtomsky noted the contrast between St Michael's dramatic slaying of the Satanic dragon and the approach embodied in the Synaxis: 'See the icon of Archistratege Michael *in monkhood* defeating the devil. This [the Synaxis] is another path — the path of attention to the Word of God and His wisdom'.[117] Here attention to God was an alternative to direct combat with Satan.

On yet another level, attention was a force for cosmic unity in devotion to God, to which Ukhtomsky referred as 'Sacred Attention' (*Sviatoe Vnimanie*). While studying Epifanovich on Maximos The Confessor's analysis of Man's intended role as unifier of the created and uncreated, Ukhtomsky (as we have seen)

114 Ukhtomsky's copy of *Dobrotoliubie*, II, pp. 228–29.
115 Ukhtomsky's copy of *Dobrotoliubie*, II, p. 99.
116 Letters from Ukhtomsky to Platonova, 27 January 1913, and 7 September 1913, St Petersburg, ARAN 749.3.69: 63rev, 68–71rev.
117 Ukhtomsky's copy of *Prolog*, ch. 1, p. 142.

summarized this in three pithy notes: 'The unification of everything in the unity of Attention', 'Sacred Attention [*Sviatoe Vnimanie*] and Unification as the means of the gravitation of existence [*bytiia*] toward God', and 'Man — master of all the created and its attention to God and deification in Him!'[118]

For Ukhtomsky, then, actively disciplined body/mind 'attention' was the means by which the individual, the Church, and the cosmos overcame the legacy of the Fall to restore harmony and communion with God.

During the decade 1911–1920 he conducted very little scientific research, initially because of his absorption in his religious quest, and then, at least in large part, because of the disruptions of the civil war. The scientific dimension of the questions he was addressing in a religious context, however, remained much on his mind, as is evident in the notebook he opened in 1916 on *Literature about Reflexes*. Amid detailed critiques of two classics of mechanistic reflexology, René Descartes' *Passions of the Soul* and Ivan Sechenov's *Reflexes of the Brain*, he commented on Sechenov's observation that, for a person with 'an ideally strong will,' it was easier to fully suppress reflexive responses than to augment them qualitatively. Ukhtomsky responded to that insight approvingly: 'It is easier to be an Ascetic than to become a new Man!' Here, he used the word 'ascetic' (*asket*) to denote those who, viewing the body as evil, limited their spiritual practice to the mere inhibition of passions; as opposed to the 'Christian ascetic' (*podvizhnik*), who, based on the 'entirely new, exclusively Christian doctrine of the body', understood mere inhibition as just one step toward enlisting body and mind in prayerful obedience and the quest for salvation.[119]

Until at least 1917, Ukhtomsky periodically wavered about his life path and considered leaving the university for full-time religious service. On such occasions Platonova reminded him that he was not seeking professional or personal advancement, but rather 'for you, the summit is theological works in which a knowledge of physiology will serve as the foundation'. Even if he left the university, he must complete his projected doctoral thesis in physiology. 'God has given you such talent [...] I believe that God has blessed you to write this cherished work'.[120]

In November 1920, the developments in Russia that Ukhtomsky had been describing in apocalyptic terms touched him directly, justifying his 'expectation of terrible trials'. The Rybinsk Soviet was confiscating private residences, and at the urging of his closest friends there he returned to save his home and its mountains of books, icons, papers and notes. He bore a letter from Petrograd University attesting to his status and supporting his claim to his home and scholarly materials.

118 Ukhtomsky's copy of Epifanovich, *Prepodobnyi Maksim*, p. 59.

119 Ukhtomsky's unpaginated notebook *Literatura o refleksakh*, in Rybinsk, Gosudarstvennyi Arkhiv Iaroslavskoi Oblasti [hereafter: Rybinsk, GAIO], 39752 DD 3414.

120 Letters from Platonova to Ukhtomsky, n. d., St Petersburg, ARAN 749.2.490: 266rev–267; 749.2.400: 89.

Upon arrival, however, he was accused of counter-revolutionary sentiments, arrested, and taken to the headquarters of the Rybinsk Cheka (Bolshevik political police), where he came harrowingly close to execution. He was informed that he was guilty of a capital offense, 'and then a stocky, elderly, and entirely gray man entered and asked, in the voice of a veteran slaughterhouse fighter, if everything was ready, and turned to me as one destined for slaughter: "Well, get going"'. Escorted to the basement, he listened as, just around the corner, other condemned prisoners were shot and their bodies buried in the courtyard. The hours spent in contemplative horror ended when a message from the Petrograd Soviet saved his life. His would-be executioners were 'unpleasantly amazed when it was decided [instead] to send me to Yaroslavl'. From there he was transferred to the Cheka's special section at the Lubyanka prison in Moscow, where he spent more than a month before the intervention of friends won his release in January 1921.[121]

This terrifying encounter with death, with the apocalyptic changes in the consciousness and behavior of people in his beloved Rybinsk — and the prospect that his manuscripts, notes, and annotated books recording decades of work might be confiscated and disappear in an instant — made a powerful impression. He would never return to Rybinsk. Yet there also followed 'such minutes of special uplift, which sometimes come to an elderly person whose earlier receptivity and impressionability return to him, with remembrance of what he faces in life! These are moments of special joy, or special grief, moments of "emotional storms"'.[122]

This psychological state brought his gradually developing ideas of the past decade about Christianity and religious experience into sharp focus and an integrated whole. Stranded in Moscow for two months after his release while awaiting permission to return to Petrograd, he opened a new diary and quickly filled some 240 pages with his thoughts. In May, still afire, he opened a new one and kept writing.

The marginalia and notes of 1911–1920 were now replaced by dense paragraphs and a synthesis. These diaries record the crystallization of his accumulated thinking in an intellectual and spiritual 'break,' a new resolution of his longstanding quest to unite the 'two paths' that would soon find its scientific expression in his concept of the dominanta.

The central moment in this break was what he termed his 'discovery' of the profound significance of 'the most essential and deliberate feature of authentic religion', which he described while inspired by the spring 1921 Easter services. His epiphany was that Christianity for the first time put Divine Truth 'outside and above man as an object of independent aspiration' and that this provided

121 Letter from Makarova to Ukhtomsky, September 14, 1920, St Petersburg, ARAN 749.2.482: 70–73rev; Letter from Petrograd University to Rybinsk Soviet, 19 November 1920, Rybinsk, GAIO, Ukhtomskii collection, DD-166 RBM #10038/9a; Letter of Petrograd Soviet, 20 November 1920, Rybinsk, MDM, RBM-33773.DS 6768; Letter from Ukhtomsky to Platonova, 30 November 1921, St Petersburg, ARAN 749.3.70: 118–119rev.

122 St. Petersburg, ARAN 749.1.90: 61rev.

a powerful impetus for 'overcoming in man the spirit of self-affirmation and utilitarian centripedality to self'.[123]

> For the Christian, Christ is truth above and beyond consciousness, independent of all things, and at the same time He is the intrinsic motive of every action and thought, for before Him [...] there is no motive or thought, for the Truth itself is Christ. This state of mind is the entirely exclusive heritage of the Christian, making him a "monk" in the midst of the rest of the world, which takes as the ultimate criterion self-satisfaction, self-affirmation, egocentrism, pagan pleasure.[124]

This 'aspirational movement' of the 'whole person' in 'all-encompassing attention' toward that highest Truth now moved decisively to the center of his reflections on the religious experience.

This, in turn, led Ukhtomsky to fundamentally reformulate his longstanding approach to the relationship between science and religion. After leaving the Theological Academy, he had been inspired by Avenarius's scientific method for studying the organism's physiological 'adaptation to reality'. He had hoped to extend that method to 'the organ' of religious experience — and so, with Avenarius's 'blessing', had entered the university and begun his studies with Vvedensky. Now, however, he found in The Book of Job a 'classical Biblical critique' of this 'utilitarian principle'. Approaching religious experience as mere 'adaptation to reality', excluding Man's inherent aspiration to an immanent higher Truth, was 'by its spirit directly contrary to Christianity'.[125] It was as if a physiologist wanting to study the brain objectively began his investigations by 'excluding his own brain!'[126] On the cover of his Russian translation of Avenarius's *Critique of Pure Experience*, Ukhtomsky now wrote in emphatic red ink the dates '1910–1919', recording his decisive departure from this perspective.[127]

Investigation of religious experience, he now concluded, must begin with recognition of this higher Truth, with Man as an active subject in united mind/body aspiration toward communion with God: 'To give a scientific account of religious experience, one must first stand on the firm and unshakable foundation of faith, or have it in front of oneself in another person, and understand how, from his point of view, the entire content of everyday experience is illuminated and developed'.[128]

---

123 St. Petersburg, ARAN 749.1.148: 23rev.

124 St. Petersburg, ARAN 749.1.148: 23rev, 74.

125 St. Petersburg, ARAN 749.1.148: 23rev–24, 32.

126 St. Petersburg, ARAN 749.1.148: 36.

127 Ukhtomsky's copy of R. Avenarius, *Kritika chistogo opyta* (St Petersburg: Izd. I. V. Shestakovskogo, 1907), St Petersburg, ORK U/2085.

128 Ukhtomsky's copy of Garal'd Gefding [Harald Hoffding], *Sovremennaia filosofiia*. D. V. Viktorov, ed.; A. P. Polivanova, trans. (Moscow: Tovarishchestvo I. D. Sytina, 1907), pp. 232–33, in Rybinsk, MDM, RBM-6945.

The 'Ariadnean thread' for understanding the relationship between the 'two paths,' then, consisted in deepening his own religious experience and understanding that of the Holy Fathers, the 'experts on human nature [...] who established bodily discipline and feat as the propaedeutics of spiritual achievement'. 'What was the secret of the discoverers of Truth, what is constant and essential in their work when they were approaching Truth? What is their method?'[129]

The ultimate goal of this religious experience, of this struggle for Truth and reworking of the human whole, was 'new experience' that transformed Man into a 'new being' (*novaia tvar'*), recovering the 'natural Adam' in whom subject and object, Man and God, merged once again. Through religious experience, attentive prayerfulness to God, the worshipper 'discovers how it is psychologically possible for man to unite in the most intimate and unbreakable fashion (even fusion) with what is for him the highest Truth of all.'[130]

Ukhtomsky had, as we have seen, reflected intensively about these subjects for a decade, but now realized with new force and clarity the centrality in religious experience of the all-encompassing 'active aspiration' toward Truth through the attentive 'governing mind' and its significance for bodily-spiritual activity. He reflected in his diary that 'In order to educate the governing mind for truth, one must begin the cultivation with psychology and physiology — the correctly pre-educated spirit will acquire Truth!'[131] Not just mind, but also body gained added significance: 'The more difficult and necessary our task in relationship to flesh and blood', he wrote, 'the more reverently we should treat it, the greater sanctity to see in it!'[132] To approach higher Truth, to maintain and develop the 'new sensation' acquired in prayer — to reform human nature — required understanding of the psycho-physiological conditions that obstruct it and the mechanisms that facilitate it, that 'educate' the body from below through habit. In his diary, he returned to Maximos The Confessor:

> The psychophysiology of bodily feat, of rule and ritual, is with particular completeness developed by Maximos The Confessor. Man is morally educated from below, by bodily habit: to that which he is accustomed in everyday bodily life, in everyday bodily expression — such ideas will subsequently develop. Initially there is assimilated the external side, the external mode of realization, the external expression of virtue (τρόπος); and afterwards internal meaning and knowledge (λόγος) is gradually comprehended. Assimilation of any virtue always begins with an external feat in order, in the end, to make the virtue an internal possession of the soul.[133]

---

129 St. Petersburg, ARAN 749.1.148: 64, 67rev–68.
130 St. Petersburg, ARAN 749.1.148: 72rev.
131 St. Petersburg, ARAN 749.1.148: 51rev.
132 St. Petersburg, ARAN 749.1.148: 116rev.
133 St. Petersburg, ARAN 749.1.148: 63rev–64.

Here Ukhtomsky finds an understanding of psycho-physiology oriented, not toward the passive adaptation to life circumstances, but rather toward attaining virtue and morality.

Reflecting on the human ability to apprehend higher Truth, he drew upon Sherrington's discussion of the eyes and ears as 'receptors at a distance,' reconceptualizing 'reception at a distance' as the function of a united, active, higher psycho-physiological 'organ' of the 'whole person' that reached across space and time to produce an integral image. This process entailed distinguishing between visual-acoustical 'phenomena' and 'the possible content (qualities) of an object in itself'. To do so, Man actively 'projects' (or 'anticipates') its content through a highly individual psycho-physiological process in which judgment and imagination play an essential role.[134] This projection of objects, events and laws of existence preceded 'direct contact'. In that sense, both scientific theory and religious faith pursued the same goal —' anticipation of the approaching reality, pre-perception at a distance'.[135]

Here Ukhtomsky emphasized the interaction of the physiological and psychological — reflexive, physiological projections of reality were altered by the drive and ability to change it: 'The reflexive understanding of thought tells us that what exists is given to us only to move on to what should be; [...] reality is always given to our receiving apparatuses in order to change it into what it should be.'[136] Reflexive responses were not passive and predictable; they were often active, anticipatory, creative acts. This ability to change existing reality — to replace it with the 'necessary', 'desirable,' and 'morally proper' — now became integral for Ukhtomsky to 'the reflexive conception of experience' as applied to 'every living being'.[137] And for the Christian, the 'morally proper' involved the transformation of reality and oneself in order that 'Christ rule within'.[138]

The 'higher truths about Christ,' then, were the 'receptors at a distance' of religious consciousness, providing foresight into objects, events and laws of existence before direct contact with them.[139] The most 'active and far-sighted' of these receptors was 'the organ of conscience' — that 'secret voice within us that collects all sources and orders of knowledge, those inherited impressions from the life of the species; and, by particular anxieties and emotions of a higher order, warns of the necessary consequences of that which is occurring before us'. This organ was especially active in 'the most extraordinary epochs of life, when something fatal is being resolved, when paths are parting, when great possibilities

134 St. Petersburg, ARAN 749.1.148: 37rev–38.
135 St. Petersburg, ARAN 749.1.148: 52rev.
136 St. Petersburg, ARAN 749.1.90: 9.
137 St. Petersburg, ARAN 749.1.148: 71.
138 St. Petersburg, ARAN 749.1.90: 5rev–6; 749.1.148: 18–19rev.
139 St. Petersburg, ARAN 749.1.90: 17rev–18.

and consequences are foreseen.'[140] In times, that is, such as those that Ukhtomsky himself was living through and had just experienced so harrowingly.

Religious and scientific experience were therefore both real parts of consciousness that sought to 'foresee real experience'. Just as scientific laws were a 'projection and construction of reality' ('an approximate construction of it') that were considered 'objective'; so the religious 'law of Retribution, law of Good and Evil, and — higher and later — law of Mercy, and law of the Interlocutor ('the hypothetical projection of the human person that I compose according to the interpolated data of experience') are considered 'true characteristics of Existence because they foresee real experience most distantly and in the greatest summation of their entire content'. The religious terms 'retribution,' 'good,' 'evil,' and 'mercy'; and the scientific terms for 'more elementary systems of experience' — terms such as 'mass,' 'force,' 'inertia,' and 'pressure' — were both distinctly 'human' (*chelovekoobraznyi*), that is, conditional, approximate, and psychological.[141]

This religious epiphany following his arrest in 1921 brought Ukhtomsky to a resolution of his longstanding quest for the 'constant psycho-physiological roots in human nature of religious knowledge and searches,' for the forces that maintained and revived that knowledge and search despite an unpropitious environment.[142] He had discovered an approach to the formation of the 'all-encompassing attention' of Man as a whole toward 'new experience,' which 'will expand and enrich the existence of Man, increase his stability and ability to respond to reality "actively," "progressively and expansively"'.[143] In so doing, he had arrived also at the framework for his principle of the dominanta.

The Dominanta as Scientific — Religious Concept

Ukhtomsky created and experimentally fortified his concept of the dominanta in 1921–1922, and explained it to the scientific community in a series of articles in 1923–1927. 'Attention' was replaced by 'dominanta,' and the tentative conclusions of his Master's thesis were now reconceptualized through his theological revelations and transformed into a scientific principle and line of investigation. As his friend and confidante A. A. Zolotarev put it, 'the scientific theory of the dominanta was at the same time religious, and even more so, Eastern Orthodox'.[144]

This is clear from the first mention of the dominanta in Ukhtomsky's extant papers, in a diary entry from August–September 1921. Here he already refers to 'my doctrine of the dominanta' (so, his first reference to it perhaps was recorded

140 St. Petersburg, ARAN 749.1.87(5) [1914?]: 34, 38; 749.1.90: 31.

141 St. Petersburg, ARAN 749.1.148: 53, 13rev–15, 38; and 749.1.90: 5rev–6.

142 St. Petersburg, ARAN 749.1.148: 27rev.

143 St. Petersburg, ARAN 749.1.148: 77rev–78rev.

144 A. A. Zolotarev, *Campo Santo moei pamiati* (St Petersburg: Rostok, 2016), pp. 687–88; St Petersburg, ARAN 749.3.90: 7.

in an earlier diary that is lost to us)[145] and clearly expresses the paradigmatic role of prayer in his thinking about it. In this entry, he ruminates about how a strong emotional state underlies, strengthens and feeds upon excitations. Drawing upon incoming stimuli, these emotions become independent 'creative principles' of human life that give rise to habit, 'the moral sense,' and 'the religious sense'. Conversely, a stimulus unconnected to such an 'emotional tone in the soul is doomed to the more or less rapid banishment from spiritual life'. The positive outcome of this process was evident in church and the negative in the theatre, where the results were 'of a lower order, in correspondence with the more primitive type of emotions and the more primitive "philosophy" by which theatre lives'.

'My doctrine of the dominanta in central nervous activity,' he continued, applied to the entire range of nervous phenomena, and in relation to higher psychic processes accorded with Freud, Bleuer, and Jung's notion of 'psychic complexes'. Here, the 'one-sided excitation' of a dominanta was 'connected and individualized' by an emotional tone that shaped a person's ideas and mentality. The dominanta also maintained its influence on the psyche when it 'descends below the level of consciousness,' as Freud demonstrated for hysteria, during which one psychic complex excluded another from the field of consciousness. Freud, Ukhtomsky observed, analyzed and treated this pathogenic complex by the 'complete abstraction of consciousness from external impressions' (the talking cure).

His own concept of the dominanta, however, was rooted in a much different, Eastern Orthodox tradition, and he had in mind a much different set of emotions, problems and treatment. For him, the paradigmatic dominanta was the attentive prayerful effort to commune with God, in which resided also the cure for harmful 'complexes' and other spiritual weaknesses:

> This is best, most seriously, accomplished during the prayerful concentration of attention, during the prayerful reading of one's soul. Look oneself in the mirror; move one's secret inner enemies into the light of consciousness; weave them into a healing, regenerative tissue! [...]
> Prayerful discipline is primarily the discipline of all-encompassing attention, illuminating all the corners and secrets of the subconscious, uniting and collecting the person [lichnost'] into one active whole, bound by a powerful emotion — the emotion of love for all Existence.[146]

He was now able to develop his ideas experimentally as a result of changes in the Russian context and his own professional status. Petrograd University was returning to life in 1920 with the winding down of the civil war. With the ailing Vvedensky's support, Ukhtomsky, now Professor of Animal and Human

145 St Petersburg, ARAN 749.1.148: 80–81. This dating accords with the recollection of Ukhtomsky's assistant M. I. Vinogradov that he first heard Ukhtomsky use the term 'dominanta' in 1921, when explaining his plans for the Peterhof lab. St Petersburg, ARAN 749.4.244: 34.

146 St Petersburg, ARAN 749.1.184: 81.

Physiology, became in 1921, a few months after his release from the Lubyanka, director of the department's small physiological facility in the Aleksandriia Park at Peterhof, on the Gulf of Finland just outside of Petrograd. Here, in 1920–1921, Ukhtomsky, his assistant M. I. Vinogradov, and several others began to investigate the nervous physiology of invertebrates; and Ukhtomsky attempted experimentally to elicit in mollusks the same 'attention of the nervous system' that he had studied one decade earlier in a swallowing cat.

They were joined in spring 1922 by a group of University students. In the dorm and lab of 'our wonderful Aleksandrija', Ukhtomsky sought to create an ideal scientific collective bound by a common mission and an ethos of mutual understanding and love — attributes that he related to Christian asceticism and the well-cultivated dominanta.[147] He prepared them for their research with a lecture course on his functionalist view of the nervous system — emphasizing the close relationship of mind and body, the active and variable quality of reflexive responses, and, with reference to John Chrysostom, the fallacy of equating reflexes with fatalism — and then set them experimenting on the mollusk and frog.[148]

During spring and summer 1922, Ukhtomsky joined three students in their investigation of mollusks and collaborated closely with another, Ida Kaplan, on frogs. His notebook on the former included vivid sketches comparing the mollusk's posture in 'a state of peaceful observation' with that of 'alert attention' elicited from the invertebrates by adding to their water weak doses of atropine and strychnine. (In another sketch, he portrayed this same attentive state in the investigators themselves).[149]

'Attention' morphed into 'dominanta' in the protocols of his and Kaplan's experiments on frogs. Applying strychnine and phenol to raise the excitability of the ventral and dorsal parts of the spine, the experimenters succeeded in creating both a sensory and motor dominanta in the rubbing reflex of the legs. As Ukhtomsky put it in the experimental protocols: 'The dominanta is comprised of two characteristics: first, the presence of a *local focus of excitation*, which attracts to itself stimuli from various receptor regions. And, second, *from the inhibition* of other reflexive acts during the time in which this dominanta developed.'[150]

---

147 Letter to I. I. Kaplan, November 12, 1922, in *Intuitsiia Sovesti*, p. 212. What Ukhtomsky refers to repeatedly as 'the human face' in his letters to Kaplan becomes 'the face of another person' beginning with his letter to E. I. Bronshtein-Shur, 6 April 1927; A. Ukhtomskii, *Intuitsiia Sovesti*, p. 252. This notion of seeing through the visible 'face' to a deeper reality is perhaps related to the worshipful engagement with the *lik* of icons.

148 For the lecture outline, St Petersburg, ARAN 749.1.151: 48rev–55. Ukhtomsky had commented on this passage of John Chrysostom in a marginal note of October 25 1920. This in his copy of *Polnoe sobranie tvorenii sviatogo ottsa nashego Ioanna Zlatousta*, vol. 6, p. 444, in St Petersburg, ORK U/1951.

149 St Petersburg, ARAN 749.1.91 (14): 16, 31rev.

150 Lab notebook for August–September 1922, Ukhtomsky papers, Rybinsk, GAIO, RBM-16811/1. DS-3716, *list* 39.

He excitedly (and somewhat ingenuously) informed the absent Vvedensky of a 'new theme that had arisen': he and Kaplan had demonstrated the possibility of creating a 'governing focus' of excitation in a nervous center, this focus 'attracts to itself' and 'responds sensitively' to 'the most distant and varied excitations,' and this changed all the involved reflex responses themselves. Reminding his mentor that this 'dominanta,' as he now termed it, was the same phenomenon he had discovered ten years earlier with respect to defecation and swallowing, he noted that research on its dynamics was based on Vvedensky's theory of parabiosis. 'This theme,' he concluded, 'interests me terribly'.[151]

Ukhtomsky's continued preoccupation with Eastern Orthodox practice and theory at the time, and its relationship through 'attention' to the dominanta, was reflected here in the term 'governing focus' (*gospodstvuiushchii ochag*), which derived from the 'governing mind' (*gospodstvuiushchii um* or *vladychestvennyi um*) through which the Christian ascetic disciplined mind/body during prayer. This term also appeared in his first publications on the dominanta.

The same preoccupation is evident in his letters to Kaplan while they conducted experiments and he prepared to present them and his new concept to the scientific community. Ukhtomsky related to Kaplan with deep affection, and wrote to her in 1922–1923 (interrupted for two or three months by his second imprisonment) about religious and spiritual issues. In one letter, he referred to the critical comment of another student-coworker, N. D. Vladimirskii, that Ukhtomsky, in his religious devotion, was 'departing from the world'. Ukhtomsky explained with citations from the Holy Fathers that one must not conflate the two different worlds in which humans reside: the conservative, dead, self-satisfied world of 'lust of the flesh, lust of the eyes, and worldly pride'; and the 'Divine world of beauty and light'. He invited Kaplan to engage with the latter through one of his icons: 'Sophia the Wisdom of God is the world [...] in aspirational movement toward the One Eternal Word of God, in attention to the One and Unity around Him'. This was the cosmos of 'progress, escape from oneself' toward God, the one in which he aspired to live and serve.[152]

In 1923–1928 Ukhtomsky published three synthetic essays on the dominanta as a psycho-physiological principle, and some ten articles on its specific nervous dynamics.[153] The rapid composition of the essays reflected the fact that he had

151 Letter of 7 September 1922 to N. E. Vvedensky, St Petersburg, ARAN 749.2.447: 13–16rev. That same religiously-resonant phrase featured in the first published definition of 'dominanta,' in Ukhtomsky and Kaplan's article 'Sensornaia i motornaia dominanta v spinnom mozgu liagushki,' *Russkii Fiziologicheskii Zhurnal* VI, 1–3 (1923), 71–88; on p. 71; and in Ukhtomsky's initial articles on the dominanta.

152 Letter to Kaplan, 19 November 1922, *Intuitsiia sovesti*, pp. 218–20.

153 'Dominanta Kak Rabochii Printsip Nervnykh Tsentrov,' *Russkii fiziologicheskii zhurnal*, VI 1–13 (1923), pp. 31–45; 'Dominanta I Integral'nyi Obraz,' *Vrachebnaia gazeta*, (1924), 26–29; 'Dominanta Kak Faktor Povedeniia,' *Vestnik Komakademii*, 22 (1927), pp. 215–41. We are citing these from Ukhtomskii, *Izbrannye Trudy*, pp. 7–19, 36–48, 63–90.

done much of the thinking behind them — in a largely religious idiom — during what seemed from the outside his scientifically fallow years of 1911–1921. The dialectic born during his simultaneous work on Church singing and his Master's thesis in 1910–1912 had deepened qualitatively, as reflected in the close relationship between his private religious ruminations and his scientific presentations. Here we can only touch briefly upon some of the basic arguments in his three synthetic essays and their underlying Eastern Orthodox framework and meaning.

In 'The Dominanta as a Working Principle of the Nervous Centers' (1923), Ukhtomsky defined this 'principle' as he had in his protocols to Kaplan's experiments and letter to Vvedensky, illustrated it with his lab's experimental findings on invertebrates, and extended it to vertebrates by the example of a cat in heat (which responded to food, not by meowing contentedly, but rather by a heightening of the symptoms associated with heat). He then moved to humans through an analysis of Tolstoy's description in *War and Peace* of Natasha Rostov's developing response to Prince Andrei Bolkonsky, and, most powerfully, by comparing the sensory receptors of mollusk and human.

When the mollusk moved through the water, it extended its tensed tentacles forward to sense through contact any threat to its main body. This working 'bodily pose', this integrated constellation of related nervous states, this dominanta — which Ukhtomsky had described and sketched in his lab notebook as 'a state of alert attention' — maintained and strengthened itself by incorporating some incoming stimuli and inhibiting reactions to others: When one ganglion was stimulated mechanically or by strychnine, all the ganglia responded; stimuli to the foot elicited, not a local response, but only strengthened locomotion and the tension of the tentacles.

In higher organisms, the contact-sensitive ganglia were replaced by brain segments that probed across space and time with what Sherrington termed 'receptors at a distance'. These receptors anticipated and reacted to dangers before direct contact, coordinating the animal's musculature 'as a whole, as a single machine'.[154] Ukhtomsky explained that this constellation of receptors:

> can be characterized as an organ with *a multiplicity of variable, extraordinarily long tentacles, from which one and then another is extended forward for the anticipation of events*; and the `experience' of the animal in the environment changes in dependence upon which tentacles it uses — that is, how differentially and how far it anticipates and projects its environment at the given moment.

The constellation's extension of some of these 'tentacles' with the inhibition of others, and the switching among them, was achieved by the formation of a

154 Ukhtomskii, 'Dominanta kak Rabochii Printsip', pp. 11, 14.

dominanta, which was thus the 'physiological basis of the act of attention and object thinking'.[155]

In his conclusion, Ukhtomsky commented briefly on various dimensions and consequences of his principle of the dominanta. He related it to his empiriocritical view that humans experience and apprehend the environment, not through basic elements or simple sensory experiences, but as packages (images) identified and created by dominanty (the plural of dominanta). When a dominanta is clear and lively, 'it holds in its power the entire field of spiritual life. Everything reminds of it and of the images and realities connected to it'. A dominanta can be positive or negative — a fruitful scientific idea or a destructive obsession. It was characterized by 'inertia' — that is, by 'the tendency to maintain and repeat itself' despite changes in the environment, including the disappearance of the stimuli that had given rise to it. Dominanty 'periodically swim into the field of spiritual work, into clear attention, and then leave the field. But they do not die'. Here, he incorporated the above-cited diary entry of two years earlier (page 159 above) regarding Freud's 'psychic complexes' (though he now omitted the religious references). Finally, he emphasized that the dominanta was not a morphological entity, a 'topographically united point of excitation,' but rather a 'complex of determined symptoms in the entire organism' — in the muscles, secretory work, and vascular activity' (for him, this complex constituted an 'organ').[156] These brief comments introduced subjects that he was preparing to elaborate in subsequent articles.

The underlying religious framework and meaning of this essay is implicit in its overall structure, which parallels that of his theological thinking as outlined in our previous section; in the relationship of attention and the dominanta, some specific language (for example, 'the *governing* focus of excitation') and some specific subjects. For example, in addressing the ability of a dominanta to maintain itself amid changing, often countervailing circumstances, Ukhtomsky, we think, had in mind his longstanding preoccupation with the difficult struggle to maintain religious understanding and faith in a world that actively dispersed attention.

He expressed this religious dimension explicitly in the extended 'conclusion' that he scribbled on his copy of the published article in *Russian Physiological Journal*. Expanding upon the above-cited religious insight of his diary of 1921 (pages 156-157 above), he elaborated on the nature and power of human receptors 'at a distance' and the integrated perceptual 'image' that the dominanta as 'organ' generated:

> Of course, the most powerful organ of contact with the environment is love, whether it be the love of the scientist for a beloved and lost idea, or the love of the Christian ascetic for lost truth and beauty, or sexual love, or the love of the mother and the child's caregiver, or, finally, the love for the face of Christ.

---

155 Ukhtomskii, 'Dominanta kak Rabochii Printsip', pp. 14–15.

156 Ukhtomskii, 'Dominanta kak Rabochii Printsip,' pp. 17–19. We have translated as 'spiritual' the word *dushevnyi*, which means literally 'of the soul'.

> That face becomes here the true organ of life. You are my organ, the organ of my life! When you are with me, my soul sings, and even more so my soul dies when you are not with me [...] And that is my happiness, but also my catastrophe. Love is the absolute enrichment of life, which however devours and depletes all other tendencies. Everything in man is depleted and inhibited except for that which has captured his image.
> But the most far-sighted of Man's organs of contact with the world is conscience. Before her and from her perspective every love is objectively re-evaluated [...]
> It is given to every intense aspiration, to every dominanta, to get its way. But this can also be disastrous for its bearer. And this is the Judgment, that is, in the highest sense, the fate of man.[157]

Ukhtomsky next elaborated his physiological and psycho-physiological concept in a talk on 'The Dominanta and the Integral Image', (1924). This was published in the form of an extended account of his remarks, which represented what Ukhtomsky termed 'only a programmatic note'.[158]

As a physiological process, the dominanta was formed, in his standard definition, when 'a sufficiently stable excitation, occurring in the centers at a given moment, acquires the significance of a governing factor in the work of other centers: accumulates in itself excitation from the most distant sources, but inhibits the ability of other centers to react to impulses having a direct relation to it.'[159] Drawing upon continuing research in his lab, he elaborated several moments in its dynamics.

Two of these moments are particularly interesting. First, in keeping with his religiously-based opposition to the notion that the organism's physiological responses were oriented toward stasis and balance, and his belief that reflexive reactions varied according to the general state of the integrated organism, he emphasized that the dominanta generated new, historically developing reflexes. For example: The cat in heat was continually subject to a mass of stimuli. The dominanta responded to these selectively, 'picking out' and strengthening only those 'in biological affinity' with the excitation of its sexual apparatus. These stimuli were thus over time incorporated into the dominanta, becoming one of its stable exciters.[160] The result was newly emergent reflexive responses, and, for Ukhtomsky, a concept of the reflex, not as mechanistic and permanent, but rather

---

157 Ukhtomsky's copy of *Russkii fiziologicheskii zhurnal*, VI, 1–3 (1923), 40–41, 43; at MDM, RBM-17175.

158 *Vrachebnaia gazeta*, 2 (1924), 26–29; Ukhtomskii, *Izbrannye Trudy*, pp. 36–48; fragment of unpublished letter to E. I. Bronshtein-Shur (no later than 3 April 1927), St Petersburg, ARAN 749.2.447: 82. He elaborated his ideas about the integral image in another letter to Bronshtein, 15 August 1928, *Intuitsiia sovesti*, pp. 290–93.

159 Ukhtomskii, 'Dominanta i Integral'nyi Obraz', p. 39.

160 Ukhtomskii, 'Dominanta i Integral'nyi Obraz', p. 43.

as a creative psycho-physiological (and often anticipatory) act developing in space and time.

A second interesting physiological note in this essay is Ukhtomsky's deployment of coworker I. A. Vetiukov's study of spinal dominanty in the frog to make the point that it was not especially *strong* stimuli that best produced a dominanta; but rather 'persistent, rare, rhythmic stimulations' that created 'an especially strong impression on the centers, facilitating their accumulation of a local, stable excitation'. Ukhtomsky noted several times in this essay that physiological and psychological analysis shared the goal of 'mastery of human experience, mastery of oneself and the behavior of those with whom one comes to live'. Given the religious roots of his investigations of the dominanta, it seems likely that, as with his interest in the effect of different kinds of Church singing, this observation reflected his vital interest in the development of a dominanta facilitating attentive prayer.[161] (In his earlier work on Church singing, he had emphasized that, not the dramatic moments in concert music, but rather the quiet rhythms of znamennyi chant best prepared the body for prayer.) In any case, the physiology of the rise, persistence, and decline of the dominanta became an ongoing line of experimental investigation, generating such articles as 'On the State of Excitation in the Dominanta' (1926), 'Parabiosis and the Dominanta (1927), and 'The Internalization of Rhythm in Light of the Doctrine of Parabiosis' (1928).[162]

In the published account of his talk, Ukhtomsky's comments about the dominanta as a psychological actor — specifically, about its role as creator of 'integral images' — were brief but pointed. 'Real and living experience deals always with integral images', and each of these was the product of a lived dominanta. As the product of human experience, science, too, was subject to these 'dominating tendencies.' He concluded with the injunction that 'In order to master human experience, to master oneself and others, to direct behavior and the most intimate life of people, one must master the physiological dominanty in oneself and the environment [...] Let only this training of humanity be executed with benevolence toward him!'.[163]

He summarized the main message of this talk more directly and fulsomely in a letter to a confidante. 'I am, you know, convinced that our organism and central nervous system normally act as a single whole. But this unity of action is not given

161 Ukhtomskii, 'Dominanta i Integral'nyi Obraz,' pp. 42 (frog), 46 (mastery). In another article, he revealed that 'the dominanta would lose all interest for me if it were reduced to the elementary difference in the strength [of competing excitations]' rather than its 'ability to strengthen (to accumulate)' excitation by incorporating and inhibiting others. This, we think, reflects his preoccupation with prayer practice, the 'elemental force' of steady, rhythmic Old Believer services, and the dynamics of that 'even wind' emphasized by the Holy Fathers. 'O sostoianii vozbuzhdeniia v dominante' (1926), in Ukhtomskii, *Sobranie Sochinenii*, vol. I (Leningrad: Leningrad University, 1950), p. 53.

162 See Ukhtomskii, *Sobranie Sochinenii*, vols I, II.

163 Ukhtomskii, 'Dominanta i Integral'nyi Obraz', pp. 46–48.

in advance as such. It is achieved by its development'. Ancient mechanisms of integrated action (for example, chain reflexes) had become hereditarily established in the spine, medulla oblongata, middle brain, and even in parts of the cerebral cortex — and these were indeed given us at birth. But:

> At the highest level, that is, in the most creative parts of the cortex, in the serious matters of individual life, this unity of action is only achieved and sought. One can say that the entire matter of so-called consciousness resides in the development of a unity of action amid a heap of impressions from ongoing life. That which is given in the lower stages of the long history of development is only sought and must be [achieved] above, in the cortex, in the activity of consciousness.
>
> Nevertheless here too everything that has more or less been achieved and is achieved in the form of "reception", "experience", "representations of discrete things" — this carries within itself the stamp of the integrative work of our nervous system. To select from the disorderly murk [mgla] the impressions of "thing", "emotional experience", or "person" is achieved by us insofar as we actively integrate ourselves. And this succeeds insofar as we are worthy, that is, insofar as we achieved in each separate case a unity of action.[164]

Here, again, Ukhtomsky expressed the distinctive combination of physiological, psychological and religious thinking in his conception of the dominanta. That by 'the serious matters of life' he refers most of all to active prayer and the quest for salvation is implicit in the structural consistency and parallels between the reasoning in this letter and his private commentaries on religious practice and ideas. It is also visible in his reference to the 'murk' of distracting sensations — a term that, as explicated by Habbakuk and Ephrem the Syrian, had played an important role in his analysis of Church singing.[165]

He wrote explicitly about this relationship in a diary entry of 1924:

> We are studying the regularities and patterns in the structure of the brain, trying to understand the laws of its work and functions. But what is the end goal and norm, the purpose and final achievement of this great mechanism with its astonishing actuators? [...] The goal and norm is achievement of that acquisition of the Holy Spirit about which the sainted Serafim Sarovskii spoke [...] This great mechanism has accomplished its purpose when Man has succeeded in attaining this summit![166]

Ukhtomsky's third, most elaborated and powerful synthetic essay, 'The Dominanta as a Factor of Behavior' (1927), originated as addresses to the student Biological-Scientific Circle and the Scientific Society of Marxists at Leningrad University, and was published in the *Herald of the Communist Academy*. These

164 Fragment of undated letter to E. I. Bronshtein-Shur, ARAN 749.2.447: 82–82rev.

165 See p. 147 above.

166 St Petersburg, ARAN 749.1.91 (17) [1924]: 16–16rev.

seemingly surprising venues resulted from the relationship between Ukhtomsky (whose religious convictions were an open secret) and some philosophically sophisticated Communist physiologists, including several in Ivan Pavlov's labs. That relationship was based in part on an important commonality between Ukhtomsky's Christian dialectics and the Communists' dialectical materialism: both sought to develop an analysis of the relationship between physiological and psychological processes that, while rooted in determinist physiology, left room for the active, independent (and heroic) role of consciousness — for what both Ukhtomsky and the Communists characterized as 'determinism of a different order'. Throughout his third essay, Ukhtomsky strikes this resonant chord with a Communist audience interested in the building of a new socialist society and a 'New Soviet Man' while himself having in mind a very different heroic endeavor — the prayerful Eastern Orthodox struggle for the 'New Being' (*Novaia Tvar'*) and deification.

His presentation of the dominanta in this essay was, then, structured around the changeability of physiological processes, the role of consciousness and the disciplined direction of human behavior. Invoking a popular metaphor ('the Soviet experiment'), he observes that we are all experimenters in life — probing, testing and projecting like a laboratory scientist — but 'with much weightier consequences'. 'The task of human culture is the development of new instincts'. And in a formulation parallel to his frequent private pronouncements that 'prayer is labor': 'Our behavior is labor'.[167]

Ukhtomsky grounds these sentiments deeply in the psycho-physiology of the dominanta. The reflex — long falsely portrayed as fixed and unchangeable — was actually active and change-oriented, highly variable and sensitive to the functional state of the organism. That functional state was defined by the dynamics of the dominanta — a center of heightened excitability with the ability to maintain itself by incorporating excitations from other regions while inhibiting others. The dominanta was both a 'law' and an 'organ' — a term that must be redefined in the light of modern science (having in mind here energetics), not as a fixed morphological entity but rather as 'any combination of forces capable of leading every time under equal conditions to identical results. An organ is most of all a mechanism with a definite singular action, [...] a dynamic, labile agent or working combination of forces.' In the higher nervous centers, the dominanta was the organ of reception at a distance, of anticipation and far-sightedness; it militated against the spinal centers' tendency toward mere habit, passivity and the path of least resistance, and 'toward activity, the enriching of the organism with new possibilities; and they reside at the basis of the formation of new reflexes'. The dominanta was 'the flywheel of our machine, helping to connect and organize

167 Ukhtomskii, 'Dominanta kak Faktor Povedeniia', pp. 83 (culture), 87–88 (experimenters, participants, labor); 'Prayer is most of all labor': for example, in his commentary on Gregory Sinaites's 'Instructions to the Prayerfully Silent' in his letter to Platonova, 25 August 1911 (p. 145 above).

experience into a single whole, but it is also an obsessive idea, a prejudice of behavior. One and the same factor breaks glass, forges steel'.[168]

He related all this to his laboratory research on the complex dynamics of nervous processes. According to Vvedensky's theory of parabiosis, excitation and inhibition were not ontologically distinct; they transformed one into the other according to the functional state of the nervous centers. So, 'one and the same reflex proceeding under our eyes can — with the very same exciters, just a little bit accelerated or enhanced — transform into phenomena of inhibition in the very same centers'. By Vvedensky's law of the 'physiological pessimum', if the nervous focus of a dominanta was excited beyond a certain maximum, it turned into its opposite and was inhibited.

The dominanta, then, was a very sensitive organ, and its management required scientifically-informed vigilance: 'If you wish to maintain a certain vector of behavior, a certain activity at one and the same level, you must very carefully consider the changing conjuncture of exciters and nervous centers,' which encompassed a multiplicity of factors such as the level of excitability of the dominating center, its relationship to neighboring centers, the frequency and strength of the exciters that continually arrived from those other centers, and so forth. To maintain a desirable dominanta, one must constantly 'educate' and 'nurture' it —' to assure that it has not become over-excited, not over-stepped a certain strength, that it constantly corresponds to ongoing conditions in the centers [...] and its environment'.[169]

The remaking of one's own behavior, therefore, required 'coercion, discipline' of self and the adjustment of the environment. 'We are not observers, but participants in existence. Our behavior is labor'. He concluded his article with the injunction to cultivate the 'dominanta of the other person' [*dominanta na litso drugogo*], meaning the drive to comprehend other people on their own terms, in terms of their own dominanty, rather than through the prism of our own.[170]

For the private Ukhtomsky, of course, that was also the essence of prayer — the banishing of one's own thoughts, emotions and images in focused attention to communion with God:

> Whenever I pray, and afterwards resort to tobacco again, I feel dizzy and poisoned. In the mundane hustle and bustle of life, this does not happen. This means that the completely distinctive tension of attention, that completely distinctive state of my organization that is formed during prayer, wrenches me abruptly from the ordinary, relaxed state. It has its own dominanta, and the ordinary inertia of life has its dominanty. The physiological state of each is completely distinctive. And that abundance of thought, the readiness to

168 Ukhtomskii, 'Dominanta kak Faktor Povedeniia', pp. 64 (reflex), 69–70 (dominanta features), 71–72 (law and organ), 81–82 (far-sightedness, new possibilities), 87 (flywheel).
169 Ukhtomskii, 'Dominanta kak Faktor Povedeniia', pp. 88–89.
170 Ukhtomskii, 'Dominanta kak Faktor Povedeniia', p. 89.

> bring people joy, the complete love for people that are given in prayer, also constitute that education of the dominanta that a person feels as the action of the Holy Spirit.[171]

Reflecting privately about Dostoevsky's *The Brothers Karamazov* in 1928–1929, he expressed the central themes of 'The Dominanta as a Factor in Behavior' in their underlying spiritual dimension: For all the differences in their personalities and expressions of spiritual pain, Dostoevsky's Fyodr, Dmitrii and Aleksei Karamazov were each trapped in the same closed circle: each created for himself the world of his perceptions, and each responded to and was shaped by that world. 'Everything confirms to a person in his inertia his beloved worldview, for he acts as he perceives the world, and feels the world as he acts [...] As are the dominanty in a person, such is his integral image; and as is his integral image, such is his behavior — and such are his happiness and unhappiness, and the face he shows to other people.' To escape that closed circle, a person must change 'his physiological perception of the world, the physiological, habitual continuity of his life. And that is very, very difficult!'

As both religious thinker and scientist, Ukhtomsky was most interested in the one character in Dostoevsky's novel who had achieved this, the Christian ascetic Zosima The Elder:

> My initial, first and last task is to understand how the cast of mind of the elder Zosima is created. I have learned that it is created by a great physical feat, by spiritual culture from others, and an attitude toward the world as a beloved, revered, intimate Companion. This is very difficult, [...] this form of perception is nurtured and maintained through great labor, with constant self-discipline and the careful guarding of conscience.[172]

Musing privately in the early 1930s about the culmination of his long effort to understand the creation of a Zosima, he reflected that his dominanta was neither theory nor hypothesis; it just 'ascertains from experience a very general principle, an empirical law like the law of gravity'.[173] Here he was referring to Newton's failure to find a satisfactory mechanism for gravity, which therefore remained a mathematical expression of observed relations in the physical world. Ukhtomsky's dominanta, as presented to the scientific community, also relied upon empirical observations and experimental findings in psycho-physiology — that is, upon the 'merely descriptive' knowledge that he considered the province of science. Yet, for him, prayer practice and the wisdom of the Holy Fathers made clear the animating cause and meaning of the dominanta — that is, they provided the

171 Diary entry of 1922–1924, St Petersburg, ARAN 749.1.145: 65rev–66.
172 St. Petersburg, ARAN 749.1.91 (18): 51rev–53rev.
173 St. Petersburg, 749.1.88 (4): 10.

kind of knowledge that he considered the province of religion. Had he lived in a different time and place, he may well have made clear that this scientific principle was the culmination of a lifelong quest: the dominanta *was* his unification of science and religion.

KOSTAS TAMPAKIS

Science as an Orthodox Weapon and the Politics of Anti-Communism in Mid-Twentieth Century Greece

▼ **ABSTRACT** This paper details how a polemic Orthodox science-and-religion venture, that of the Christian Union of Scientists and its magazine *Rays*, paved the way for one of the most intense efforts of anticommunist propaganda in 1950s Greece, that of the Greek Light organization. In Greece, science was seen as an important component of communist ideology, by proponents and enemies alike. Thus, science became a contested battleground between communists and non-communists, and later, a resource for institutionalized anticommunist rhetoric. In the end, this paper argues that, while the communism/anti–communism struggle has been recognized as a crucial facet of European and global twentieth century history, the fact that science and religion narratives were crucial in such battles has remained relatively unexplored. Greece, which ended up on the Western side of the Iron Curtain, but was also Orthodox and had a communist party with a strong presence, is an especially informative such case.

Introduction — A Furious Summons

During the first days of 1947, Nikos Zachariadis (1903–1973), the leader of the Communist Party of Greece (KKE) hastily summoned several communist and leftist intellectuals in a theater in central Athens, appropriately named the Central Theater. Zachariadis had a lot to deal with, in the first months of 1947. The first phase of the Greek Civil War (*Emfylios*) was in full swing. In March 1946, armed bands of veteran members of the recently disbanded Greek People's Liberation Army (ELAS), who had fought Nazi occupation in Greece for several years

Kostas Tampakis • National Hellenic Research Foundation

Orthodox Christianity and the Study of Nature, ed. by Kostas Tampakis and Ronald Numbers, Science and Orthodox Christianity, 4 (Turnhout: Brepols, 2024), pp. 185–211
BREPOLS ❦ PUBLISHERS 10.1484/M.SOC-EB.5.141384

under the auspices of KKE, attacked a police outpost and killed the policemen there. This military action was a direct consequence of the 1945 Varkiza Treaty, according to which KKE had disbanded ELAS. From 1944, there had been sporadic fighting between ELAS and the British-backed government forces, in which ELAS was often treated as a hostile occupying force, rather than the liberating army its members saw it to be. Many of the government forces and officials had until recently collaborated with the occupying German forces, and members of ELAS feared that the Varkiza Treaty, which was supposed to safeguard a peaceful transition to amnesty and general elections, would be a ruse to render ELAS powerless. They were proven right by the wave of persecutions of demobilized ELAS members that followed after Varkiza. As a result, the Democratic Army of Greece (DSE) was created by former ELAS members. In December 1946, more than 15,000 armed men operated under DSE, with the blessing and guidance of KKE. The Greek Civil War would escalate to full military confrontations until 1949, where a total defeat of DSE left tens of thousands of DSE fighters and sympathizers executed, exiled or jailed. More than 30,000 children were sent abroad to sympathetic communist countries. Another 25,000 children, mostly orphans or with parents affiliated with the DSE were placed in involuntary foster care in specially prepared 'child towns' (*paidoupoleis*), under the direct supervision of Queen Frederica. The Greek Civil War, nowadays seen as the one of the first proxy wars of the upcoming Cold War, would remain a deeply divisive national event, which polarized and shaped cultural, social and intellectual Greek life until well into the 1970s.[1]

Zachariadis certainly had a lot to talk about with other leftists. At the time of Zachariadis' speech, the Greek monarchy had just been reinstated through a dubious referendum, DSE had successfully attacked military and police outposts and had taken over the town of Naousa in Macedonia and the death penalty has been enacted for jailed communist. A prominent communist journalist, Kostas Vidalis, had been kidnapped off a train in Thessalia and murdered in broad daylight.[2]

---

1 The Greek Civil War is a matter of contention among historians. It is not the purpose of this paper to provide a concise treatment of the relevant extensive historiography. A starting point for non-Greek speakers would be the collection of essays Ph. Carabott and Th. Sfikas, eds, *The Greek Civil War. Essays on a Conflict of Exceptionalism and Silences.* (London: Routledge, 2004); A. Gerolymatos, *An International Civil War: Greece, 1943–1949.* (New Haven: Yale University Press, 2016) and M. Mazower, ed., *After the War was Over. Reconstructing the Family, Nation and State in Greece, 1943–1960* (Princeton: Princeton University Press, 2000). For specific dimensions of the Greek Civil War, see P. Voglis, *Becoming a Subject: Political Prisoners during the Greek Civil War* (Oxford: Berghahn Books, 2002) and L. M. Danforth and R. Van Boeschoten, eds, *Children of the Greek Civil War: Refugees and the Politics of Memory* (Chicago: University of Chicago Press, 2019).

2 The various episodes mentioned here are discussed in greater detail in L. Bærentzen, J. O. Iatrides, and O. Langwitz Smith, eds, *Studies in the History of the Greek Civil War, 1945–1949.* (Museum Tusculanum Press, 1987), especially Ole Smith's essay on Self-defense Communist policy, pp. 159–78 and Lars Baerentzen's paper on children, pp. 127–58.

Nevertheless, rather than focusing on any of these events, the topic of his furious speech was the recently published *Manifesto of the Christian Union of Scientists*, written collectively by the homonymous Christian Union of Scientists (CEE), which was itself affiliated with the para-ecclesiastical religious brotherhood *Zoe* (Life).[3] Zachariadis berated his audience for having slept 'like the Foolish Virgins' (of the New Testament) while their enemies were working under their very shade against the foundations of KKE's existence as a party.[4] Why would then Zachariadis choose to focus on the *Manifesto* and describe it in such dire terms?

An attempt to answer this question forms the backbone of my paper. The confluence of communism, religion, materialism and science, taken in pairs, is of course not a novel historiographic consideration. The history of early modern and modern Natural Theology is the historiographical exemplar for how the study of nature can be used for the glorification of religion.[5] In the same vein, historians of nineteenth century German science have discussed the role of scientific materialism, and accounts of the effect of communism in twentieth century Soviet Russian science are by now numerous.[6] In contrast, while the communism/anti-communism struggle has been recognized as a crucial facet of European and global twentieth-century history, the fact that science and religion narratives were crucial in such battles has remained relatively unexplored. In fact, the historiographical lure of Soviet science and Soviet religious history has obscured the borderline cases of NATO–allied countries which had their own bitter internal strife. The historiographical partitioning of spaces into communist and anti-communist, or of their science as materialist or anti-materialist, obscures the different type of engagement in countries like Spain, Italy and Greece. These are spaces which ended up on the Western side of the Iron Curtain, but which also had a communist party with a strong presence for the whole of the twentieth century. Science was seen as an important component of communist ideology, by proponents and enemies alike. Thus, a different kind of discourse on religion and science emerged, one that was politically charged, socially sensitive and often punishable by exile, imprisonment or execution. In Greece, science was contested territory, over which communists and Marxists fought conservative Orthodox scholars. As the Greek communist leader Zachariadis recognized early on, ideological and intellectual battles are fought alongside military battles. Actions beget

---

3 *Διακύρηξις της Χριστιανικής Ένωσης Επιστημόνων, Για κάθε Έλληνα* (Athens, 1946). In subsequent contemporary English publications, CEE appears as the *Christian Union of Scientists and Professional Men*.

4 N. Zachariadis,'Το ιδεολογικό μέτωπο', *Kommounistiki Epitheorisi*, 2 (1 February 1947), 53–56. See D. Constantelos, *The Zoe Movement in Greece* (New York: St Vladimir's Seminary Press, 1959), pp. 5–6.

5 J. H. Brooke and G. Cantor, eds, *Reconstructing Nature: The Engagement of Science and Religion* (Oxford University Press, 1998), pp. 141–76.

6 See for example F. Gregory, *Scientific Materialism in Nineteenth Century Germany*. (Dordrecht: D. Reidel, 1977); H. Kragh, 'Science and Ideology: the Case of Cosmology in the Soviet Union, 1947–1963,' *Acta Baltica Historiae et Philosophiae Scientiarum*, 1.1 (2013), 35–58.

reactions. In my paper, I intend to approach the issue of science, communism, and religion through a doubly lateral historical repositioning. I aim to focus on a special case of an Orthodox appropriation of science, that of the Greek *Christian Union of Scientists* and its journal *Rays (Aktines)*, active in Greece from the 1930s onwards. I want to engage with the issue of religion, science and communism, not as a 'communist' history, but from the other end, as a history of Orthodox anti-communism and anti-materialism. I will discuss CEE's history and intellectual genealogy and investigate the first forty years of its existence. I also aim to show the political uses of Orthodox anti-communism. It should be noted that CEE is still in existence, active, and still publishing *Rays*. So is the *Zoe* brotherhood and the constellation of religious organizations around it, which we will have cause to discuss later on. All in all, CEE and *Rays* represents one of the most enduring, and greater in scope, apologetic endeavors in Orthodox Christianity, using science and scientific and literary rhetoric to promote a very specific religious view. It is to the genealogy of this endeavor I will now turn.

Genealogies of Christian Unions

At first glance, the beginning of CEE is easy to trace. It was founded in 1937 by Alexandros Tsirintanis (1903–1977), the charismatic Professor of Insurance Law at the University of Athens and by Archimandrite Serafeim Papakostas (1892–1954), the elected leader of the religious brotherhood *Zoe*, itself founded in 1907. Since the role of *Zoe* in twentieth century Greece has been the focus of both Greek and international historiography, it is tempting to consider CEE as a mere offspring of *Zoe*.[7] In fact, most historians seem to take it for granted that CEE and *Rays* were just one of the many satellite organizations around *Zoe*, since both contemporary and later commentators imply that this is the case.[8] However, such a view would be misleading. The genealogy of CEE must be expanded to include not only *Zoe*, but also the historiographically less visible religious association *Anaplassis* (Renewal), which was founded in 1886, with goals quite similar to

7 See for example Ch. Maczewski, *Die Zoi–Bewegung Griechenlands: ein Beitrag zum Traditionsproblem der Ostkirche* (Göttingen; Vandenhoeck u. Ruprecht, 1970), translated in Greek and published by Armos in 2002; V. Makrides, 'Orthodoxy in the Service of Anticommunism: The Religious Organization Zoë during the Greek Civil War', in *The Greek Civil War. Essays on a Conflict of Exceptionalism and Silences*, ed. by Ph. Carabott and Th. Sfikas. (London: Routledge, 2004). 159–74; A. Logotheti, 'The Brotherhood of Theologians Zoe and Its Influence on Twentieth-Century Greece', in *Orthodox Christian Renewal Movements in Eastern Europe* ed. by Milovanović, Aleksandra Djurić, and R. Radić (Palgrave Macmillan, 2017), 285–302 and A. Logotheti, 'The Religious Justification of Anti-Communist Persecutions in Greece (1920–1949)', in *The Palgrave Handbook of Anti-Communist Persecutions* ed. by Ch. Gerlach and Cl. Six. (Springer, 2020), 329–45; Also A. Gousidis, *Οι χριστιανικές οργανώσεις. Η περίπτωση της αδελφότητας Θεολόγων "Η ΖΩΗ" — κοινωνιολογική προσέγγιση* (Thessaloniki; Pournara, 1993).

8 See for example Makridis (2004).

both *Zoe* and CEE. CEE's beginnings — as those of *Anaplassis* — should also be traced to the work of the disgraced religious leader Apostolos Makrakis (1831–1905).[9] This network of scholars, clergy men and institutions did not always see eye-to-eye, but their activities nevertheless form a lineage of anti-materialism, and later anti-communism in Greece, which I argue CEE drew from and expanded upon.

Apostolos Makrakis was born on the island of Sifnos but was educated in the Phanar Greek Orthodox College (also known as the Great School of the Nation) in Istanbul. In 1852, he started teaching in several schools in Istanbul, then became the secretary of the Lesvos Bishopric until 1862, where he left for France as the private tutor of a rich banker's children. Makrakis came back in 1865 and he settled in Athens in 1866.[10] Makrakis was a fierce and charismatic preacher, who attacked head-on what he saw as the corruption of the Church and the evils of the foreign-born monarchies of Greece. He has been rightly seen as the heir of several anti-Western populist religious figures, such as Kosmas Flamiatos (1786–1852) and Christoforos Papoulakos (1785–1861).[11]While Flamiatos and Papoulakos often called for, and participated, in armed confrontations with the Greek state, Makrakis instead founded newspapers, such as *Logos*, and religious associations and schools, such as the 'School of the Logos' (1873), 'John the Baptist' (1877), 'Constantine the Great' (1879), 'John Theologian' (1884) and 'Plato' (1901). His great popularity initially attracted the praise of the Church. In 1876, however, Makrakis accused specific Ministers and Bishops of buying their way to their seats. He later started teaching a version of Orthodox theology of his own making, while vehemently attacking the State, the University of Athens, even the Greek Church's hierarchy. After some attempts at reconciliation, the Church denounced him in 1879, and Makrakis was convicted to two years in prison in 1880. By that time, Makrakis was denouncing everything he saw as being against his vision of Christianity and Hellenism. This vision included, among other things, the provision that to be Greek and not Orthodox is something alien, degenerate and monstrous.[12] Abandoned by many of his former students, but still very popular, Makrakis died in 1905.

Makrakis combined pre-existing Greek Orthodox strands of anti-Westernism and anti-modernity with a missionary 's zeal for fighting intellectual battles,

---

9 I have opted to translate *Anaplassis* as Renewal. Other scholars, such as Eftymios Nicolaidis have opted for term New Creation. See E. Nicolaidis, *Science and Eastern Orthodoxy: From the Greek Fathers to the Age of Globalization* (Baltimore: John Hopkins University Press, 2011), p. 184.

10 Biographical information on Makrakis can be found in E. Karageorgoudis, *Χριστιανισμός και πολιτική κατά τον Απόστολο Μακράκη* (Athens: Vriennios, 2005).

11 G. Metallinos, 'Απόστολος Μακράκης (1831–1905). Φιλόσοφος και ηθικολόγος, λαϊκός ιεροκήρυκας με πλατιά ακτινοβολία και ηγέτης αφυπνιστικού κινήματος στη μετεπαναστατική Ελλάδα,' *New Sociology*, 27 (1999), 8–11.

12 For a full discussion from a historical and theological perspective, see L. Brang, *Το μέλλον του ελληνισμού στον ιδεολογικό κόσμο του Απόστολου Μακράκη* (Athens: Armos, 1997).

using Christian associations and journals as weapons. Many of his students, such as the founder of *Zoe* Efsevios Matthopoulos (1849–1929), the future Bishop of Patra Hierotheos Mitropoulos (1840–1903), and the scholars Konstantinos Dialesmas (1855–1921) and Michael Galanos (1862–1948) took this lesson to heart and carried out this Makrakian mode of public engagement, even after they distanced themselves from his teaching.[13] Even before his fall from grace, a number of Makrakis' students, such as Dialesmas and Galanos, founded the para-ecclesiastical association *Anaplassis* in 1886, and a year later, published the homonymous journal. *Anaplassis* had a tighter focus than Makrakis' teachings. According to the founding charter of *Anaplassis*, its members were to be 'men of science and art' that would work towards 'understanding the Great truths of Christianity' and who must strive to apply these great truths in their everyday life. *Anaplassis* should show that the perceived opposition between modern science and Christianity was a misrepresentation. To that end, its members should follow the international debate taking place in all Christian countries and translate into Greek the most important relevant publications. Finally, and tellingly, the Association would found and operate Sunday and evening schools for the people, in which to spread its message.[14] The echoes of Makrakis' mode of operation are clear. The Association and its journal, however, distanced themselves from Makrakis after his disgrace. Makrakis, in turn, felt that his intellectual property was being stolen by his former students and responded with vehement vitriolic articles. The *Anaplassis* response was less vitriolic, but no less firm, and extended to theological criticism of Makrakis' work.[15] *Anaplassis*, perhaps because it displayed all the missionary zeal of Makrakis while denouncing his heretical theology and political engagement, soon received the favor of the Church. An Encyclical of the Holy Synod of the Church of the Greece in February 1893 commended *Anaplassis* to all clergy men, promoted its dissemination, declared it the 'only religious journal of its kind' and underlined that it operated under the aegis of both the Archbishop of Athens and of the heir to the Greek throne, Prince Konstantinos.[16] *Anaplassis* was a successful journal, averaging around 3000 copies and even reaching 7000 in 1895. It was also the first journal to carry out a sustained attack against what its editors saw as a burgeoning tide of materialism within Greek society. Makrakis had, even before *Anaplassis*, carried out several

---

13 See Ch. Mpontzidis, *Λειτουργική αναγέννηση και Χριστιανικές Αδελφότητες κατά τους 19° και 20° αιώνες* (PhD Thesis, University of Athens, 2017), p. 12 for a very illuminating depiction of the network.

14 *Καταστατικόν Του Συλλόγου «Η Ανάπλασις»* (Athens, 1886).

15 The only study of Anaplassis is the glorifying doctoral thesis of the Archimandrite S. S. Fakas, *Σύλλογος «Ανάπλασις» και η προσφορά του, 19ος — αρχές 20ου αιώνα* (PhD Thesis, University of Thessaloniki, 2013). Due to his sympathetic viewpoint, Fakas concisely details the clashes between Anaplassis and Makrakis. See especially pp. 37–44, 138–48.

16 *Περί συστάσεως του θρησκευτικού περιοδικού της Αναπλάσεως*, Encyclical of the Holy Synod of the Church of the Greece, 78/174, 18 February 1893.

attacks against modernist theories, but they were sporadic, and mostly based on his general views of Christianity and theology. In contrast, *Anaplassis* was at the forefront of the denunciation of Darwinian evolution and Heackelian and Buchnerian materialism for several years. The journal translated relevant French and German articles in Greek, and published articles written by senior members of the association, often to specifically attack Greek proponents of evolution and materialism.[17] However, denunciations of materialistic feminism and of masonic infiltrations were also common. The lawyer, lawmaker and member of the parliament Ioannis Skaltsounis (1824–1905) was especially active in these debates, and in fact assumed the editorship of the journal in 1895. While his work has received scant historiographical attention, he was seen by many of his contemporaries, as well as from the next generation of Orthodox scholars, as one of the forefathers of the Orthodox critique of materialism.[18]

However, the most influential religious organization, even if it was neither the first, nor at times the most populous, was the religious brotherhood of theologians *Zoe*. Its story is intimately tied to CEE. *Zoe* was founded in Athens in 1907 by another former student of Apostolos Makrakis, the Archimandite Efsevios Matthopoulos (1849–1929). A fearsome and very charismatic preacher, Matthopoulos met Makrakis when he was a monk in Megalon Spilaion in Kalavrita. From 1876, Matthopoulos became one of the confessors in the 'School of the Logos' and was thus convicted and sentenced to exile for eight years in 1879. In 1882, Matthopoulos was allowed to return to Athens, and he denounced his former teacher publicly in 1884. After twelve years of travelling as a preacher all over Greece, and in many, at the time, Ottoman-occupied territories, Matthopoulos came back to Athens to found *Zoe*.[19] Alongside him, were some of his closest students, such as the theologian and apologist Panagiotis Trempelas (1886–1977), Dimitris Panagiotopoulos (1885–1967?), Ioannis Kolliopoulos (1884–1963) and the noted Archimandrite and preacher Dionysios Farazoulis (1882–1920).

Zoe occupies a curious lacuna within Greek historiography. Despite the books and articles that have been devoted to its history, either directly or indirectly, the actual primary sources are rather few.[20] There is of course the journal *Zoe* itself,

---

17 For the history of Darwinism in Greece and the role of *Anaplassis*, see Nikolaidis (2011), pp. 184–87. On Darwinism in general, see M. Zarimis, *Darwin's Footprint: Cultural Perspectives on Evolution in Greece (1880–1930s)* (Budapest: Central European University Press, 2015) and K. Krimpas, *Ο δαρβινισμός στην Ελλάδα* (Athens: National Hellenic Research Foundation, 2017).

18 A notable exception, again from a theological perspective, is I. Panagiotou, *Ο Ιωάννης Σκαλτσούνης ως απολογητής του Χριστιανισμού* (PhD thesis, University of Thessaloniki, 2018).

19 S. Papakostas, *Ευσέβιος Ματθόπουλος (βιογραφία)* (Athens: Brotherhood of Theologians Zoe, 1930).

20 The actions of Zoe are discussed in E. Gazi, *«Πατρίς θρησκεία, οικογένεια». Ιστορία ενός συνθήματος. 1880–1930.* (Athens:Polis, 2011); P. Karamouzis, *Κράτος, εκκλησία και εθνική ιδεολογία στη νεώτερη Ελλάδα (κλήρος, θεολόγοι και θρησκευτικές οργανώσεις στο μεσοπόλεμο)* (PhD

which has been in constant circulation since 1911, and comprises an impressive corpus of sources. However, the inner workings of *Zoe* are still largely unknown. The brotherhood itself reportedly issued records of its meetings, but they were distributed only to its members. That leaves either autobiographical records of members and former members, articles and books written by protagonists of the organization, or second-hand accounts of contemporaries. None of these accounts, however, can be considered impartial, since active members glorify the work of *Zoe*, while former members and intellectual opponents tend to vilify it.[21] Nevertheless, some facts are well established. *Zoe* was founded to be a semi-monastic brotherhood of theologians and associated professionals. Its members took some of the vows of monks (obedience, celibacy, poverty) and lived communally in the brotherhood's main building in the center of Athens. Matthopoulos, in founding *Zoe*, had been explicitly inspired by the polemical and missionary zeal of Makrakis. The members of *Zoe* aimed to improve themselves morally and spiritually, but also to help answer Greece's grave political and social problems with a recourse to Orthodox Christianity.

For the first two decades of its existence, *Zoe* placed more emphasis on moral conduct and ethics, and less to theological matters or sociopolitical ones.[22] Perhaps that explains why *Zoe*'s initial impact was limited. However, in 1927, Seraphim Papakostas (1892–1954), a young and energetic Archimandrite and the personal choice of Matthopoulos, was elected director of *Zoe*. A year earlier, in 1926, *Zoe* had organized its first seven Sunday schools. From 1929 to 1946, under the sure guidance of Papakostas, a number of organizations sprung from, or were associated with, *Zoe*. Under the general name of 'The cooperating Christian Societies of St. Paul', the Brothers of *Zoe* and the Sisters of *Efsevia* (the respective female sisterhood), founded and assumed key positions in the *Academic Social Association* (1933) (later renamed as Christian Student Union), the *Christian Union of Educators*, the *Christian Union of Working Youth* (1945) and of course, the *Christian Union of Scientists* (1937). Each of these cooperating unions and associations published its own journal, often with significant, for the era, circulation. During the same period, the attendance of *Zoe*-supervised Sunday schools rose to the tens of thousands. In 1949, *Zoe* had organized and run 500 Sunday schools all over Greece, with a total of 60,000 students. In 1959, the number of Sunday schools had risen to 2200, which was the greatest number ever reached.[23] As an organization aimed primarily at an urban population, *Zoe* was seen as filling the gap left by a more traditional, village-oriented Greek Church. Its success

---

thesis. Panteion University, 2004); Th. Tsironis, *Εκκλησία πολιτευομένη. Ο πολιτικός λόγος της Εκκλησίας της Ελλάδος (1913–1941)* (Athens: Epikentro, 2010).

21 A notable example is the autobiographical book by notable Orthodox theologian C. Yannaras, *Καταφύγιο ιδεών* (Athens: Ikaros, 2000). Yannaras was initially a member of Zoe, but later became one of its strongest theological opponents.

22 Makrides (2004), p. 160.

23 Logotheti (2017), p. 292.

prompted the Greek Church to launch its own Sunday school program, and many powerful clergymen rose from the ranks of *Zoe* to occupy key positions within the Church hierarchy.

The Emergence of the Christian Union of Scientists and the Creation of *Aktines*

The network of *Zoe* members appearing in different unions, and in CEE itself, lends itself to the conclusion that such unions and associations were directly controlled by *Zoe*. Yannaras, for example, remembers submitting articles to *Anaplassis* in the 1950s, which must be noted was not formally affiliated with *Zoe*, while also working as a clerk at the headquarters of *Zoe*, writing a column at *Aktines*, and publishing the journal of the *Student Union*.[24] Moreover, many of the key figures of *Zoe* were explicit in the fact that each union and association was more or less a soldier in a common cause, using his specialized weapon — chemistry, shoemaking or trading — to fight the greater battle alongside his peers.[25] These interrelations are the source of the misconception that CEE and its journal *Aktines* was just an appendage of *Zoe*. In fact, there were different and distinct endeavors of Orthodox engagement with politics, modernity and science. *Aktines* had much more in common with *Anaplassis* than with *Zoe*. Vice-versa, the group of scholars that founded and operated *Zoe* acted less as intellectual appendages and more like symbiotes. This in turn critically affected how they weaponized science in their pursuit of a specific vision of Orthodox Greece, creating a militant discourse on science and religion that was inherently political.

The masterminds behind the official creation of CEE in 1937 were Seraphim Papakostas, the director of *Zoe* and Alexandros Tsirintanis. Contrary to many accounts, Tsirintanis was not a *Zoe* pawn, but rather the head of a circle of intellectual associates, who had been childhood friends while growing up in Piraeus, and who went on to have academic careers. They had their own involvement with Christian Unions apart from *Zoe*, having worked with *Anaplassis* at a young age, and having created their own association, also called *Apostolos Pavlos (St. Paul)*. The company included Alexandros Tsirintanis, the controversial future Archbishop of Athens and initially devoted *Zoe* member Hieronymos Kotsonis (1905–1988), and the psychiatrist Aristos Aspiotis (1910–1983). When CEE was founded, the team was expanded to include the future Professor of Astronomy Dimitrios Kotsakis (1909–1986), the poet Alexander Gkialas (nom de plume G. Veritis) (1915–1948), a noted Christian scholar with strong anti-communist ideas, and some of the founders and pillars of *Zoe*, such as the theologian Panagio-

---

24 Yannaras (2000).

25 E. Mastrogiannopoulos, *Προσανατολισμοί. Ομιλίαι προς χριστιανικά σωματεία* (Athens: Zoe Publications, 1962), p. 39.

tis Trempelas.[26] The scholars of CEE stood somewhat apart from *Zoe* members. While most Brothers were theologians, educated in Greece and with no studies abroad, Tsirintanis and his circle were cosmopolitans, with international contacts and with active and prestigious career outside CEE. Logotheti (2017) is correct in noting that the Tsirintanis circle was approached by *Zoe* to help address a more educated and international audience. Tsirintanis went on to play a major role within the internal politics of *Zoe*, and became a kind of 'spiritus rector' for *Zoe* brothers.[27] Yannaras, a young aspiring Brother in the 1950s the time, remembers[28]

> 'The staff of *Aktines* was much more imposing. There, the presence of Tsirintanis, the second demigod of the movement, was dominant. An aristocrat with elegant manners…he had the grace of the cosmopolitan intellectual. Next to him was his colleague and Law Professor Georgios Rammos… a clever person with some sense of humor. Third was the Archimandrite Hieronymos Kotsonis, a simple and sweet man with an irresistible boyish smile. Very close was also the psychiatrist and son in law of Tsirintanis, Aristos Aspiotis, very polite, with a grave and imposing voice and a patina of nobility in his every movement […]
> There was a very obvious difference between the social and intellectual level of the leadership of *Zoe* and of the staff of *Aktines*. And you could tell right away that both sides knew very well that difference — the former, a sense of inferiority, the latter, a sense of superiority. The oldsters of *Zoe* had the imprint of their provincial upbringing […] On the contrary, the people of *Aktines* had studied abroad, had academic careers, and came across as people with social experience and at ease when moving in upper social circles'.

The main instrument of CEE was the journal *Aktines*, and in the eyes of most, including Yannaras, the CEE *was Aktines*. The journal started in 1937 but it restarted its circulation in 1938. It was a bimonthly journal until 1946, when it became a ten-times per year monthly publication. In 1939, it averaged 48 pages per issue, up from initially 16, and it was split into three parts. The first and longest part was devoted to a main article, usually written by Tsirintanis, and then to articles on Theology, Philosophy, Science and general social and educational issues. The second part was devoted to letters sent to the journal, comments, news and book reviews, and the third contained indexes and tables. For many years, authors did not sign the articles but instead used their initials or pseudonyms drawn from the names of first Christians. Tsirintanis used the name Melitis for the main articles and Zinas the Lawyer when commenting in the News section, Pothos was Klonaris and Gkialas was Veritis or Lamprinos. *Aktines* was scheduled to publish a thousand copies, but the pre-orders for the first issue were twice as

26 A. Stathakis, '"Ακτίνες" 1938–1997', in V. Charalampopoulos, ed., *Ευρετήριο 'Ακτίνων', Τόμοι: 1938–1997* (Athens: Aktines, 1998), pp. 7–20, on pp. 8–11.
27 Logotheti (2020), p. 340.
28 Yannaras (2000), p. 90. (translation is mine).

much, with a demand for 1500 more after that. *Aktines* was widely and wildly successful, and by 1949, it sold 40.000 copies, which, as the people of *Aktines* proudly note, was the same number of copies that the communist journal *Kommounistiki Epitheorisis* (Communist Overview) sold.[29] Again in 1949, CEE numbered more than a thousand members, but only a small number actually wrote in its journal. The success of *Aktines* allowed Tsirintanis, which was recognized as the prime mover of the endeavor, to express his views on Orthodoxy, Hellenism and their relation to a wide audience, and as a result leave a serious mark in the discussions of his era.

Science as the Denial of Denial

When *Aktines* was first published, Orthodox scholarship against undesirable modernistic and scientific theories was not something new. *Anaplassis* was well into its fifth decade, and prominent *Zoe* members, such as Trempelas, had published such treatises for years. What *Aktines* brought to the fore, was a cosmopolitan critique with the full weight of the academic credentials of its editorial team, combined with organic ties to existing Orthodox ideas. These credentials were in the fields labeled in Greek as *επιστήμες*, which denotes not only the Natural Sciences, but every academic discipline. Thus, more wissenschaft than science. As a result, when an article in *Aktines* said that science and Orthodoxy are not in opposition, its author referred not only to Physics or Chemistry, but also to Law, Psychology, History, Medicine or Archaeology. The enemy against who *Aktines* rallied is spelled out in the first page of the inaugural 1938 first issue, with the headline 'Denial': It is the denial of Christian values and ideas by modern thinkers. Tsirintanis spent several pages describing why and how such a denial would be catastrophic for the Greek nation, if it came to fruition. In an ironic, and probably unintentional, twist of the Hegelian negation of the negation, Tsirintanis concluded that just denying denial is not enough, and that a positive, constructive goal is necessary.[30] In one article, *Aktines* acquired an enemy and a goal.

Who the proponents of denial were became very clear in the next issues, and was a constant theme in almost all scientific, pedagogical or philosophical articles for the next twenty years: they were the materialists and the 'so-called' rationalists. It is no accident that these were exactly the philosophical commitments that the rising Greek Communist Party had adopted, and in which its critique of Christianity and conservative ideas of Hellenism was based upon. This was also the era of the Ioannis Metaxas' military dictatorship in Greece, whose main ideological

29 I. Spiropoulos, '"Ακτίνες" 60 χρόνια', in Charalampopoulos (1998), pp. 448–77, on p. 449. *Kommounistiki Epitheorisis* soon stopped circulating, due to a government ban on communist texts.

30 'Η άρνησις', *Aktines*, 1 (1938), 1-2.

constituents were anti-communism, anti-materialism and Orthodoxy.[31] But until the first skirmishes of the Greek Civil War, the connections were rarely made explicit. Instead, the editors of the journal devoted article after article to attacking any progressive or even slightly materialistic idea. In the same 1938 issue, Aspiotis, under the pseudonym Psychiatros (Psychiatrist), wrote a piece on the unity of psychological functions, thus decrying any materialistic or biological explanations of psychological illnesses. It is only when the body and the soul are treated as one that the patient can heal.[32] A year later, in the 1939(7) issue, Aspiotis tackled Freudian psychoanalysis, which he described as an impressive, fashionable but ultimately flawed theory, which fails because it is materialistic at heart.[33] Another article, in the 1938(1) issue, painted Voltaire, considered the founder of modern rationalism, as a deeply immoral person.[34]

The critique and use of contemporary science in *Aktines* is very uneven. At its core, the CEE was headed by a Professor of Law, a psychiatrist, two theologians and an astronomer. When the articles were epistemically close to these disciplines, the argumentation, while often flawed and misleading, was quite detailed and complex. When, however, Darwinism was attacked — and it is attacked at least once every two or three issues, in the first years — the arguments reverted back to apologetics already used by Sklavounis 50 years ago, or to outright intellectual sleight-of hand. An example of epistemically-proximate critique can be found in the 1938(2) issue, an article devoted to the recently published book by the physicist Pascual Jordan, titled 'Physics of Twentieth Century'.[35] Many of the editorial team could read German and French and Kotsakis could read English, a rare skill at the time. The main point of interest in the book, according to the reviewer, was that matter is no longer considered the primary substance by Physics, and a specific reference to a page in the German publication is given. As such, any materialist science and philosopher should apologize for believing so. However, Jordan says nothing of the sort in his book, which instead discusses ancient Greek materialism, the notion of Democritean atom and whether modern physics is a refinement of it.[36] Tellingly enough, the book contained a section on religion, which does not get mentioned at all. The same treatment is reserved some pages later to Max Planck's 'Religion and Physics' lecture, given in 1937 and published some months earlier in 1938. According to the *Aktines* review, Planck concedes that there is a Higher Power acting in the world, while in fact Planck talks about positivism and the necessity of physical laws as a proxy for higher

31 For the period see the Special Issue: Fascism in Interwar Greece, *Fascism*, 11.2 (2022), ed. by A. I. Kallis and G. Souvlis.

32 Psychiatros,'Η ενότης των ψυχικών λειτουργιών', *Aktines*, 1 (1938), 8-12.

33 Psychiatros,'Από τη θεωρία του Φρόϋντ', *Aktines*, 7 (1939), 19-27

34 Z. N., 'Βολταίρος', *Aktines*, 1 (1938), 12-17.

35 'Η στροφή', *Aktines*, 2 (1938), 74-77.

36 P. Jordan, *Physics of the Twentieth Century* (New York: Philosophical Library: 1944), pp. 144–46.

power.[37] The trend continued for many issues. In the article on matter on the 1938 (3) issue, modern Physics is said to have proven that matter does not exist, only energy.[38] Similar is the paper by Dimitrios Kotsakis in the 1939(8) issue, on whether other planets are habitable. Writing about a subject in which he has excellent academic qualifications, and with references to Eddington, Jeans and Lunmark, Kotsakis declares at the end of the article that the uninhabitability of other planets is a serious problem for materialism, since either the laws of nature must be different elsewhere and thus preclude life on other planets, or Earth is special, and thus something more than materialism is involved.[39]

Critique was not so sharp when it comes to Darwinism. None of the editors were biologists, and they seem unaware of the great strides taking place in the 1930s towards the Modern Darwinian Synthesis. Instead, taking issue 1938(3) as an example, the readers of *Aktines* are informed about the discovery of the Peking Man and assured that is not the missing link in human evolution, the legendary ape-man, but rather a man.[40] The fact that it was the subspecies *Homo erectus* and in fact indeed a link in human evolution is not mentioned at all, while the objections of Teilhard de Chardin are duly noted, glossing over the fact that he was in fact instrumental in the finding's classification. In the issue 1938(5), there is a scornful short story where a materialist and Christian discuss human evolution. The knockout arguments of the Christian were the differences between the ape and human skull and the absence of a nation of surviving apemen, which are more or less the same arguments that Skaltsounis was using when writing in *Anaplassis* in the late nineteenth century.[41] In issue 1939(7), the Presidential Address given by the fervent evolutionist Sir Ed. Poulton in the British Association for the Advancement of Science a year earlier, which in fact describes the history of evolutionary ideas, is presented as an argument for the incompleteness of evolution in general. In the same issue, contemporary scientists who were known even at their time for being extremely religious and obstinate anti-evolutionists, such as Freiher von Huene (1875–1969) and Hans Driech (1867–1941) are presented as world-leading experts.[42] And so on, and so forth.

Aktines also weighed on social issues. Feminism is discussed in an article by Tsirintanis in the 1938(7) issue, titled 'The *clever woman* and her guides'. Tsirintanis, at the time 35 years old, contrasts the Christian vision of women,

37 M. Planck, 'Science and Religion' in *Scientific Autobiography: and Other Papers* (London; William and Northgate, 1950), pp. 182–84.

38 'Υπάρχει ύλη;', *Aktines*, 3 (1938), 89-94.

39 D. K.,'Κατοικούνται οι αστέρες;', *Aktines*, 8 (1939), 75-80.

40 'Ο άνθρωπος του Πεκίνου;', *Aktines*, 3 (1938), 98-99.

41 See for example, the post–mortem republication of Ioannis Skaltsounis earlier article, I. Skaltsounis, 'Πάντα τα γνωστά γεγονότα διαψεύδουν τη δαρβίνειο θεωρία', *Anaplassis*, 49 (1936), 398–400. Tellingly, the same issue contains quotes from Ioannis Metaxas, who was the head of the fascist military junta in power at the time, and a devoted anti-communist himself.

42 D. K.,'Η εξέλιξης της θεωρίας της εξελίξεως;' *Aktines*, 7 (1939), 12-15.

in which they are glorified and considered to have the same immortal soul as a man, with the materialistic vision of the female. If no soul exists, then women are left with a weaker body. While their beauty lasts, they smoke and 'cross their legs', but when their beauty fades, then they are left alone, with no husband, father, or children. All in all, Tsirintanis clearly implies, it is only the Christian immortal soul that makes women equal to men. In the 1938(4) of the same year, an article by V. Ploumidis titled 'The troubles of a people' discusses the plight of Jewish people all over Europe. However, after a brief sorrowful description, the Jewish are unfavorably compared with the Greek nation and the article ends with the conclusion that the Jewish troubles are a direct result of the emergence from its ranks of a specific kind of rational materialistic philosophy. The spiritual emptiness that followed led to its woes.[43] The editors of *Aktines* were also ready to defend, and in a way declare, their intellectual forefathers. In the same 1938(4) issue, there is a nine-page (out of a total of 48), article titled *Ο υλισμός εν Ελλάδι και ο Ιωάννης Σκαλτσούνης* (Materialism in Greece and Ioannis Skaltsounis) which hails Skaltsounis as a hero and an intellectual giant. In 1947, another article by I. Kassianos hails the work of Kosmas Flamiatos, the nineteenth century Orthodox monk, as a precursor to the Declaration of the CEE. Flamiatos, it should be noted, tried to topple the Greek government of his time and impose a thoroughly Christian nation in its stead, and was also the inspiration for Papoulakos, whose movement the Greek army was called to forcibly put down.

As a final note, *Aktines* seemed to have an impact early on, which allowed them to make their mission clear. Already in its fourth issue, *Aktines* devoted one and half page to an 'angry letter', accusing the journal of one-sidedness. The editor's answer is clear:

> 'Our journal is a journal of principles. Principles which have been attacked with great ferocity over several decades, using verbal and written speech, and when possible, by the sword. These attacks had only one goal. The obstinate denial [...]
>
> Is it then wrong to want to present the truth? Is it one-sidedness to have principles and use them to guide his life in all its facets?'[44]

Aktines was a journal of principles, and that is that. In the service of these principles, *Aktines* published 658 articles devoted to the Natural Sciences over the course of 46 years, from 1938 to 1974. This averages out to 18 articles per year, or three articles every two issues.

43 V. Ploumidis, 'Το δράμα ενός λαού', *Aktines*, 4 (1938), 155–158.
44 'Μονομέρεια', *Aktines*, 4 (1938), 133–34.

The Manifesto and the Declaration of the Christian Union of Scientists

The *Christian Union of Scientists* (CEE) and *Aktines* had a steady rise, which went hand in hand with the rising influence of *Zoe*. However, it became a household name with the publication of *Manifesto of the Christian Union of Scientists, for every Greek* (1946) and its accompanying *Declaration*. The Manifesto was a 200–page book, that tackled everything the team of *Aktines* had so far worked on. This can be seen in its main chapters: *Current reality and its creators, The Denial that ruled, Science and Religion* (with sub-chapters on the creation of the world and the Natural Sciences, Freudian theory, the Immortality of the soul and others), *What History says* (with a subchapter on feminism, among two others) and *What Faith says*. For the long-time readers of *Aktines*, many of the topics and arguments mentioned in the *Manifesto* would seem familiar, and intentionally so. It was an apologetic venture similar to what Trempelas had been doing for decades, but polished by Tsirintanis, who was informally recognized as the mastermind behind the publication. Of equal importance was the appendix of just twelve pages, which contained the *Declaration of Greek Scientists and Artists*, signed by initially more than 180 Greek University Professors and Artists, and later on by forty more. The *Declaration*, which first appeared as an autonomous text in the November–December of 1946 issue of *Aktines*, proclaimed that from 'the objective and impartial study of science', as well as from 'literature, art and culture in general, and from the observation of human experience, especially of recent years' one must conclude that the future of humanity, and Greece in general rests on using Christianity as the spiritual foundation of civilization. Not only science and religion are not at odds, and not only science cannot be used to refute religion, but in fact, progress and social justice require a Christian morality and a Christian education. All that, we should re-emphasize, are direct results of the objective and impartial study of science. While the *Manifesto* was Tsirintanis's brainchild, the *Declaration* seems to be the work of the astronomer Dimitrios Kotsakis. Three years later, in 1949, Kotsakis wrote the introduction to the book 'Science speaks', in which he expanded the *Declaration* to include the endorsements of foreign scientists who were approached for that specific purpose, but also texts from notable scientists that purport to show that their work is also influenced by religion.[45]

The impact of this two-fold publication was tremendous, in both sides of the ideological fence.[46] The *Manifesto* sold more than 150,000 copies, an extraordinary amount. Yannaras, at the time 11 years old but already in orbit

---

45 *Η επιστήμη ομιλεί* (Athens: Brotherhood of Theologians Zoe, 1949).

46 For a first attempt to discuss EAM's response, see D. Skordos and C. Skordoulis, '1946 Declaration of Christian Association of Scientists as viewed by EAM's (National Liberation Front) Press. Examining the controversy over the role of science and the relation to religion within the context of the Greek Civil War', *International Conference on Science and Religion* (Athens, 3–5 September 2015).

around *Zoe,* and well acquainted with its internal workings, remembers the wider network of *Zoe* visiting 'every nook and cranny' to find literate Greeks who would buy it.[47] It provided a solid foundation for a type of conservative Christian anti-communism which would become dominant in the next few years, as the Greek Civil War became a reality. Most of the common people had little understanding or need for the grandiose apologetics of the *Manifesto,* but they could understand and evaluate the significance of more than two hundred esteemed professors and artists declaring that science was on their side. What is noteworthy is that many of the signatories were not in fact especially conservative. This was noticed by members at the time. Yannaras again says:

> [...]for the intellectuals of the era, signing or not the Declaration of CEE must have been a crucial dilemma of openly declaring political allegiance. This is why, within the 220 signatures there were names of people openly uninterested in Christian faith or even people which had advocated the exact opposite ideas than those of the Declaration in their whole life — and continued to do so afterwards.
>
> In any case, for the collection of the signatures, all the forces of the "movement", but also "friends", were mobilized and networked [...] And it was systematically glossed over that the specific text was to accompany the Declaration [...]
>
> In the end, we cannot know how many were naively taken by the idealism of the text and how many signed from fear of being branded as supporters of the "army of atheism and materialism".[48]

In fact, on the same day that Zachariadis gave his fiery response to the *Manifesto,* and probably not coincidentally, three signatories wrote to the liberal journal *Elefthera Grammata* (Free letters) to say that they were duped, and would like to withdraw their support.[49]Zachariadis, far from being a disinterested observer, mentions that 'the reactionary spiritual sermons ... affect people within the popular democratic camp and of course within our lines, while scholars that ... gave in to the pressure and bullying of CEE look like the children and the fools that, frightened by the frogs, leap into the river'.[50] In fact, even conservative Christians noticed the disparity, as when the arch-conservative priest Augustine Kantiotis (1907–2010) portrayed in his journal *Christianiki Spitha* (Christian Spark) the liberal Professor of Viniculture Vasileios Krimpas (1889–1965) as the devil adding his signature to the *Declaration.*[51]

---

47 Yannaras (2000), p. 45.

48 Yannaras (2000), p. 44. (translation is mine).

49 A. Giannopoulos, S. Doukas and S. Papaloukas, 'Δήλωσις' και αντιδηλώσεις', *Elefthera Grammata,* 59 (7–1–1947), 31.

50 Zachariadis (1947), p. 54.

51 S. Bournazos, 'Το κράτος των εθνικοφρόνων: αντικομμουνιστικός λόγος και πρακτικές', in *Ιστορία της Ελλάδας του 20ού αιώνα,* Vol D2, ed. by Ch. Chatziiosif (Athens: Vivliorama, 2009), pp. 9–49.

Whatever the case, the publishing success of *Manifesto* was seen by communists and anti-communists alike as part of the emerging Civil War, and intimately tied to it. In 1946, apart from the developments already mentioned in the introduction of the paper, the short-lived Tsaldaris government had voted the infamous Ninth Resolution on the 'Purge of Public Services', which led to 18 professors losing their position due to their communist sympathies. In 1947, Dimitris Hondros, the leading Professor of Physics in the University of Athens and the Secretary of the University, wrote and helped pass the 'Resolution of the general assembly of the board of professors: On the subject of the struggle carried out by the nations against bandit bands supported from abroad'. In it, the University condemned — in four languages, Greek, English, French and Italian — the DSA fighters as lawless bandits working under the auspices of international communism and called for international intervention to help Greece. The *Manifesto* was a strike from the intellectual side of the same front, and the response, by the taken unaware communists, had to be swift. Zachariadis, in his 1947 paper was very clear: This was an attack on the People by one of many reactionary forces , and thus part of the ideological war that accompanies class struggle. This had become, as the title of his article says, an ideological battlefront, and intellectuals should better sit up and fight back.[52] And indeed, a flurry of different kinds of publications shortly followed.

The first such response, some days after the initial publication of the *Declaration* in *Aktines*, was by Professor Charis Theodoridis, in the journal *Elefthera Grammata*.[53] In a single page, he accuses the signatories of the *Declaration* of hypocrisy, saying that they are known to act very differently in real life than what they profess, and that they could have shown their Christian charity towards their colleagues who had lost their position due to their leftist ideas. The official newspaper of the Greek Communist Party, *Rizospastis*, took a much more severe tone in its 7th and 8th of January 1947 issues. In a series of articles titled 'The Declaration and those who declare' (Η Δήλωσις και οι δηλωσίες), *Rizospastis* personally attacked the signatories as unworthy scientists and immoral people.[54] A week later, in the 15th of January, another signatory is mocked for putting up an ad for marriage.[55] And again, in the following issue of January 16th, 1947, the Christian Union of Scientists is derided for remaining silent about the 'bloodbath'

52 Zachariadis idem, p. 53–53. He also cites S. Myrivilis (1890–1969), a famous novelist which at the time did anti-communist radio shows, and an article in the respected literary journal *Nea Estia* as part of the same attack. See P. Charis, 'Ελεύθεροι πνευματικοί άνθρωποι', *Nea Estia*, 468 (1947), 5–7.

53 Th. Charis, 'Διαβάζοντας μια δήλωση', *Elefthera Grammata*, 57 (15 Dec 1946), 367.

54 It should be noted here that 'δηλωσίας' was a byname for the traitors of the communist cause during the Metaxas era. The name stems from the "Declarations of remorse and disavowal of communism" that the Metaxas regime forced communists to sign, as a way to alienate them from the Communist Party.

55 A. Spilios, 'Αι μικραί αγγελίαι', *Rizospastis*, (15–1–1947), 1.

against the Greek people, all the while preaching about Christian values. Or as the author writes: 'Our people have understood that twenty centuries of Christian faith, with which our nation's history has been intertwined, have not made things better, and have no patience for twenty centuries more'.[56] The pioneer Marxist philosopher Yannis Imvriotis — who had lost his University position in the Purge resolution — joined the fray in February, 1947, with an article in *Elefthera Grammata*. He accused the CEE of being dishonest people who, under the guise of a pseudoscientific morality, in fact reproduced the darkest hours of fanatical theological conservatism. One of the final paragraphs sums it up: 'Here is a book which claims to do objective, patriotic and Christian research, while being in radical opposition to science, patriotism and Christianity'.[57] More sustained and elaborate, but no less vitriolic, were the responses in the *Socialist Review*, the official journal of the leftist Popular Democracy Union (ELD) party. In a series of articles, Kostas Vrettos attacked the implicit ideas, implementation and general viewpoint of the *Manifesto*, finding in it a modern revision of the theocratic Dark Ages, and thus, deeply anti-Christian in its essence.[58] Helias Sarantos, in the engineering leftist journal *Antaios*, focused his critique on the biological arguments of the *Manifesto* and from there, on what he saw as an attempt to 'attack the scientific thought which made man recognize his value and strength'.[59] However, the most magisterial defense of dialectical materialism *vis-a-vis* the Manifesto was the 1947 book *The Philosophy of modern Physics*, by Nikos Kitsikis (1887–1978).[60] Considered the father of modern Statics in Greece, Kitsikis was at times the Rector of the National Metsovian Polytechnic in Athens,, the President of the Technical Chamber of Athens and a member of the Parliament. In 1947, he had lost his position due to his communist sympathies and wrote in *Antaios*, which he had co-founded. The *Philosophy* is now considered one of the inaugural books for Greek philosophy of science, which inspired a generation of leftist intellectuals.[61] Despite its theoretical elegance and its obvious mastery of modern physics, the author intended the book to partly be a response to the Manifesto. As Kitsikis says in the chapter 'The importance of Physics for Philosophy':

> 'A heated battle is taking place today in the theoretical scientific level, with grave impact in the social, the religious, the political level. Scientific theories of modern physics are being exploited, which defiles the purity of scientific

56 M. Avgeris, 'Χριστιανική ρητορεία', *Rizospastis*, (16–1–1947), 1.

57 Y. Imvriotis, 'Οι αρνητές της επιστήμης', *Elefthera Grammata*, 60 (1947), 36.

58 K. Vrettos, 'Χριστιανική απολογητική της καπιταλιστικής παρακμής', *Sosialistiki Epitheorisi*, 2 (1947), 67–70 and 4 (1947), 126–30, and 'Επιστήμη και θρησκεία', *Sosialistiki Epitheorisi*, 6–7 (1947), 224–30.

59 H. Sarantos, 'Οι επιστήμονες και τα προβλήματα της ζωής χωρίς μεταφυσική', *Antaios*, 2: 5–6, (1947), 169.

60 N. Kitsikis, *Η φιλοσοφία της νεώτερης Φυσικής* (Athens: Papazisis, 1947).

61 See the 'Introduction' of E. Bitsakis on the 1989 re–issue of Kitsikis' book (Athens: Gutenberg, 1989).

> thought. Professors and writers are being subjected to never-before-seen blackmail, reminiscent of the Catholic Middle Ages, in order to sign manifesta and to confess to beliefs contrary to their scientific dignity, laborers of science and art are being conscripted to create an ideological battle-front, which has as a goal to deflect the fell danger posed by scientific truth to the regime of social injustice and obscurantism.'[62]

Kitsikis here does not mince words and cites directly both the Manifesto and the various responses to it. His writing echoes word for word the article of Zachariadis on the ideological battlefront. Despite its erudition, however, and its powerful exposition of modern scientific theories — in fact quite advanced relative to other expositions of relativity and quantum mechanics available in Greece at this period — the book was becoming irrelevant at the battle it wanted to fight.[63] By the end of 1947, the Civil War would be in full swing, and arguments made by both sides would be accepted or rejected according to one's political stance.

Leftist scholars and scientists were not the only ones who objected to the *Manifesto* and the *Declaration*. Some of the criticism came from Christian and conservative circles. The rapid rise of *Zoe* and its burgeoning influence within the Church was met with alarm by some theologians and clergymen. In two separate times, in 1914 and in 1924, *Zoe* had been brought before the Holy Synod of the Church of Greece, under the charge that it had circumvented Church hierarchy. Both times, *Zoe* was exonerated, and even received a commendation instead of a condemnation.[64] Another kind of attack came from established theologians, like Dimitrios Mpalanos (1877–1959). Mpalanos was a prominent Professor of Patristics and Dogmatics, a former Minister in the Greek government, and a member of the Academy of Greece since 1931. He took it upon himself to oppose *Aktines*. Mpalanos had clashed with CEE members before, especially when Hieronimos Kotsonis had submitted a dissertation in 1945 as part of his bid for a university position. *Zoe* theologians had a dismissive attitude towards University theologians and rumors abounded about Trempelas and his circle causing problems within the School of Theology of the University of Athens.[65] For whatever reason, Mpalanos devoted his April of 1947 lecture in the Academy to *Aktines* and its *Manifesto*.[66] The *Manifesto*, for Mpalanos, was a work of 'one-sided religiosity, which devolves to hysteria' (p. 3). Greeks had not lost their faith, as the *Manifesto* claimed, and the bleak pessimism of CEE was misplaced (p. 4 and 5). The *Manifesto* was so biased that it even made excuses for the Inquisition. The

---

62 Kitsikis (1947), pp. 4–5.

63 See K. Tampakis and C. Skordoulis, 'The History of Teaching Quantum Mechanics in Greece', *Science & Education*, 16.3 (2007), 371–91.

64 Karamouzis (2011), pp. 330–31.

65 Yannaras (2000), p. 165.

66 D. Mpalanos, *Η θρησκευτικότης του ελληνικού λαού* (Athens: Aetos, 1947).

Manifesto should follow its own advice and not mix Religion with the Natural Sciences:

> 'As we have the very reasonable claim that the natural sciences should not mingle with metaphysical matters, and (should not) oppose science to religion, in the exact same way it is reasonable (to support) the claim of the natural sciences that obstacles should not be put forth in their research and progress, in the so-called name of religion [...]'.[67]

And so on and so forth. The answer was swift by Mpalanos' adversary, Kotsonis, in a text almost as long as the original Mpalanos lecture, which was fully endorsed by the *Aktines* staff. Kotsonis, among other accusations and attempted refutations, tacitly accused Mpalanos of being on the side of materialist communism.[68] Mpalanos responded even more furiously in a letter in the very next issue, declaring Kotsonis' article to be a 'libel'. Mpalanos also decried that the *Declaration* was appended to the Manifesto without the knowledge of the signatories, and that 'if my detractor wants the children to be taught as dogma that the world was created in six days, let him do so. Without any doubts whatsoever, I believe that this is a wound to our religiosity'.[69] The second answer of Kotsonis is even longer and, if possible, more dismissive. Among other things, Kotsonis says 'We had no need to resort to artifice for CEE to declare that Aktines comes from more than a thousand scientists for the simple reason that ...the Union (has) more than a thousand members...' and that if Mpalanos finds the temples full, it is because the people of the Manifesto have worked for that.[70] Mpalanos retired the next year, and it seems that no further debates took place, at least publicly.

Another attack closer to home came from Michael Galanos (1868–1948), in a book written in 1947 but published at the year of his death, in 1948.[71] If Mpalanos was part of the theological establishment, Galanos was a brother in arms for the people in *Zoe* and *Aktines*. A former editor of *Anaplassis*, a friend and co-worker of Ioannis Skaltsounis and another former student of Makrakis, Galanos had been at the trenches fighting against materialism for more than sixty years. He was part of the people who had sued the leftist pedagogue Alexandros Delmouzos for his teachings on religious grounds and had signed the Church excommunication against the then Prime Minister Eleftherios Venizelos in 1917.[72] Galanos has been in the game too long to mind the youngsters like Tsirintanis and Trempelas. In fact, he goes at pains to say that he remembers them as his students, and that

---

67 Mpalanos (1947), p. 9.

68 I. Kotsonis, 'Από τα απίστευτα', *Aktines*, 71 (August 1947), 378–84.

69 D. Mpalanos, 'Από τα απίστευτα', *Aktines*, 72 (September 1947), 425–27.

70 I. Kotsonis, 'Οφειλόμενη απάντησις', *Aktines*, 72 (September 1947), 427–32.

71 M. Galanos, *Η γνώμη μου περί του όλου έργου του Αρχ. Σεραφείμ Παπακώστα*, (Athens, 1948).

72 An account of the excommunication is presented by none other than J. G. Frazer in 1917, see J. G. Frazer, 'The cursing of Venizelos' *Folklore*, 28.2 (1917), 13–140.

he does not understand why they portray themselves as neophytes.[73] Galanos considers only the head of *Zoe* to be a suitable target, and in fact accuses him of selective memory, bad scholarship, and of heretical and dictatorial behavior not suitable for a senior clergyman.[74] He sees with alarm the meteoric rise of *Zoe*, which takes credit for organizing Sunday Christian schools, for publishing Orthodox thought abroad or for fighting materialism, when in fact Galanos had done so for decades.[75] Galanos accuses Papakostas and his circle of specifically obfuscating the accomplishments of Galanos' and others from the old *Anaplassis* guard (the Archimandrite Efsevios Matthopoulos, and the theologians Gregorios Papamichael and Panagiotis Mpratsiotis) to highlight their own.[76] The tone is obvious: The old veteran is being sidelined by his younger protégés, who pay no respect to him right at the time when his sixty-year old fight finally becomes fashionable. Nevertheless, the *Aktines* circle of scholars did not seem to want to pick a fight with an old grognard, especially since he died the same year.

Finally, an unlikely source of criticism came from Theodoros Lanaras, a successful textiles industrialist in Naousa. Lanaras had developed a circle of followers and friends with whom he discussed and taught blatantly heretical views of Christianity, which he insisted he could debate with anyone in front of a recording machine or audience. The Brotherhood of Zoe and the church had tried at first to bring him into the fold — he was a very visible and rich industrialist, after all — but they were not successful, and eventually Lanaras was excommunicated. This did not stop him, and he continued to write books and articles addressed to the Church and to the general public. Lanaras' unsolicited and surprising response to the *Manifesto* and the attached *Declaration* appears in January 1947, only months after its publication.[77] Lanaras, as was his style, took issue with CEE's apologetics and theology, because they did not coincide with his own. He recounted his own conservative position, found numerous problems with the facts presented in the *Declaration* and in the *Manifesto* and took issue with even the possibility of science being helpful to Christian faith (p. 64–66). He also considered the signatories to be 'talking puppets', moving and talking under the direction of others (p. 6) and sarcastically notes that he saw very few actual signatories from most Universities, Polytechnic Schools and the Geoponic departments. 'That means that most do not agree with your Declaration and did not sign it' (p. 147). *Aktines* and *Zoe* did take issue with the 180-page long pamphlet and devoted two articles in the March 1947 *Zoe* issue, to take their turn in mockery, noting that they have been attacked by both communists and industrialists '[The premium Communist journal] Rizospastis is being assisted by the [industrialist] Theod. Lanaras and

73 Galanos (1948), pp. 28–29.
74 Galanos (1948), pp. 12–16, 20–24, 34–38.
75 Galanos (1948), pp. 28–32.
76 Galanos (1948), pp. 34–36.
77 Th. Lanaras, *Απάντησις κριτική επί της Διακηρύξεως των υπερχιλίων επιστημόνων της Χριστιανικής Ένωσεως* (Athens, 1947).

Theod. Lanaras walks hand in hand with Rizospastis. Let them rejoice and be proud of each other as allies'.[78] With one stroke, the Brothers of *Zoe* had managed to insult both their detractors. That of course brought about another long and rambling response by Lanaras, penned within three days, who apparently needed thirty pages to answer a two-page article. Lanaras was rather indignant at being labeled a communist and took the opportunity to describe his theses once more.[79] But the confrontation ended there.

Why was the *Declaration* and the *Manifesto* so successful? It is tantalizing to consider its success as an offshoot of their novelty. However, it has not so far been observed that a venture similar to the *Declaration* had been attempted a decade before, in 1937, by the *Akadimaikos Koinonikos Syndesmos* (Academic Social Association), an earlier enterprise of some of the members of *Aktines*. A booklet called 'The Wise on God and Religion', probably authored by Kotsakis, was a collection by quotes of famous scientists that showed their belief in God, and also contained small texts by Greek academics solicited for that exact purpose.[80] That was the format that the *Declaration* and the expanded *Science speaks* book followed later on. As the 1938 booklet says in its third page, these quotes had been selected from the earlier, 1916 Greek translation of P. Svjetloff's *Scientists and Religion* and from the 1932 Greek translation of Robert de Flers *Les sentiment religieux et la science*. Other references cited include the 1913 book *Science and Religion* by Michael Galanos and the *Zoe* journal. Clearly, this approach was not new. Nor was the *Manifesto*'s apologetics a novelty, since Trempelas, Galanos and Papamichael had been doing something similar for decades. It would be thus helpful to inverse the relationship between intellectual foreground and political background: It was not the success of the *Manifesto* and the *Declaration* that contributed to an ideological and intellectual war. Instead, the emergence of an ideological and intellectual war led to the success of the *Manifesto* and of the *Declaration*. It was the right publication at the right time, and it was promoted as such.

Aktines in the Service of State Anti-Communism: The Rise and Fall of the *Elliniko Fos* Organization

The success of the *Manifesto* and of the *Declaration* buoyed Tsirintanis and his circle in the higher echelons of Greek politics. From 1949, Hieronimos became the Archpriest of the Greek Palace — since Greece at the time was a constitutional monarchy, at least in name — and after a short time, the personal priest of the Royal Family. Soon after, the Greek Civil War would end, with DSE stopping all

---

78 'Ενας συνοδοιπόρος' and 'Εξωφρενικόν, αλλ΄ουχί ανεξήγητον', *Zoe*, 1569 (23–4–1947), 112.

79 Th. Lanaras,*Ανοιχτή επιστολή πρός τον πρόεδρον της Αδελφότητος 'Ζωή' Αρχιμ.Κον Σεραφείμ Παπακώσταν και τους Θεολόγους της Ζωής* (Athens, 29/4/47), p. 17.

80 Akadimaikos Koinonikos Syndesmos, *Οι σοφοί περί Θεού και θρησκείας* (Athens, 1938), pp. 37–62.

hostile activities. However, its aftermath would be felt for decades, especially since not all fighting units would concede at the same time. The official Greek state propaganda against the Communist party would intensify in the next decades, even as the Cold War picked up steam internationally. The Greek State would vehemently deny that this had been a Civil War, and instead would consider it a military action by bandits, directed by Slavs and Russians. Communism was a lie and a disease in the body of the nation and the fight against the communist bandits was a holy war of Christianity against atheism.[81] In that era, Tsirintanis, who had developed excellent relations with the King Pavlos II, proposed that a new organization should be instituted, to provide the intellectual weapons and tools needed in this fight. Its name was to be *Elliniko Fos* (Greek light) and would be a voluntary action, headed by the King himself. The King wholeheartedly agreed, and placed the organization under the aegis of Tsirintanis.[82]

Elliniko Fos was wildly successful, and in fact was the main intellectual anti-communist organization of the period 1949 to 1955.[83] Its goal was 'the general uplifiting of the intellectual and moral level of the Greek people' and especially of the younger generation 'according to the national ideals and traditions'. Three basic principles would nominally guide the effort: the cultivation of the love for the country, a clear and consistent position towards Christian religion and the Christian Church and a complete separation of the organization's efforts from politics.[84] In actual practice, this amounted to a continuing propaganda worth an average of 40 to 50 million drachmas each year. The organization ran Sunday schools, organized series of lectures for Army officers, gave out pamphlets, donated toys to the kids of child-towns, and translated articles and books to be sent abroad. It also hosted radio shows, often ran advertisements in newspapers, found jobs for family members of soldiers and facilitated hospital care for families of proper ideological orientation. Its main pamphlets, 'Για μια νέα Ελλάδα' (For a new Greece), 'Εις τον αγώνα του Έθνους' (In the nation's struggle) and 'Διατί αγωνιζόμεθα' (Why we fight) circulated in the tens and hundreds of thousands.[85] As a way of comparison, according to the 1951 Census, Greece at the time had a population of 7 million, of which 5 million were adults. 35% of this adult population was completely illiterate and 55% had only finished primary education. The enormous circulation was made possible by the institutional and probably financial support of the state, in

81 For the anti-communist rhetoric of the era, see Bournazos 2009, and also S. Bournazos, 'Ο αναμορφωτικός λόγος των νικητών στη Μακρόνησο: η ένταξη του Εμφυλίου στην προαιώνια ιστορία της φυλής, ο «διηθητός ιός» του κομμουνισμού και ο ρόλος της «αναμόρφωσης»', *Dokimes* 6 (1997), 101–34.

82 Yannaras (1990), p. 101.

83 Bournazos (2009), pp. 32–34.

84 Elliniko Fos, *Σύλλογος «Το Ελληνικό Φως». Κατευθύνσεις και καταστατικό* (Athens, 1948), pp. 3–4.

85 Elliniko Fos, *Σύλλογος «Το Ελληνικό Φως», Απολογισμός μιας πενταετίας* (Athens, 1954), pp. 48–50 and Yannaras (1990), pp. 102–04.

which embassies requested copies of its pamphlets, generals ordered army units to receive volunteers of *Elliniko Fos* for lectures and foreign journalists directly received information about its activities.[86]

However, the anti-communist orientation of *Elliniko Fos* and the indirect leadership of Tsirintanis does not directly imply some relation to *Aktines* and its weaponization of science and religion discourse. To show the existence of such an umbilical cord, one has to look at the inner operation of *Elliniko Fos*. The first such indication is the Executive Board itself, in which we find not only Alexandros Tsirintanis, but also Theodoros Mertikopoulos (1895–1981), another founding members of *Aktines*, as well as Plakides and Kotsakis. Of the eight board members, four were founding members of *Aktines*, and two more were from *Zoe*. More importantly, Tsirintanis himself took pains to prop up *Elliniko Fos* during its first months of operation, by underlining its connection to *Aktines* and its *Manifesto* and *Declaration*. To show the necessity of *Fos*'s actions, Tsirintanis, who wrote all *Fos*' major programmatic essays, says:

> '[…] the General Secretary of KKE (the Greek Communist Party) attacked with real fury the famous Declaration of the Christian Union of Scientists […]
> The spiritual strength, that flows from Christianity, communism forever fears and fights everywhere […] Because it knows well that "since (communism) is based on the worldview of materialism, it cannot win the battle unless it defeats the only, truly alive, spiritual worldview, which is Christianity"'.[87]

The quotes here are by the author himself, and are from a 1949 article in *Aktines*, titled 'In the struggle of the nation'. In fact, as Tsirintanis underlines, all of the main pamphlets of *Elliniko Fos*, which would be later circulated in the tens of thousands, had initially been published in *Aktines*, in the March and April issues of 1949, and would already be familiar to long time readers of the magazine. Or, as once again Tsirintanis says, in the same programmatic essay for *Elliniko Fos*, 'For the theoretical foundation of its efforts, "Elliniko Fos" used the relevant to the national struggle articles that *Aktines* offered to the Greek nation'.[88] Later on, one of the first 'Officers' weeks', in which *Elliniko Fos* members lectured Greek Army officers, is described in detail. The titles of the lectures for the Week — 'Overcoming materialism in the natural sciences' or 'The denial in medical science' — would have sounded very familiar to readers of *Aktines*.[89] Finally, the connection between the *Aktines*' discourse on science and religion and the anti-communist propaganda of *Elliniko Fos* becomes even more clear in an impressive, and somewhat chilling, document, titled 'Speeches of the National

86 All data are collected from Elliniko Fos (1954), pp. 80–150.

87 Elliniko Fos, *Ελληνικό Φως, Εις την πνευματική συμπαράσταση του εθνικού αγώνος, Έκθεσις πεπραγμένων (Απρίλιος — Σεπτέμβριος 1949)*, (Athens, 1949a), p. 7.

88 Elliniko Fos (Athens, 1949a), pp. 9–10.

89 Elliniko Fos (1949a), pp. 22–26.

Spiritual Struggle', published in 1949. In contrast to other publications, this is an internal document, aimed at speakers who would tour the countryside and deliver 45-minute-long lectures to a general, and usually non-rural, audience. The document contains blueprints for six speeches, whose topics range from 'The struggle above all' to 'Social justice and communism'. The document states clearly the goal of each speech, presents in detail eight to ten arguments, and offers specific rhetoric devices that can be used, with a clarity that many pedagogical textbooks would envy. Each lecture is also accompanied by helpful handouts and reading material. Tellingly, for the four of the six lectures, the suggested reading material are articles from *Aktines*.[90] The work that had been done for *Aktines* was now the main material for the anti-communist propaganda of *Elliniko Fos*, not only for urban populations and for Army officers, but for rural populations also.

The 1950s were the apex for Tsirintanis and his circle. They were the rising stars within the increasingly powerful *Zoe* Brotherhood, but also boasted strong ties to the Greek Monarchy. There were even whispers and innuendos that Tsirintanis had been offered thrice an unconditional Presidency, which was a euphemism for a dictatorship.[91] However, the political atmosphere at the time was not ripe for such actions, and such rumors probably stemmed from palatial political intrigue. Nevertheless, the work that *Aktines* had begun twenty years ago was at its height, right at the forefront of the ideological battlefront against communism, and with the *Aktines* team at the peak of their scientific careers. Tsir-intantis' programmatic text 'Για έναν χριστιανικό πολιτισμό' (Towards a Christian civilization), was translated into English in 1950 and received positive reviews from none other than the pioneer Orthodox theologian G. Florovsky and later by the Calvinist theologian J. K. S. Reid.[92] But the 1960s would prove more troublesome. In 1960, a crisis of leadership within *Zoe* led to a splintering of the powerful organization. As a result, several of the old guard, such as P. Trempelas, Ioannis Kolliopoulos and sixty other members, left *Zoe* to found the Brotherhood *Sotir* (Saviour). Tsirintanis, also a loser in the struggle, founded the *Enosi Ellinikos Politismos* (Greek Civilization Union) and the journal *Sizitisis* (Conversation). Tsirintanis was later seen as having tacitly supported the Greek military Junta of 1967, which irregularly installed his old friend Hieronymos as the Archbishop of Athens and thus, as the Head of the Orthodox Church of Greece. The Greek Junta of 1967, it must be said, was brought about by the Army officers Georgios Papadopoulos, Stylianos Patakos and Nikolaos Makarezos ostensibly to combat communism, and much of its rhetoric was reminiscent of the *Fos* and *Zoe* rhetoric.

---

90 Elliniko Fos, *Ομιλίαι εθνικού πνευματικού αγώνος* (Athens, 1949b), pp. 1–3, 7, 11, 29.

91 G. Karagiannis, *Η Εκκλησία από την κατοχή στον Εμφύλιο* (Athens: Proskinio, 2003), pp. 123–24.

92 J. K. S Reid, 'Review of "Towards a Christian Civilization. A Draft Issued by the Christian" Union of Professional Men of Greece." "Damascus" Publications, Athens, 1950', *Scottish Journal of Theology* 6 (1953), pp. 220–24 and G. Florovsky 'Towards a Christian Civilization: A Draft Issued by The Christian Union of Professional Men in Greece, by A. N. Tsirintanis. 270 pp. (Athens, Damascus Publications, 1950)', *Theology Today*, 8.2 (1951), 279–81.

By 1974, the Greek Junta had lost most of its international support — it was initially sponsored by the US government — and fell after a series of catastrophic political blunders. In the eyes of most people, the association of *Aktines* and *Zoe* with the Junta quickly rendered the old rhetoric of science and Orthodoxy as part of a Greek Christian civilization not only obsolete, but abhorrent. Even on a purely intellectual level, it was getting harder to defend an old school anti-materialist stance in the age of molecular genetics, nuclear reactors and worse, the Beatles. Tsirintanis himself, in a 1972 article published five years before his death, seems to look back longingly to the old days. He almost praises the older atheism for being at least original and intellectually robust. He seems lost on how to handle his younger contemporaries:

> '[...] In the end, to be an atheist because you got misled by Darwin, even a misunderstood Darwin, is a destructive error. But to be an atheist, because Lennon or Brigitte Bardot said so, that is a shameful comedown.'[93]

It was a losing battle. *Aktines,* and CEE, are still active, but are shadows of their former glory days. Tsirintanis died in 1977 and was remembered by his colleagues, and enemies, for several years on. Aristos Aspiotis died five years later, but Kotsakis and Plakides would hold on for a decade more. Their erstwhile opponent, Nikos Zachariadis, had been exiled 17 years earlier in Siberia by his own Party, for opposing the ascension of Nikita Khrushchev, and committed suicide in 1973.

Conclusion

What is then the conclusion to be drawn by looking into the history of CEE and *Aktines*? The point of this paper is that the work and role of CEE in mid- twentieth century Greece brings to the fore a so far neglected component of science and religion interactions. In specific spaces, in which Greece is but one, the intellectual and cultural power of the Natural Sciences made them a crucial component in ideological battles that have so far been historiographically neglected. Greece stands out as the only space that is at the same time part of the Western bloc, Orthodox, while hosting a strong communist party. The work of *Aktines* enables us to see how the Orthodox intellectual space reacts to materialism by marshalling science to its defense, while being an heir to a century-old Orthodox Christian apologetics scholarship. It also shows how inherently political and ideological such endeavors are. The communist/anti-communist struggle, and its intellectual corollary, the materialist/anti-materialist debate, was not an abstract argument taking place within academic halls. It involved people dying in the Greek

93 A. Tsirintantis, *Εμφάνισις και αντιμετώπισις της συγχρόνου αθεϊας* (Athens: Apostoliki Diakonia, 1972).

mountains, families being split and children getting raised like orphans in special towns. In all these events, a weaponization of science and religion narrative took place by both sides, with the more powerful state-sponsored institution bringing in the most resources. Articles about Darwin or quantum mechanics in a journal about Orthodoxy and the sciences contributed, indirectly but specifically, to a five-year anti-communist campaign that cost millions of drachmas and took place in tens of towns, as part of the bitter aftermath of a cruel civil war. The authors and publishers of these articles, initially a group of Orthodox conservative scholars, end up becoming the most prominent intellectuals of Greek anti-communism. In the end, however, the world of *Aktines* ended not with a bang, but with a whimper: As I was preparing this article, I systematically hunted books by the astronomer Dimitris Kotsakis in the University of Athens' library. They had been gifted to the Library of the School of the Natural Sciences by Kotsakis himself. In the last fifty years, I was apparently the first to request them. What an incongruous fate for the astronomer who penned most of the *Declaration*.

CHRISTOPHER W. HOWELL

The Holiness of Creation

Philip Sherrard and the Climate Apocalypse

▼ **ABSTRACT** Philip Sherrard is little known outside of Greece, but he exerted a significant influence on modern Orthodoxy. Translation was his greatest career success, and Sherrard introduced the world to Greek poets and the *Philokalia*. But Sherrard's later interests focused more on modern science and he was deeply concerned about climate change. Though environmental destruction has often been interpreted in socio-economic terms, Sherrard went a step further and faulted science itself for paving the way for devastation. It was science, he argued, that was behind the dehumanization of humanity and the desecration of the cosmos. Consequently, humanity could not use the tools of science to engineer a way out of it. The crisis was, at bottom, a spiritual one, and so required a spiritual solution.

Introduction

In June of 1956, some noteworthy names gathered to take communion in a small Athenian church at the edge of the Pedion Areos park. John Romanides, the eminent theologian, celebrated the liturgy with three devotees of the faith, among them the icon painter Fotis Kontoglou and the former Catholic Demetrios Koutroubis. The third figure present was an Anglo-Irish convert named Philip Sherrard (1922–1995), in church for his first communion as a Greek Orthodox communicant.[1] A poet, lay theologian, and radical environmentalist, Sherrard is little known outside of Orthodox circles, but had an outsized influence in translating and introducing the English-speaking world to modern Greek poetry, as well

1 A. Louth, *Modern Orthodox Thinkers: From the Philokalia to the Present* (Downers Grove, IL: InterVarsity Press, 2015), pp. 231–46.

Christopher W. Howell • Duke University

Orthodox Christianity and the Study of Nature, ed. by Kostas Tampakis and Ronald Numbers, Science and Orthodox Christianity, 4 (Turnhout: Brepols, 2024), pp. 213–240
BREPOLS PUBLISHERS 10.1484/M.SOC-EB.5.141385

as translating the Philokalia. But as Christopher Knight observes, 'the reaction to Sherrard among his fellow-Orthodox has, however, often been somewhat ambivalent'.[2] His personal philosophy as well as his vibrant life experience, however, formed the basis of his views on science, views which are both complicated and challenging. Despairing of the modern world, of capitalism, and of technologies of war, violence, and death, Sherrard rejected science as it was practiced in the twentieth century. He felt physics to be puerile and stunted, and he saw evolution as a failed attempt to restore meaning to a fundamentally pointless world. Though he was not at all a creationist (or even an intelligent design advocate of the kind that rose to prominence around the time of his late writing), Sherrard's rejection of evolution was at least, if not more, emphatic. As Andrew Louth observes, when it came to evolutionists, creationists, and intelligent design theorists, 'he would have regarded all of these positions as not radical enough'.[3] His views went so far against the contemporary grain that they are hard to categorize. Crucially, however, Sherrard's perspective on science enabled him to see the coming climate catastrophe before many others. As John Chryssavgis and Bruce Foltz write, 'Sherrard was the first Orthodox theologian to articulate a systematic approach to the ecological crisis, doubtless drawing upon his experience of monasticism and his exposure to the classic texts of the *Philokalia*'.[4] Sherrard's career and thought, then, raises uncomfortable but important questions about what humanity must do if science, even right science, can become a destructive force, and whether it can be redeemed.

Life and Conversion

The circumstances of Philip Sherrard's birth and early life might invite an observer to accept a strong form of predestination, for it seems he was marked out for a Greek-inspired rejection of modern science, technology, and industry. He was born on September 23, 1922, one day after the end of the Great Fire of Smyrna, the conflagration in which all of Greece's modern irredentist dreams — its 'Great Idea' and Romantic self-adoration — were burned up. His mother, Brynhild Olivier, was from the progressive Olivier family and associated with the Bloomsbury Group. Her parents had close ties with William Morris, and the name Brynhild was taken from the heroine queen in Morris's *Sigurd the Volsung*.

2 C. C. Knight, 'The Christian Tradition and Faiths of the World: Some Aspects of the Thought of Philip Sherrard', *Theology*, vol. 111.836 (), 336–45.

3 A. Louth, 'Philip Sherrard on Evolution: A Note', (unpublished manuscript, Dec 30, 2021), typescript.

4 J. Chryssavgis and B. V. Foltz, 'Introduction. "The Sweetness of Heaven Overflows onto the Earth": Orthodox Christianity and Environmental Thought', in *Toward an Ecology of Transfiguration: Orthodox Christian Perspectives on Environment, Nature, and Creation*, ed. by J. Chryssavgis and B. V. Foltz (New York: Fordham University Press, 2013), pp. 1–6.

Brynhild herself worked at one point as a jewelry apprentice for Henry Wilson, one of Morris's associates. After her first marriage to Arthur 'Hugh' Popham broke down, Brynhild married Raymond Sherrard and had three children with him, including Philip, who was born before Brynhild's divorce was realized. The fallout from the scandal led to a period of sustained poverty, as the Sherrards struggled to eke out an existence in the English countryside. Brynhild eventually became seriously ill and died in 1935 at the age of 47.[5] Despite the hardship, Philip Sherrard recalled his childhood fondly. As Kallistos Ware related, Sherrard's daughter Liadain remembered him speaking with 'sheer delight and nostalgia about his farm upbringing, the early rising, the milking, the harvest in an England still rural and to a large extent unmechanized'. But it would be in Greece that Sherrard would find his calling and realize his own Arcadia. In 1942, Sherrard joined the British army; he declined a commission but was eventually promoted to captain in the Royal Artillery. Although he was a soldier, he was committed to non-violence and so carried around a wooden pistol instead of a real one. After first serving in Italy and Austria, Sherrard was sent to Greece in 1946, and it was there that, according to Ware, 'marked the beginning of a life-long love of the Greek land, of its history, culture, and religious faith'.[6]

It began with a chance encounter in an Athenian bookstore. A bookshop worker named Anna Mirodia recommended Sherrard a French translation of George Seferis's poetry. Upon reading, the 23-year-old Sherrard wrote an enthusiastic letter to Seferis, praising the poet and asking if he could translate his poems into English. Seferis did not reply.[7] Sherrard nevertheless persevered with his own translation of Seferis's 'Last Stop', and sent another letter before, on Christmas 1948, Seferis finally offered a belated response, thanking Sherrard and commenting on the translation. Perhaps trying to make up for previous lack of engagement, Seferis also added, 'By the way, I do not know how it happened that we have never met.'[8] What followed was an eight-year correspondence and discussion between the two (though one cannot help but notice that Sherrard wrote to Seferis far more often than Seferis wrote back), before the volume and frequency of letters greatly diminished after the spring of 1956, when Sherrard converted to Orthodoxy (this was also a year after Seferis published *Logbook III*).[9] Anna Mirodia became Sherrard's first wife and the mother of his two daughters. Anna was descended from a legendary Greek family on her mother's side — the

---

5 S. Watling, *The Olivier Sisters: A Biography* (New York: Oxford UP, 2019).

6 K. Ware, 'Foreword', in *Christianity: Lineaments of a Sacred Tradition*, by Philip Sherrard (Brookline, MA: Holy Cross Orthodox Press, 1998), pp. IX–XLV.

7 Philip Sherrard to George Seferis, Feb 21, 1947, in *This Dialectic of Blood and Light: George Seferis — Philip Sherrard an Exchange: 1947–1971* (Limni, Evia, Greece: Denise Harvey Publisher, 2015), pp. 54–55.

8 George Seferis to Philip Sherrard, Dec 25, 1948, *This Dialectic of Blood and Light*, p. 78.

9 Andrew Louth remarked that Seferis was a bit alarmed by Sherrard's turn towards Orthodoxy. See Louth, *Modern Orthodox Thinkers*, p. 235.

Mavromichalis Maniot clan — and because she was fluent in English, French, and German, she assisted Sherrard with his poetry translations.[10]

It was something of a surprise that Seferis's work led Sherrard closer to Orthodoxy, given the poet's ambiguous relation to the Church.[11] But Sherrard detected an undercurrent of Christian theology and insight within Seferis's work, a thread he followed until it brought him to conversion. In Seferis, as well as other Greek poets like Dionysios Solomos and Angelos Sikelianos, Sherrard saw a holistic co-inherence of nature and supernature, a mingling of the two hemispheres of being that had been broken apart by modernity, especially by modern science and its attendant naturalist metaphysics. In an introduction to Seferis's poetry, Sherrard explained that he came from a rationalist, evolutionary, and progressive society, and that he had lived in a disenchanted world. In Seferis, and in Greece in general, he found what he called 'the other mind' of Europe, an alternative to the coldly mechanistic philosophy of industrial England and the broader West. 'It defeated', he wrote, 'my frame of references, challenged, as it were, the whole ground on which I stood, demanded of me that I should step through a door which I didn't even know existed.' That door turned out to lead to the 'Greek Byzantine world', one reflected in Seferis's poetry — though, perhaps, without even Seferis fully realizing it. There was, Sherrard believed, a sensuous element in Seferis's poetry that he linked to its Orthodox heritage. Greece was 'heir to an older, universal tradition', he continued, 'and the consciousness, the norms of thought, imagination and feeling enshrined in this tradition were very different from those reflected in the modern western European paradigm which I was presented with.'[12] It is telling that Sherrard used the phrase 'other mind of Europe', which, as the annotations in *This Dialectic of Blood and Light* point out, he borrowed from a 1984 book about Goethe and science. The author, J. P. S. Uberoi, attempted a recovery of Goethe as a scientist who held to a 'non-standard' account of the scientific method drawn from Paracelsus.[13] While Uberoi failed to convince, Sherrard nevertheless adopted the phrase and applied it to Greece, seeing the country, the people, and their traditions as an alternative to the stultifying effects of modernity, science, and capitalism. For Sherrard, Greece was a challenge and rebuttal, and to understand his perspective on science one must understand his perspective on Greek poetry, tradition, and religion.

---

10 Denise Harvey, email message to author, July 23, 2024.

11 I have written elsewhere on Seferis's religious views. See 'George Seferis and the Freedom of Tradition,' *Public Orthodoxy*, August 31, 2020, <https://publicorthodoxy.org/2020/08/31/george-seferis-freedom-tradition/> [accessed 27 December 2022].

12 P. Sherrard, 'The Other Mind of Europe', pp. 37–42, in *This Dialectic of Blood and Light*, pp. 37–50.

13 J. P. S. Uberoi, *The Other Mind of Europe: Goethe as Scientist* (New York: Oxford UP, 1984).

The Other Mind of Europe

The Search for Tradition

Sherrard would not be alone in turning to Greece for salvation from modernity. Following on the heels of Henry Miller and *The Colossus of Maroussi* (1940) a wave of 'new philhellenism', in David Roessel's phrase, washed over the English-speaking world — comprised of writers such as Miller, Patrick Leigh Fermor, Lawrence Durrell, Edmund Keeley, Kevin Andrews, and Sherrard himself. Miller's depiction of George Katsimbalis made the latter man an emblem of modern Greek life and masculinity (a Zorba before *Zorba the Greek*), and his rapturous depiction of the poets Seferis and Dimitris Antoniou brought the vibrancy of modern Greek poetry to Anglophone readers.[14] Beyond the poetry, though, the landscape of Greece fascinated and intoxicated these new philhellenes. As Fermor wrote in *Mani*, the characteristics of the Greek countryside 'have a strange effect on the Greek landscape. Nature becomes supernatural; the frontier between physical and metaphysical is confounded',[15] Miller observed that 'God's magic is still at work and, no matter what the race of man may do or try to do, Greece is still a sacred precinct.'[16] And Durrell recounted the wisdom of a shepherd in his poem 'To Argos', writing, 'Only the shepherd in his cowl / Who walks upon them really knows / The natural history of a sacred place.'[17] If the world had been disenchanted, the toxins of modernity had not yet spread everywhere. Greece, for a time at least, was still an enchanted holdout, a place where the land and the sun and the sky were still suffused with sacramentality.

Sherrard shared this sacred sense of Greece, a place capable of healing modernity's breach between nature and spirit. The country had retained a sense of supernatural haunting, which could be discovered through poetry. 'The poet is a messenger between the supernatural and the natural world', he wrote, and the poet's tasks are to 'reveal to us the great underlying fabric of the universe, its unalterable principles'.[18] According to John Chryssavgis, Sherrard hoped to 'reinstate a sense of poetry in the aridity of life, and to restore a sense of sacred cosmology to the world'.[19] In a deracinated world, drained of life by industry, consumerism, and the exploitation of nature's inert matter, Sherrard sought to

---

14 D. Roessel, *In Byron's Shadow: Modern Greece in the English & American Imagination* (New York: Oxford UP, 2002), pp. 252–84.

15 P. Leigh Fermor, *Mani: Travels in the Southern Peloponnese* (New York: New York Review of Books 2006), p. 323.

16 H. Miller, *The Colossus of Maroussi* (New York: New Directions, 2010), p. 15.

17 Quoted in Roessel, *In Byron's Shadow*, p. 266.

18 Ph. Sherrard, *The Marble Threshing Floor: Studies in Modern Greek Poetry* (London: Valentine, Mitchell & Co., 1956), pp. 239–41.

19 J. Chryssavgis, 'Essay Review: A Tribute to Philip Sherrard (1922–1995),' *Journal of Modern Greek Studies*, Vol. 14, Issue 2 (1996), 345–54.

revive a pre-modern, sacramental view of the cosmos. In rural Greece and in modern Greek poetry, he found a haven spared the 'the debauchery of rationalism of which the modern world is the product'.[20] And so, in addition to writing his own poetry, he collaborated with Edmund Keeley throughout his career to translate and introduce modern Greek poetry to the English world.

The sacramentality of nature is key to Sherrard's reading of the Greek poets. From 1947–48, Sherrard was Reader of modern Greek literature at King's College, London, where he worked on his doctoral dissertation that was eventually published as *The Marble Threshing Floor*. The work comprises a literary biography of five Greek poets — Dionysios Solomos, Kostis Palamas, Constantine Cavafy, Angelos Sikelianos, and George Seferis. Throughout, as Ware wrote, Sherrard 'treats them as witnesses, bearing testimony to a reality greater than themselves'. This reality — these 'Great Realities', as he describes them in his chapter on Solomos — exemplify an 'attitude towards art that is neither "classic" nor "romantic" but "traditional"'.[21] He is noticeably harsher on Solomos, an Italian-educated and Italian-speaking Greek from Zakynthos heavily influenced by the Romantic poets, who is most known now for the 'Hymn to Liberty' that has become Greece's national anthem, as well as his apotheosis of Byron after the latter's death at Missolonghi. In Solomos, Sherrard criticized the split between the natural and supernatural that he saw throughout his work, writing that 'such an attitude not only ignores the bonds that link these two aspects of reality, but also, and as a consequence, threatens the very heart of life itself'. Despite this, however, Sherrard approved of the intimations of spiritual reality in Solomos's largely unfinished oeuvre, signs of hidden spiritual meaning that 'reveal [themselves] to whoever has the creative capacity to look penetratingly and deep'.[22] In Kostis Palamas, however, and especially Angelos Sikelianos and George Seferis, Sherrard found welcome discussion of the holiness and sacrality of the natural world. In Palamas, especially in his most famous work 'The Dodecalogue of the Gypsy', 'the ideal world is not irreconcilably separated from the natural world; it grows out of it, is part of it, its fulfillment but not its denial'.[23]

Sikelianos, the Dionysian celebrant and friend of Nikos Kazantzakis, comes across as potentially Sherrard's favorite. Sherrard saw in his poetry a constant 'lyric assertion of the natural world' as well as a reciprocal view of 'the austere vision of the seer who knows the natural world is full of tragedy and suffering and that the true center of man's life lies elsewhere'.[24] Sherrard documented a 'mythological attitude to life' in Sikelianos's early work, like 'The Visionary':

20 Ph. Sherrard, *The Wound of Greece: Studies in Neo-Hellenism* (London: Rex Collings, 1978), p. 61.

21 Ware, 'Foreword,' *Christianity*, pp. XIII, XX–XXI; Sherrard, *The Marble Threshing Floor*, p. 20.

22 Sherrard, *The Marble Threshing Floor*, p. 37.

23 Sherrard, *The Marble Threshing Floor*, p. 75.

24 Sherrard, *The Marble Threshing Floor*, p. 125.

> 'And when your grasp is firm / upon the sacred earth, / in triumph and in deliverance / will I forge wings for you / which the sun cannot destroy, / that you ascend and before it / raise up my indomitable / heart among the stars.'

In his treatment of nature, Sikelianos 'brings together the worlds which have fallen apart, he restores their original unity'. Sherrard saw in Sikelianos a belief that the traditions of the Greek peasants and villages preserved an ancient sensibility — a retention of belief in the Eternal Wisdom as the source of the world. And when the intellectuals have foundered, 'then it is precisely in the beliefs, art, and customs of the illiterate peasantry that may be found, in however adulterate a form, the material through which the wisdom of a former age can be recovered'.[25]

And lastly, as Sherrard reads Seferis, the sense of nature's intermingling with the supernatural — the forgotten enchantment of the world — is pronounced. Drawing from Seferis's 'Mythistorima', Sherrard wrote, 'Into this world [...] penetrate, as Plato divined, certain shoots and rays from the other, the lost world'.[26] Seferis's poetry captured this, in Sherrard's eyes, as did the Greek Orthodox character of much Greece's literature. 'This understanding of Plato's', continued Sherrard:

> 'That the natural world of the senses is rooted in the supernatural world; that the sight of beauty in the world leads directly back to the world of ideas — this understanding has been preserved in the Greek East in a way that it has not been in the West'.[27]

It should come as no surprise that Sherrard converted to Orthodoxy in 1956, the same year *The Marble Threshing Floor* was published. As Ware wrote, not only was this, for Sherrard, an embrace of the truest form of Christianity, it was also an entry into a Church that preserved 'an organic and life-giving connection between human beings and the natural world'.[28]

In all these poets, Sherrard found an enchanted world. The basis of his qualm with modern science is already evident there, long before he ever turned to writing on science directly. But it must also be added that Sherrard's view of creation was far afield from the common antievolutionary stance. Instead, Sherrard's emphasized the concept of Wisdom, or Sophia — which took the form in these poems as Wisdom herself, the Eternal Feminine, and often the Virgin Mary. In

---

25 Sherrard, *The Marble Threshing Floor*, pp. 126–29.

26 Sherrard's reading of Plato resists the accusations that the great philosopher was too otherworldly or that he hated the physical realm. Elsewhere, Sherrard — who despised the dualist bifurcation of reality into separate spheres of matter and mind — argued that there are *two* Platos, the otherworldly, gnostic one, and the one that attempts to 'express the immanence of reality.' This latter Plato is more evident in the later dialogues, and not the most famous ones like *Symposium* or *Republic*. See Ph. Sherrard, *Human Image: World Image: The Death and Resurrection of Sacred Cosmology* (Limni, Evia, Greece: Denise Harvey, 2015), pp. 12–13.

27 Sherrard, *The Marble Threshing Floor*, pp. 196–97.

28 Ware, 'Foreword,' *Christianity*, p. XII.

examining Solomos's 'myth', one finds 'at its heart is the imagining of a divine presence, of a feminine, maternal power, who seems to possess all the qualities of an all-seeing, eternal Wisdom'. In temptation and Fall, as in the story of Adam and Eve, humanity is sundered from its relationship to the 'Great Realities', but fallen humanity can regain this connection again through 'the birth of the divine Word, the Logos, through Mary'. In Palamas, likewise, 'woman plays a central role', both as the 'Eve-element' but also in terms of redemption — 'we might even say that, in Palamas's myth, woman represents the cross. She has a redemptive function. Through her the individual enters not a kind of pre-natal state from which his ordinary, waking consciousness, that of his selfhood, separates him'. Sikelianos, too:

> 'represents this fall [of humanity] as the loss by man of contact with that transcendent and universal wisdom which he symbolizes as a feminine power, as a Divine Mother, on whose sustenance depends the vitality of man's life and the truth of his understanding.'

Drawing all these observations together into Seferis's work, Sherrard opined that 'if [woman] is the daughter of Eve, she is also the image of a greater power, of a Magna Mater who is at the same time the Mater Dei'. And in summation, in Sherrard's own words, 'Thus, we return to the mysterious presence that so haunts the imagination of these poets, to this Divine Nature, Sophia, Creatrix...This is that unites and completes the scattered and partial forms of created and uncreated worlds'. In Sophia is the fullness and wholeness of the world, the unity of supernature and nature.[29] This supratemporal, Sophianic conception of creation, fall, and cosmic reality parallels, in some respects, Sergei Bulgakov's, but while it is plausible that Sherrard read him, he never cited him. Regardless, it is clear he divined these views from the Greek poets, as well as the Russian Sophiologists Vladimir Solovyov and Pavel Florensky (whom he did cite).[30] In this discussion, one sees explicitly developed the sense of nature's sacredness and wholeness — fractured and broken in the modern world — that Sherrard sought to restore, both by translating and introducing Greek poetry, but also in joining the Orthodox Church that he felt preserved these vestiges of ancient wisdom.

---

29 Sherrard, *The Marble Threshing Floor*, pp. 242–55.

30 Sherrard drew on Solovyov heavily for his book *Christianity and Eros: Essays on the Theme of Sexual Love* (Limni, Evia, Greece: Denise Harvey, 1995). He thanked both Solovyov and Florensky in a rather eclectic acknowledgement section in *Human Image: World Image*. For more on Bulgakov and Sherrard, see A. Louth, 'Theology of Creation in Orthodoxy', *International Journal of Orthodox Theology*, 8. 3 (2017), 51–76.

René Guénon, Perennialism, and Traditionalism

We must take a short detour here to address the question of perennialism, especially the influence of the French Muslim esotericist René Guénon, both to understand Sherrard better and to frame his later writings on science. The books Sherrard read and the influences he cites sometimes evince a perennialist or even Traditionalist School angle. One need only look at the acknowledgements of *Human Image: World Image* to find a cohort of Traditionalists present, among them some from Guénon's circle like Titus Burkhardt, Marco Pallis, and Ananda Coomaraswamy (at the other end of the spectrum, he also thanks C.S. Lewis, who decided that Guénon was 'as obvious a quack as I ever smelled out' after being gifted his books by their mutual student Martin Lings).[31]

Traditionalism is a strange subject not only because of its pervasive influence but also its secretive nature. Principally founded by Guénon, the 'Traditionalist School's' primary figures since then have been figures as disparate as the Swiss mystical philosopher Frithjof Schuon, the religious studies titan Mircea Eliade, and the fascist 'baron' Julius Evola. It has counted among its sympathizers an even wider array of figures, from King Charles III to Huston Smith to Thomas Merton and to even the 'small is beautiful' economist E.F. Schumacher. In his history of Traditionalism, Mark Sedgwick writes that 'at times it felt like I was uncovering the secret intellectual history of the twentieth century'.[32] Traditionalism is a twentieth century, Advaita-inspired interpretation of the perennial philosophy, and holds that behind all religious traditions there is a 'transcendent unity'. At root of Traditionalism is a rejection of modernity, a belief that the modern world has abandoned tradition and so anyone searching for ancient wisdom is *de facto* banished to the fringe of society. As Traditionalist Harry Oldmeadow writes, the 'spiritual crisis of modernity' has one cause, 'the betrayal of tradition'. This betrayal has led to 'ecological catastrophe, a material sign of the rupture between Heaven and Earth; a rampant materialism and consumerism', and much more. 'No amount of gilded rhetoric', he continues, about science, technology, or democracy 'can hide the fact that our age is tyrannized by an outlook inimical to our most fundamental needs'.[33] Guénon provided the template for a solution. Inspired by Advaita Vedanta, he abandoned his French Catholicism, moved to Cairo, converted to Sufi Islam, and took the name Abd al-Wahid Yahya. Though he practiced as a Sufi, however, his belief system was animated by the Traditionalist concept of a single, underlying metaphysical oneness behind authentic religious 'tradition', and

---

31 *The Collected Letters of C. S. Lewis*, Volume 2, ed. Walter Hooper (San Francisco: HarperOne, 2004), p. 90.

32 M. Sedgwick, *Against the Modern World: Traditionalism and the Secret Intellectual History of the Twentieth Century* (New York: Oxford University Press, 2003), p. 15.

33 H. Oldmeadow, 'Introduction: Signs of the Times and the Light of Tradition', in *The Betrayal of Tradition: Essays on the Spiritual Crisis of Modernity*, ed. by H.Y Oldmeadow (Bloomington, IN: World Wisdom, 2005), p. xi.

so there was as split — as is common in Traditionalism — with his 'exoteric' practice and his 'esoteric' spirituality.[34]

Philip Sherrard's affiliation with Traditionalism has made his place in modern Eastern Orthodoxy somewhat controversial. He appears in Sedgwick's narrative only once, described (rather inaccurately) as an 'apparently Schuonian Traditionalist' who helped the poet and William Blake scholar Kathleen Raine found the Temenos Journal, which eventually became the Temenos Academy, in 1981.[35] Temenos, which is ancient Greek for 'sacred precinct', is an 'educational charity which offers education in philosophy and the arts in the light of the sacred traditions of East and West'.[36] Due to the illusory nature of a lot of Traditionalist belief, the depth of Sherrard's commitment to Orthodoxy-as-Orthodoxy has been the subject of some dispute. Kallistos Ware anticipated some of this criticism, writing that 'Philip was genuinely Orthodox, but his vision of Orthodox Christianity was generous and wide-ranging not defensive or timidly parochial'.[37] Critics have not been as favorable. James L. Kelley writes that Sherrard 'succumbed to the heretical teaching that God has saved and continues to save men through non-Christian religions'. Kelley then put him in a list of 'Christian and quasi-Christian' historians like Christopher Dawson, Eric Voegelin, and Herman Dooyeweerd.[38] Somewhere in the middle is Christopher Knight, who writes that though Sherrard's views often sound like a familiar kind of Christian inclusivism, it is in reality more 'subtle', and that he instead promotes 'a position in which, at least implicitly, the possibility of faiths other than Christianity being fuller expressions of the divine reality cannot be excluded'.[39]

Was Sherrard a Traditionalist? It is worth looking closely, as Guénon likely influenced Sherrard's scientific views and most certainly his religious views — though Sherrard also departed from him strongly. In a 1953 letter to Seferis (before his conversion), Sherrard acknowledged the teachings of his 'masters' (quotes original) and used explicitly Traditionalist language to describe the 'dissolution of our cycle' (the cyclical view of history being one of the key stances of Guénon and Traditionalism). He then asked Seferis if he had ever read Guénon,

34 Sedgwick, *Against the Modern World*, pp. 208–212. One can find Traditionalist influence scattered about modern Orthodoxy. Seraphim Rose, for instance, was first introduced to Orthodoxy through reading Guénon (whose books had been given to him by Alan Watts). See D. Christensen, *Not of this World* (Forestville, CA: Seraphim Rose Foundation, 1993), pp. 60–65.

35 Sedgwick, *Against the Modern World*, p. 214.

36 See 'Temenos Academy,' *Temenos Academy*, <https://www.temenosacademy.org/> [accessed 27 December 2022].

37 Ware, 'Foreword,' p. XIX.

38 J. L. Kelley, 'Philip Sherrard: Orthodox Theosophy and the Reign of Quantity', *Romeosyne 'Breaks and Links Series' No. 1* (Norman, OK: Romanity Press, 2015), 12–13.

39 Knight, 'The Christian Tradition and the Faiths of the World', p. 341.

and praised the Frenchman's criticism of modernity, *The Reign of Quantity*.[40] Traditionalist verbiage appears in his subsequent work, notably in *The Greek East and the Latin West* (1959), wherein he argues that one must take other religions into consideration, but that this does not 'admit a limitless relativism into matters of religion'. Rather, one recognizes that there is a 'transcendent unity, or inward identity, of all religions [which] is matched by a corresponding distinction of the outward forms and dogmas'. This leads to outward contradictions between faiths, making it 'only possible for one who has passed beyond forms altogether and has realized the Truth in its universal nature to discern their inward harmony'.[41] Beyond the Traditionalist hallmark of 'transcendent unity,' Sherrard also repeated a Traditionalist staple in drawing a distinction between the esoteric tradition (sacraments) of Christianity and the exoteric tradition, as well as observing the need for 'initiation' into a genuine religious tradition. Importantly, as Kallistos Ware noted, this emphasis 'is less evident in his later writing'.[42] In his early work, however, his localization of tradition, and also of reason, led him to see in Christianity the solution to modernity's malaise, for it is 'the spiritual form *of Western society* [emphasis added]'.[43]

The transcendent unity of religions invites a problem, however. If they are all exoteric manifestations of an underlying esoteric oneness, why do they differ from each other so? The reason, for Sherrard, is that humanity is limited. God self-discloses, through revelation, in ways contingent upon our understanding. These contradictions owe to:

> '…the difference in aptitudes, capacities, and temperaments of the various groups of humanity to which each particular form is addressed. Truth is one; but in expressing Itself in a way which is accessible to the human intelligence, It has to take account not only of the limitations of the human state as such, but also of the various, though relative, divisions within mankind that are themselves expressive of various facets of the divine plenitude'.[44]

Like Moses on Mt. Sinai, we can only see God's back. If this is true, though, what do we make of all the differences between religions? And how do they all relate to each other? In an unpublished fragment entitled 'Exclusivity or Inclusivity: A Christian Dilemma,' Sherrard wrestled with this problem. He lamented that often Christians viewed 'non-Christian religious experience as so obviously suspect as to be either too dangerous to study or else not worth the

40 Philip Sherrard to George Seferis, September 13, 1953, in *This Dialectic of Blood and Light*, p. 255.

41 Ph. Sherrard, *The Greek East and the Latin West: A Study in the Christian Tradition* (Limni, Evia, Greece: Denise Harvey, 2017), p. 53.

42 Sherrard, *The Greek East and Latin West*, p. 94; Ware, 'Foreword,' *Christianity*, p. xxii.

43 Sherrard, *The Greek East and Latin West*, p. 194.

44 Ph. Sherrard, *Christianity: Lineaments of a Sacred Tradition* (Brookline, MA: Holy Cross Orthodox Press, 1998), pp. 62–63.

trouble of being studied'. The most damaging aspect of this was the historicization of ecclesiology, such that all true religion was a one-to-one identification with Church membership, and that 'salvation is an exclusive prerogative of Christians'. He suggested that there was more to Christianity than this, and that the Apostle Paul and Church Fathers such as Justin Martyr suggested that unknowing pagans could have achieved salvation as well. But this more open approach — based on the 'universal Logos inherent in the creation He has brought into existence' — was defeated by a historicized version of the faith with sharp boundaries between the saved and unsaved as those inside and outside the church walls.[45]

If the matter were left there, perhaps one would identify Sherrard as a perennialist or Traditionalist. But that is not the whole story, for Sherrard also critiqued aspects of perennialism and disputed Guénon's stance on Christianity. First, Sherrard rejected the syncretic approach characteristic of Schuonian Traditionalism. Whereas Guénon advocated for intense commitment to a single exoteric expression (in his case, Islam), the Swiss metaphysician Frithjof Schuon, based in his mysterious compound in Bloomington, Indiana, was more eclectic and drew off multiple religious traditions.[46] Kathleen Raine favored this approach, but Sherrard emphatically did not. As Louth observes, this put Sherrard more in the Guénonian camp. The Frenchman preferred rigorous commitment to a single tradition one had been initiated into, and not an adventurous syncretism.[47] Furthermore, despite his interest in transcendental unity, Sherrard was concerned to defend uniquely Christian concepts, especially the Trinity. In a 1974 article, Sherrard argued that 'while it is one thing to say that all traditional forms ultimately express the same universal truth, it is quite another to say that they all express it to the same degree'.[48] Traditionalist Harry Oldmeadow disagreed with Sherrard on this, offering a more culturally relativist contention that, say, Islam does not *better* represent the Truth than Buddhism, but 'it does so for Muslims only'. He then accused Sherrard of not understanding the esoteric/exoteric distinction.[49] Such criticism indicates a gap between Sherrard and Traditionalism.

His adherence to Christianity meant that Sherrard broke significantly with Guénon, who had no confidence in the religion as an authentic tradition (Guénon remained curiously ignorant of Orthodox Christianity, instead basing his rejection on the supposed western sterility of Protestantism and Catholicism).[50] In terms

---

45 Katounia, Philip Sherrard: Archives, MS 2b.

46 Sedgwick, *Against the Modern World*, pp. 170–77. Schuonian Traditionalists emphatically reject that their approach is syncretic, arguing instead that it is 'synthetic,' but it is hard to be persuaded by such hairsplitting. See K. Oldmeadow, *Traditionalism: Religion in Light of the Perennial Philosophy* (San Rafael, CA: Sophia Perennis, 2011), p. 16.

47 Louth, *Modern Orthodox Thinkers*, p. 245.

48 Ph. Sherrard, 'The Tradition and the Traditions: The Confrontation of Religious Doctrines,' *Religious Studies*, 10.4 (1974), 407–17, on p. 409.

49 Oldmeadow, *Traditionalism*, pp. 184–85.

50 Oldmeadow, *Traditionalism*, pp. 13–14.

of religious practice, Sherrard resisted the pessimistic elitism that characterized Guénon's thought. Whereas Guénon abandoned all hope in the common folk holding on to religious tradition, instead advocating for a small, cloistered set of elites to practice their esoteric faith in secret, Sherrard found great hope in the people — especially the peasants of Greece.[51] In this, he followed Angelos Sikelianos, who perceived in the rural Greek peasantry's language, customs, songs, and dances, 'reflections of a genuine autonomous spiritual tradition'.[52] But Sikelianos failed to link this to Christianity, instead turning further East and looking to the Upanishads. Sherrard did not believe this move was helpful, writing that there was no practical way to 'incarnate' such religious truths in the modern Western world, unless they be incarnated in the dogmas, practices, and revelation of Christianity.[53] Sherrard's perspective is nuanced and subtle, and can sometimes be hard to parse. What is clear is that he carved out his own path, taking from Guénon a commitment to a single tradition, from Sikelianos a predilection for Greek rural life, and then moving beyond both by resolving that Eastern Orthodoxy was a legitimate tradition.

It was not just an issue of practice; metaphysics was a crucial part of the disagreement too. In his posthumous *Christianity: Lineaments of a Sacred Tradition*, Sherrard praised Guénon's work but argued against the metaphysician's rejection of Christianity as an authentic 'traditional' religion.[54] Sherrard argued — in both *Christianity* as well as in his unpublished notes — that Guénon's metaphysic was not as apophatic as he believed, and that his conception of God that he believed was purely apophatic ended up smuggling back in a positive articulation of doctrine (in his case, the Truth was really Advaita Vedanta). In rigorously, and logically, laying out a metaphysics of the absolute Oneness of God, Guénon made cataphatic what should have been apophatic. The 'holy agnosia' of Orthodox Christianity, however, which embraces seemingly anti-rational paradoxes like the Trinity, was a better approach to apophatic theology.[55] Not only was Guénon articulating a specific ultimate doctrine, he also arbitrarily walled off other religious expressions. After all, adopting Guénon's metaphysics would make it impossible to 'embrace that reality which is revealed in the incarnation of the Universal Logos in the person of Jesus Christ'.[56] The locus of Sherrard's objections can be seen here — in articulating such a metaphysics, Guénon privileged Advaita Vedanta above all, blocked other religious expressions (especially those that depended on Christ's incarnation), and most importantly, created a breach between God and the world that could never be bridged. Without the essence-energies distinction characteristic of Orthodox thought (which, Sherrard noted emphatically,

51 Sedgwick, *Against the Modern World*, p. 26.
52 Sherrard, *The Wound of Greece*, p. 78.
53 Sherrard, *The Greek East and the Latin West*, pp. 193–94.
54 Sherrard, *Christianity*, pp. 76–113.
55 Katounia, Philip Sherrard: Archives, MS 11 f.
56 Katounia, Philip Sherrard: Archives, MS 21a.

conceives the energies as not 'any less real or less Absolute than the Essence'), Guénon's thought introduced 'a total hiatus between the Infinite and the Absolute and the manifest sensible world, so much so that one might say that the really unaccountable thing is the existence of the world'.[57] Creation then becomes inexplicable.

Because of his concern about creation, these are not minor quibbles on Sherrard's part. The endpoint of Guénon's approach is 'rejection of creation' and an overall 'anti-Incarnational character, a way of thinking that has as its inevitable counterpart a de-incarnational view of spiritual realization'.[58] This metaphysics results in an abolition of the world's sacrality along with the world itself. 'Creation is fundamentally deprived of all reality and has, and can have, no relationship of any kind at all with the Absolute,' observed Sherrard in his notes, 'that is to say, [creation is] ultimately without significance, and is a mere shadow and illusion'. An Orthodox Christian cannot abide this view, and Sherrard concluded that the Orthodox attitude to creation is 'essentially positive' because the creature and creation can (and do) 'participate in the eternal reality of the Divine'. In hewing too close to Guénon, one would end up affirming 'a false understanding of the Divine and hence of the relationship between God and creation, the Absolute and its manifest forms'.[59] Ironically, this chasm between God and creation was a mistake Sherrard believed was also made by the Schoolmen as well as modern scientists. Though Guénon criticized both, he inadvertently made the same mistakes. At bottom, it was Sherrard's desire to understand and protect creation that led him to turn away from both Guénon and modern science.

Sherrard and Science

Modern Science and Modern Capitalism

Reality has become bifurcated. This split between reason and revelation is the key thesis for Sherrard on science. However, reality was cloven in twain long before the modern period; instead, Sherrard turns to medieval scholastics and even to the *filioque*. As he wrote in *Greek East and Latin West*, there are two different 'implicit' worldviews in the question of the *filioque*, and the acceptance of it by the West allowed for 'the conception, and setting up, some thousand years later, of such an organization as that of the United Nations', as the UN was the inevitable outgrowth of a broadly secularist, rationalist, denatured liberal world.[60] As Sherrard saw it, the *filioque* led to a conception of God as absolute

57 Katounia, Philip Sherrard: Archives, MS 21b, and MS 11 f.
58 Katounia, Philip Sherrard: Archives, MS 11 f.
59 Katounia, Philip Sherrard: Archives, MS 21b.
60 Sherrard, *The Greek East and Latin West*, p. VIII.

simplicity, an ontological principle of Pure Being, which 'made it impossible for the Latins to admit the Trinitarian doctrine maintained by the Greeks'. The difference resulted in a God ontologically removed from the world, and not the panentheist coinherence of God with creation that Sherrard favored (this critique is not that far removed from his qualms with Guénon). Even in Thomas Aquinas, for instance, one finds a latent empiricism that results from sealing God off from nature, in which humanity can only learn anything 'by a process of abstraction from sensible objects'. Such a 'secondary beatitude' was 'imperfect' and consisted only in the 'study of speculative sciences'. But Thomas wanted to safeguard Revelation, and so introduced a regrettable breach between reason and revelation, wherein the Logos discloses revealed truth that cannot be known from nature. Scholasticism, then, assisted in dividing faith from philosophy (and revelatory truth from science). A hermetic seal between faith and science resulted.[61] Andrew Louth notes that *Greek East and Latin West* was not received well, but it highlights that Sherrard's thoughts on science were already formed as early as the 1950s, and he would continue to return to theme of reason's proper place in the investigation of reality.[62]

Sherrard held to a rather low view of human reason — or 'the reason' as he often put it. As we have already seen, he chided Guénon for relying too heavily on reason in the construction of his metaphysics, but he was equally critical of its almost omnipotent role in the sciences. In an unpublished dialogue between a scientist and a philosopher, Sherrard has a philosopher subject a scientist to an interrogation that reveals the entire scientific pursuit is a house built on sand. The philosopher contests science's claims to objectivity because all investigation must be filtered through the faulty tools of the human senses. One cannot stand outside oneself and achieve an objective view of nature; even empirical evidence fails to overcome the problem of filtered observation. 'As a scientist', argues the philosopher, 'you can only ever acquire a relative knowledge of a tree [for example] — a knowledge which is liable to change and may turn out to not be knowledge at all'. The conclusion is that the scientist can:

> '[...] never acquire more than a relative or partial knowledge of anything, and that you can never know how relative or partial your present knowledge is, and that you can never know that it may not be entirely false and so not knowledge at all'.

The scientist, forced to agree, is asked why he continues his work and replies, 'I'll have to think about it'.[63] A flawed view of reason meant that science was heading down the wrong path from the very outset of its investigation.

As is evident in this conversation, a point to which Sherrard continually returned was the importance of *a priori* principles. The methodological naturalism

61 Sherrard, *The Greek East and Latin West*, pp. 146–53.
62 Andrew Louth, *Modern Orthodox Thinkers*, p. 239.
63 Katounia, Philip Sherrard: Archives, MS 60c.

that constitutes science's best practice had hypertrophied into a metaphysics; scientists unwittingly denied the existence of any plane beyond the material. As such, a critique of science cannot be written by a scientist, because 'few contemporary scientists appear to be aware' of the way that their worldview 'in fact presupposes and implements, in its every aspect, theoretical and practical, a metaphysical or philosophical view of things that is anything but neutral, self-evident, self-proven or a matter of common sense'.[64] The consequence of this blindness, however, was that anything that was intangible, any value that could not be reduced to 'the world of quantity' (note the Guénonian phrase), was banished from the realm of reality and reduced to subjective opinion. Art, in particular, became functionally meaningless. Nothing existed except that which could be measured.[65] Reason, the guiding light of science, depended on a 'framework of conceptions' to be rational, but these were themselves not rational discoveries, they were simply brute facts, givens that must be accepted in order to make science function, *ipso facto*.[66] In terms of the philosophical origins of science, Sherrard owed much to E.A. Burtt's classic but underappreciated study, though he did not 'share the author's evaluation of what he analyzes with such clarity'.[67]

Science's perversion of reality manifests in two ways, according to Sherrard: a 'fetish' for mathematics and a fixation on linear time. Though he felt that most scientists, even the founders of modern science like Newton and Galileo, 'were not great thinkers', they nevertheless 'were brilliant mathematicians'. One of the main assumptions of modern science is that truth about nature 'must be expressed in mathematical language, because is their mathematical qualities that constitute the reality of these facts'. And further, 'It is quantitative combinations that constitute the reality of the sensible world'. The converse, qualitative aspects of reality, are entirely foreign — 'nature as such knows nothing of qualities and is entirely void of them'. Such things are subjective, personal, and non-scientific: in a word, an 'illusion.' Following, but not citing, Guénon, Sherrard argued, 'The world of nature is entirely under the reign of quantity: its total reality is quantitative, and can be expressed only in mathematical terms'. So much for art, under these circumstances, which could not communicate any immortal truths, instead

64 Ph. Sherrard, *The Rape of Man and Nature: An Enquiry into the Origins and Consequences of Modern Science* (Colombo, Sri Lanka: The Sri Lanka Institute of Traditional Studies, 1987), p. 3. Sherrard is far from unique in making this argument. Locating a critique of modern science on the shift of methodological naturalism to metaphysical naturalism is a hallmark of writing on the topic, ranging from intelligent design advocates such as Phillip Johnson — see *Darwin on Trial* (Downers Grove, IL: Intervarsity Press, 1991) — or Orthodox theologians who reject intelligent design and creationism such as D. Bentley Hart — see 'Where the Consonance Really Lies,' in *Theological Territories: A David Bentley Hart Digest* (South Bend, IN: University of Notre Dame Press, 2020), pp. 123–37.

65 Sherrard, *The Rape of Man and Nature*, p. 69.

66 Sherrard, *The Rape of Man and Nature*, pp. 81–82.

67 Sherrard, *Human Image: World Image*, p. 34. See also E. A. Burtt, *The Metaphysical Foundations of Modern Science* (London: Kegan Paul, Trench, Trubner, 1924).

merely indicating an internalized preference on behalf of the viewer or artist (if the artist was sufficiently and unfashionably didactic). The world was thus conceived as a material, mathematically understandable 'object', outside initially of the deist God and then subsequently seen atheistically as running along on its own without outside interference. Thus, the total mechanization of humanity and reality had been achieved, and 'the whole universe could be viewed as a vast despiritualized, mathematically computable system of masses moving in absolute space and time'.[68]

Coterminous to this revolution in thought was a conception of time as an unstoppable, autonomous arrow — proceeding ever forward. This sin of historicism had religious origins and effects, as it affected Christian ecclesiology and missiology — causing the Church and Christendom as a whole to see itself as a progressive 'series or succession of salvation events designed to culminate in the appearance of Christ as the end of the history of the Old Covenant and as the end of human history', This meant, as we have already seen above, that dialogue and understanding with other faiths became impossible.[69] Science though, played a key role too (perhaps adapting this conception of time from its Christian background). Sherrard argued that under a properly spiritual outlook, time was not that significant, but in the secularized scientific world, we witness 'a veritable apotheosis of time', a view that 'involves a radically new view of the universe'. Time becomes a mathematical quantity and not something lived in or truly experienced by humans, other than illusorily. 'This notion of time', wrote Sherrard, 'as a measurable continuum presenting a moving mathematical limit between past and future means that we as human beings are effectively excluded from it'. First, God was removed from the cosmos, then humanity follows along the divine coattails in being abstracted out of existence. 'Man has no part in this world', and all human 'purposes, feelings, loves, strivings, and practically everything else characterizes his being and makes him what he is, are all non-quantifiable qualities'. As such, they must not exist. The conclusion is that the material world is chaotic and void — it is nihilistic, fundamentally without purpose. And here, Sherrard argued, is where evolution comes into the picture, as a last-ditch attempt to restore meaning to a cosmos that now languishes without one.[70]

Evolution

Together with the 'apotheosis of time' came the 'bogey' of evolution, to which Sherrard devoted a chapter in *Human Image: World Image*, his most sustained look at evolution.[71] As we have seen, Sherrard took a cosmic reading of creation, fall,

68 Sherrard, *Human Image: World Image*, pp. 35–42.
69 Sherrard, *Christianity*, p. 59.
70 Sherrard, *Human Image: World Image*, pp. 62–66.
71 Sherrard, *Human Image: World Image*, pp. 56–75.

and history — his view of sacred history was supratemporal.[72] His qualms with evolution, then, had nothing to do with readings of Genesis or protecting biblical authority. He was instead concerned about the disappearance of teleology. The platonic and Aristotelian concepts of organic evolution (shared with some saints, for instance Augustine) were not evolutionary in the modern sense, according to Sherrard, but rather 'mutual exfoliation and infoliation that takes place between each thing and its eternal archetype'.[73] They were explicitly teleological. Unlike creationists, however, who tend to see evolution itself as pointless, Sherrard identified evolution as a vehicle for meaning, a way to inscribe the randomness of reality with a purpose. Sherrard argued that evolution was actually not random at all; it was not 'haphazard and arbitrary' but was instead 'invested with an internal logic…operating according to some intelligible principle or principles'. It was a smuggling of telos in through the back door, and a way to resolve the Cartesian split between body and mind by 'introducing this notion of life as a third principle mediating between them.' When Darwin spoke of natural selection, his 'language is loaded with teleological implications…[but it] is still regarded as a blind process'.[74] This understanding of evolution is far afield from many creationist critics, and is actually fairly close to what many Darwinists would say about it. Darwin himself was interested in the way his American colleague (and Christian) Asa Gray highlighted this same teleological point. In an article for Nature in 1874, Gray wrote, 'let us recognise Darwin's great service to Natural Science in bringing back to it Teleology: so that, instead of Morphology *versus* Teleology, we shall have Morphology wedded to Teleology'.[75] Darwin replied in a letter, 'What you say about Teleology pleases me especially, and I do not think any one else has ever noticed the point'.[76] Likewise, even today, Richard Dawkins has expressed consternation at antievolutionists who argue that it is purely random ('Darwin set no store by randomness').[77]

Sherrard believed that evolution was not random even if it was blind. It was, rather, an attempt to reanimate the cosmos, a desperate act of resistance against the mechanization of reality and the degradation of humanity. The meaning it offered, though, based on a linear view of time that had humanity at the end point, was ultimately a false promise, for there was no reason inherent in evolution to say that just because humanity is here now that we have the ability to adjudicate

72 Sherrard, *The Greek East and Latin West*, pp. 41–43; Ph. Sherrard, *The Sacred in Life and Art* (Limni, Evia, Greece: Denise Harvey Publisher, 2004), p. 87.

73 Sherrard, *Human Image: World Image*, p. 59.

74 Sherrard, *Human Image: World Image*, p. 68–70.

75 A. Gray, 'Scientific Worthies: Charles Darwin,' *Nature*, 10 (June 4) (1874), 79–81.

76 Charles Darwin to Asa Gray, June 5, 1874, in *Darwin Correspondence Project*, ed. F. Burckhardt, and others, <https://www.darwinproject.ac.uk/letter/?docId=letters/DCP-LETT-9483.xml> [accessed 27 December 2022].

77 R. Dawkins, 'Inferior Design,' *New York Times*, 1 July 2007, <https://www.nytimes.com/2007/07/01/books/review/Dawkins-t.html> [accessed 27 December 2022].

metaphysical truth. 'By definition', he argued in a vein similar to C.S. Lewis's argument from reason, 'the human mind still in the process of evolving cannot be a truth-knowing faculty'.[78] And if evolution is not over, then what purpose can it offer those of us now, who simply occupy a mid-point between ape and, say, Stanley Kubrick's post-human Star Child floating in the abyss?

Evolution bore consequences for art, too. Art, especially sacred art, long fascinated Sherrard (and it proved one of his more Traditionalist associations, as he often drew off the thought of Traditionalist art historian Ananda Coomaraswamy). Sherrard saw evolution's influence manifest in modern art, especially the avant-garde, and he focused particularly on Herbert Read and his Whiteheadian notion that 'art was a natural organic process of unfolding', making the artist more than a 'passive recorder of changing reality' and instead 'instrumental in the very process of evolution itself'.[79] In the absence of sacred art, revealing immortal and transcendent truths, art instead must be new, progressive, changing. It evolves like nature and its movement even indicates cultural progress. By being 'cutting edge', an artist is then a prophet of progressive change. Because there is no inherent moral in evolutionary progress, the only respite is the faint belief that something is superior because it comes later. 'The new,' he wrote, 'becomes a value in its own right and there can by definition be no authentic creativity without innovation'.[80] We see, then, an immense social pressure on artists to be 'innovative' or 'original,' with a paradoxical reluctance to criticize the results, because to do so would be to invoke an artistic standard. Style innovation is the only aspect of art left standing, and so 'formal originality' becomes a goal simply for its own sake. The evolutionary conception of reality, then, has destroyed the metaphysical and religious purposes of art. The idea that 'man is no more than an inescapably time-bound, evolution-conditioned being' means that art cannot take on any significance beyond 'hedonistic ("I know what I like"), sociological, scientific or psychological terms'.[81]

Ultimately, in Sherrard's view, there could be no harmonization with modern evolutionary theory and traditional religion. The methodological naturalism necessary for science's method, and the metaphysical naturalism that it logically induced, meant that religion was at bottom incompatible with it. 'Every attempt to fit these theories in to the context of metaphysical doctrine has meant', wrote Sherrard, 'buckling metaphysical doctrine so that it can accommodate what are in fact, from the metaphysical point of view, pure lies'.[82] Proposals that religion inhabits the sphere of morality while science the sphere of nature (like Stephen Jay Gould's NOMA, though Sherrard does not cite it) are non-starters, as it results in

78 Sherrard, *Human Image: World Image*, p. 73.
79 Sherrard, *The Sacred in Life and Art*, p. 44.
80 Sherrard, *The Sacred in Life and Art*, p. 52.
81 Sherrard, *The Sacred in Life and Art*, p. 53.
82 Sherrard, *The Sacred in Life and Art*, p. 153.

the very same bifurcation of reality that Sherrard had all along decried.[83] Synthetic attempts, like those promoted by Pierre Teilhard de Chardin (or by Teilhard's few scientist followers, such as Theodosius Dobzhansky) were likewise dead ends. Teilhard accepted 'the modern scientific conception of time or, rather, of space-time,' and he made Christ 'subordinate' to evolution.[84] An evolving Christ, one which culminates in the Omega Point at the end of cosmic history, 'stands Christian doctrine on its head'. Teilhard had no doctrine of the Fall (and, for Sherrard, the Fall and evolution were incompatible at every level). This held true even for those who held to supratemporal accounts of the Fall (though Sherrard did not critique Solovyov's attempted marriage of the Fall and evolution).[85]

Sherrard understood that evolution by natural selection is not, strictly speaking, a 'random' process, even if it is random ontologically, and he did not base his critiques on biblical interpretation. Nevertheless, he accused modern science of destabilizing all of reality and of perverting human self-understanding, and he faulted evolution for trying, and failing, to provide meaning in this materialist world. Such claims go wildly against the status quo. But Sherrard had one proof on which he pinned his entire outlook: 'ye shall know them by their fruits'. And what fruit did the modern scientific world bear? Apocalypse.

Climate Change

The preservation of nature, and humanity's relation to it, became increasingly important to Sherrard as he pondered his career. He was frustrated with his scholarly life and felt, as Kallistos Ware described, 'not totally at home in academia'. In response, Sherrard departed from the academy and retreated to a haven he created on the island of Evia. In 1959, he had purchased a defunct (and therefore relatively cheap) magnesite mine near Limni, and over time he transformed it into his permanent residence. From 1977 on, after Sherrard left his post in academia, he and his second wife, Denise Harvey, lived out an 'almost monastic' life there, wrote Ware. 'They had no telephone and no electricity', he continued, 'and heating was provided by firewood'.[86] They built an Orthodox chapel on the property with their hands (along with many other workers), using 'building techniques and architectural styles that were strictly traditional.' Patrick Leigh Fermor, who stayed there for a time, described it as 'half-logger's camp, half-Arcady' (and of

83 Sherrard, *The Sacred in Life and Art*, p. 154.

84 Sherrard, *Human Image: World Image*, pp. 106, 119.

85 Sherrard, *Human Image: World Image*, pp. 122–26. One wonders what Sherrard would have thought of Bulgakov's synthesis of the supratemporal Fall and evolutionary theory. See S. Bulgakov, *The Bride of the Lamb* (Grand Rapids, MI: William B. Eerdmans Publishing Company, 2002).

86 Ware, 'Foreword,' *Christianity*, pp. xiv.

course Fermor built his own Hellenic refuge in the Mani outside Kardamyli).[87] In an interview, Ware commented on the hospitality at the heart of Sherrard's retreat, welcoming guests like Fermor, Ware himself, or the English composer and Orthodox convert John Tavener.[88] It was in this refuge from the denatured, deracinated, and disenchanted world that Sherrard began writing a series of books on the environmental crisis.[89]

Sherrard's love and appreciation for nature, especially Greek nature, meant that as the twentieth century progressed, he would perhaps inevitably become disconsolate at the way Greece and the world were modernizing. Kallistos Ware recounted that one of Sherrard's early expressions of this frustration was an essay on the roads of Mount Athos. In 'The Paths of Athos,' Sherrard expressed his dismay at the way the monks had started going from place to place via car, causing the ancient pathways between the monasteries on the Athonite peninsula to become overgrown.[90] As Ware described it, Sherrard 'saw this as a symptom of grave spiritual sickness, indicating as it did the tragic loss of living communion with the world of nature'.[91] Furthermore, there is always a tension in the way adventurous writers such as the 'new philhellenes' would write about unspoiled, enchanted Greece, for the very publicization of such places would lead to an increase in tourism and a corresponding despoliation of the fragile enchantment they had found still lingering in modernity. Patrick Leigh Fermor wrote of the paradox in hoping that Greece's struggling economy could be boosted by tourism, because 'this gregarious passion [...] destroys the object of its love.' Tourists flock to hidden inns and tavernas. Fermor urged, without irony, that 'the works of writers mentioning these places by name should be publicly burnt by the common hangman'. 'Greece is suffering its most dangerous invasion since the time of Xerxes', and the power of capital may well finally eradicate Greece's natural wonder. 'Bad money may drive out good,' wrote Fermor, 'but good money, in this case, drives out everything'.[92] Sherrard was likewise horrified by the influx of destructive tourist kitsch, rendering the island of Corfu, so beloved by Edward Lear, Henry Miller, and Lawrence Durrell, unrecognizable. He wrote to Edmund Keeley that the coast was:

---

87 Ware, 'Foreword,' *Christianity*, pp. xiii–xv.

88 'Philip Sherrard: His Life & Work — The Thoughts of Bishop Kallistos Ware,' online video recording, YouTube, 14 June 2014, <https://www.youtube.com/watch?v=J-QNPTJE4ng> [accessed 27 December 2022]; 'Sounding Icons,' part 1, Peter Levi, Channel 4 Television, 1989, online video recording, YouTube, 7 December 2011, <https://www.youtube.com/watch?v=hIo-XnrOImk> [accessed 27 December 2022].

89 Ware, 'Foreword,' p. xiv.

90 Ph. Sherrard, 'The Paths of Athos,' *Eastern Churches Review*, 9 (1977), 100–07.

91 Ware, 'Foreword,' p. xv.

92 P. Leigh Fermor, *Roumeli: Travels in Northern Greece* (New York: New York Review of Books, 2006), p. 126.

> '[...]So immersed and desolated beneath ferro-concrete hideosity of hotel and boarding-house, discotheque, bar, cafeteria and chop-house, and wave after wave of pink-faced, white-bodied neo-Visigoths that summer-long blotch and bespatter its beaches, with the accompanying raucousness of motorcar, motorbike, motorscooter, transistor radio, motorboat, and the other gimcrackery and detritus (plastic and mineral) of mass tourism, that one searches in vain, across the wreckage of this dishallowed world, for the virginal loveliness that confronted Lear at every footstep'.[93]

One can see the same kind of transformation today at Santorini — the ancient island of Thira, famed for its mysterious connection to Atlantis, its volcanic crater, and its colonial odyssey in the story of King Grinus, reduced now to a Disneyland-esque simulacrum of Greek life: the island's steep cliffs studded with cruise ships, resorts, and overpriced gyros.

However, Sherrard did not end his critique of modernity's encroachments with tourism — he saw, rather, that the entire world had succumbed to fidelity towards consumerism and industrial exploitation. Crucially, however, this turn was caused by the modern scientific outlook and its dehumanizing consequences. Unlike an easier criticism of capitalism, perhaps one that lays fault for the ravages of the Earth at the feet of CEOs, entrepreneurs, bankers, energy tycoons, or all together, and sees science as simply a pawn that is used by avaricious capital, Sherrard instead faulted modern science as the foundation of dehumanization. As radical as this thesis is, Sherrard put forward a proof: judge by our fruit. Look upon the works of humanity, ye mighty, and despair. We believe we 'have the right, if not the duty, to invade and ransack the physical world at any and every point — to split atoms, dissect living animals, defile the sky, experiment with gene and foetus'. Micro and macro, we put nature under our boot. 'We will find,' when we pay attention, 'that we have despoiled our inheritance in the name of a science that exists and can exist only through the appalling negation of all wisdom and true philosophy'.[94] Borrowing language from Guénon again, modern science was an 'inversion' of truth — an inversion all the way down to the base level of reality, and which created a crisis in humanity and the environment.[95]

For Sherrard, the image of humanity was reflected in the image of the world (and vice versa). We moderns have completed a 'desanctification of nature', and because nature had been desanctified, then humanity has likewise been desanctified.[96] Most prominently in *The Rape of Man and Nature*, Sherrard traced the origins of this desanctification to the seeds of the scientific revolution sown during the Renaissance. The 'dehumanization' of humanity, and then destruction

93 E. Keeley, *Inventing Paradise: The Greek Journey, 1937–47* (New York: Farrar, Strauss, & Giroux, 1999), p. 34.

94 Sherrard, *The Sacred in Life and Art*, p. 11.

95 Sherrard, *Human Image: World Image*, p. 55.

96 Sherrard, *The Rape of Man and Nature*, pp. 90–112.

of nature, 'is inevitable once the norms of the modern scientific outlook begin to be applied in the sphere of human and social relationships for…an inhumanity is built into the very presuppositions on which this outlook is based'. Once the Earth, and humanity with it, has been 'severed' from the divine, from the sacred, this process can undergo. He singled out Descartes for blame, writing that the French philosopher broke 'the last tenuous links between God and the world He has created'. The thinkers of the scientific revolution thereby evacuated the soul from the world and the spark of the Logos from all living things. 'For Newton', he wrote, 'the celestial spheres are a machine, for Descartes, animals are machines, for Hobbes, society is a machine, for La Mettrie, the human body is a machine'. Then the marriage of scientific measurement and managerial capitalism can be made, and the Earth can be finally offered up for total sacrifice.[97] As Sherrard wrote, modern science has:

> 'Licensed the technical mind to desecrate the whole social context, the entire planet, and to send out interminable squads of scientific-technical experts to chart, dissect, ransack, deface and ravage, dispassionately, on the basis of empirical evidence and experiment, and if possible by the intervention of mathematics or other specialist methodology, the total fabric of human and cosmic life.'[98]

Science, for Sherrard, is totalizing, and it has been totalizing in the effect it has had on nature and humanity.

Forgetting the holiness of both humanity and of nature meant that both could be ravaged without the pinprick of a guilty conscience. Upon the decline of sacramental imagination, there began the 'utilization of machines and gadgets in order to manipulate nature and to produce concrete results of a quantitative kind', something that used to be seen as 'beneath the dignity of man'. Now, a human is 'a two-legged animal whose destiny and needs can best be fulfilled through the pursuit of social, political, and economic self-interest'. A pact was struck between 'hard-headed industrialists and bankers possessed by a single-minded devotion to making money no matter what devastation they produced', and they were aided by scientists turning 'their expertise to the practical exploitation of the world's natural resources'.[99] An alliance between capital and technology, shorn of any sacramental fidelity for the world, meant that widespread devastation was inevitable. We do not respect nature, because we do not respect ourselves.[100]

This devastation manifests itself most visibly in the environment. Forgetting that humanity is the image of God, and nature the image of humankind, we then embarked upon a destruction of the desanctified planet. 'We are in the throes of a crisis of the most appalling dimensions', wrote Sherrard, 'we tend to call this crisis

97 Sherrard, *The Rape of Man and Nature*, pp. 63–89.
98 Sherrard, *The Rape of Man and Nature*, pp. 70–71.
99 Sherrard, *The Rape of Man and Nature*, pp. 63–70.
100 Sherrard, *Human Image: World Image*, p, 9.

the ecological crisis [...] our entire way of life is humanly and environmentally suicidal, and unless we change it radically there is no way in which we can avoid cosmic catastrophe'. The extension of the 'empire of sterilized artificiality and specialist methodology' leads us to 'computerized or electronic wilderness, devising bigger and better banking systems, manipulating the natural reproductive processes of plants, animals, and human beings'. And we are ransacking nature's resources 'at a speed which defies belief or understanding'. It is a 'psychosis' that has resulted in an 'industrial and technological inferno'. Nature is 'seen as an impersonal commodity, a soulless source of food, raw materials, wealth, power, and so on'.[101] In his last book, published posthumously, Sherrard likewise wrote that we have built 'a type of society that desecrates and mutilates human and natural life in all its aspects'. The rampant consumption of resources cannot be stopped either, so long 'as the means through which [resistance] is waged involve the same "logic of production", the same free-market selling techniques and the same ruthless competitive exploitation that have produced these ecological and social catastrophes in the first place'.[102] In the novel and movie *The Martian*, the solution to an environmental crisis (being abandoned on the red planet) is 'to science the shit out of this'.[103] Sherrard might have said in response: isn't it pretty to think so?

Perhaps one finds this extreme. We could point to all the advancements of science, especially in medicine and technology, to counter Sherrard's argument. He anticipated this objection and tried to meet it. 'It might be claimed', he wrote, 'that the gains in the field of medicine are so vast that these alone justify the whole scientific experiment of the last few centuries'. Not so fast — one cannot separate these things out, he replied, they are bound together. If you want modern medicine, you have to take the whole modern apparatus with it. The same goes with the perks of modern technology. If you want the ease and convenience of a car (or you are forced to drive one because your city has no public transport), then you must also accept 'all the rest as well, from the dereliction of the oil-rigs, the refineries and the motorways down to the lead poisoning, the carbon monoxide and the noise that ruins the life of our cities'.[104] Novocain is nice, but clean water is nicer. For Sherrard, we have to choose.

As Kallistos Ware observed, it was difficult for Sherrard to find a publisher for these works, as 'his ecological standpoint was not popular' (he ended up publishing them through Golgonooza and The Sri Lankan Institute for Traditional Studies, both Traditionalist presses, though it must be noted that his final book *Christianity* was published by Holy Cross Orthodox Press). While it is true that Sherrard's social, cultural, and economic critiques were not so unusual — one can find similar arguments given eloquent voice by Lewis Mumford as early as

101 Sherrard, *Human Image: World Image*, p. 1.
102 Sherrard, *Christianity*, p. 1.
103 *The Martian*, directed by Ridley Scott (2015; 20th Century Fox).
104 Sherrard, *The Rape of Man and Nature*, p. 89.

the 1930s — the spiritual element of Sherrard's critique, and his bedrock hostility to modern science, made his critique stand out.[105] Deep down, Ware clarified, the ecological crisis was just a symptom of what, for Sherrard, was a greater problem — a spiritual problem.[106] The real issue, for Sherrard, the root of the crisis, was that we have lost 'the recognition and knowledge of the realities of the spiritual realm, and hence of communion with them'.[107] All of this was done through the clear light of day and the clear eye of reason. Nothing has happened without our explicit approval and guidance, even if the most severe environmental consequences were unanticipated. Be that as it may, if we do go extinct, he argued, it is because we made the decision to do so.[108]

Sherrard did not want to end these observations mired in hopelessness. Spiritual crises need spiritual solutions, and Christianity still offers the right medicine: *metanoia*.[109] A total about face was needed, a literal turning around. Just going 'back to nature' is not enough. That is a materialist solution too, seeing nature as some 'thing' out there, and this mindset contains 'a loss of the sense of the divine in nature — the sense that the very stuff of the universe has a sacred quality'.[110] We need instead to renew our imaginations and see nature again as a sacramental creation, and most certainly not as a soulless machine. And if, after all, how we see ourselves is how we see nature, then we must return to our Christian roots and reconceive nature and humanity the way the Church Fathers conceived of it. Sherrard followed Maximos the Confessor, one of his favorites, in seeing humanity as a microcosm of the cosmos and the cosmos as a macrocosm of humanity. In recovering this sacred anthropology, one will also recover a sacred ecology, one still animated by a conviction that the world was sacred — an image of both God and the Godman.[111] Perhaps a new science, a science founded on sacred metaphysics and not unacknowledged naturalist ones, a science that, in Kallistos Ware's words, 'respects the sacred character' of the world, could then arise.[112]

Conclusion

Sherrard's legacy within Orthodoxy is significant, even if it sometimes comes across as a bit subterranean. In Greece, his work on translations opened the door for modern Greek poets to find an audience outside their country, and two of

105 See L. Mumford, *Technics and Civilization* (Chicago: University of Chicago Press, 2010).
106 Ware, 'Foreword,' p. xxxvii.
107 Sherrard, *Christianity*, p. 3.
108 Sherrard, *The Rape of Man and Nature*, p. 121.
109 Sherrard, *The Rape of Man and Nature*, p. 122.
110 Sherrard, *The Rape of Man and Nature*, p. 94.
111 Sherrard, *The Rape of Man and Nature*, pp. 18–37, 55–61.
112 Ware, 'Foreword,' *Christianity*, p. xli.

Sherrard's poets — George Seferis and Odysseas Elytis — became the first, and only, Greeks to win the Nobel Prize for literature (1963 and 1979 respectively). Within Orthodoxy, his partnership with Kallistos Ware and G. E. H. Palmer to translate the *Philokalia* into English has probably spread his name further than anything else. But his work as a theologian should not be forgotten. One must not forget, either, the way Sherrard's wife Denise Harvey has not only sustained and maintained his work, but also rescued, published, and promoted the work of dozens of Greek and philhellenic writers through her publishing house.[113] In many ways, Sherrard's legacy is also hers — and vice versa. But this chapter has been focused on Sherrard's views on science and ecology. As Chryssavgis and Foltz mentioned above, Sherrard was the first Orthodox figure to draw attention to the threat of climate change, and to characterize it as a deeper spiritual rather than simply environmental crisis. It is there that, while his ideas are bracing and difficult, Sherrard proved prophetic.

Sherrard at one point called the world we live in an 'inferno'.[114] Perhaps he did not mean it so literally when he wrote, but in Greece and elsewhere around the world, reckoning with actual fire is becoming more and more immediate as climate change intensifies wildfires, mostly through extreme weather and dry conditions caused by drought.[115] In 2018, an out-of-control fire raged through Mati, in Attica, killing 103 people and making it the second-deadliest fire in the twenty-first century. I remember because I was there. I was in Mati that morning, and a few hours later, from the Eleftherios Venizelos airport, I saw thick, black smoke like out of a blacksmith's billow, rising perilously close. I knew that there were already fires in Kineta from earlier, but this was different, and we were fortunate to have left Mati mere hours before the flames swept down. I think now of the hotel my wife, cousin, and I stayed in the night prior; about the pizzeria where we had dinner, lodged in the narrow, crowded streets of the neighborhood, streets which would trap and choke those fleeing a day later; of the English room service worker who was telling us about her upcoming marriage to a Greek man; of the American vacationer who dropped his bags behind us at check-in, ready to stay for two weeks. Most of it burned down. I do not know what happened to these people.

Sherrard's haven on Evia has been threatened repeatedly by such summer fires. In 2016, they came right up against the border of the property. In 2021, they returned and devastated northwest Evia (though the flames did not come so close to Katounia as they did in 2016). The municipality around Limni lost 750,000 fruit trees, mostly olives, 3000 animals, and 4000 beehives, along with much of the forest. Denise Harvey still lives on the island, and so does Sherrard's daughter

113 'Denise Harvey — Publisher,' *Denise Harvey Publisher* <https://deniseharveypublisher.gr/> [accessed 27 December 2022].

114 Sherrard, *Christianity*, p. 23.

115 J. Goodman and J. Horton, 'Wildfires: How Are They Linked to Climate Change?' *BBC*, 11 August 2021, <https://www.bbc.com/news/58159451> [accessed 10 January 2022].

Liadain. They elected to stay, according to the poet A.E. Stallings, though 'they did descend to the safety of the beach. The danger with forest fires, they said, was flare-ups, so they circled the property with buckets, dousing the little fires that would blossom again out of the ashes'.[116] Having prepared defenses after the 2016 fires, they were able to prevent Sherrard's home from being destroyed. The forest further north, however, took the brunt of it. As Andrew Louth told me, when recollecting about his visit not long after, 'much of it now is covered (still) by rank on rank of blackened pines, like an army standing in ashes, denuded of foliage and branches. It is very eerie: homesteads have been burnt down and livelihoods destroyed. It will take decades for anything like normality to be restored'.[117]

It is hard not to feel a special kind of bitterness at the irony. Philip Sherrard wrote of the dangers of climate change almost forty years ago, but his concerns, and the concerns of many others, have often gone unheeded. Perhaps it was because his message was contained within a broader stance of anti-modernism that is simply too radical to gain much support. I find myself frankly disturbed by the intensity of his positions — his rejection of modern science, of evolutionary theory, of the benefits of technology — when it is combined with his prophetic apocalypticism regarding the fate of the planet. He offered his proof — we shall know them by their fruits. Thousands of fruit-bearing trees died. Modern science, in at least the standard popular conception, might be the most powerful tool we have for instigating nature and learning the truth about ourselves and our cosmos. It also, if Sherrard is to be believed, might kill us before too long. It is easy to submit to fatalism in this regard.

I do not believe that Sherrard would permit such a turn, however. He resisted fatalism and maintained hope in the restoration of a sacral understanding of nature and of the cosmos. The first version of this chapter was animated by a more pessimistic conclusion, but I was fortunate to visit Evia in the middle of it, meet Denise Harvey as well as Liadain Sherrard, and see Philip Sherrard's home. The trip through Evia to Katounia was an odd one — both beautiful and unnerving. The mountainous countryside and fields of verdant flowers bear the scars of the fires, and the blackened tree trunks and remnants of soot are visible. But the flowers still bloom. From the front porch of Sherrard's house one can see the legacy of the 2016 fire that came close to destroying Katounia, but the view from the porch is still beautiful despite this sober reminder. And eventually, it starts to get better; the trees and plants grow back. If anything, this experience communicated that, as dire as things are, nature is still capable of renewal, and so there is ultimately still time. Thousands of trees died in the recent fires, but many thousands more still live. The world is vivacious and stubborn and insists on continuing to survive even as it is wrecked. Like Gerard Manley Hopkins testified,

---

116 A. E. Stallings, 'The Poetry of Earth: A visit to Euboea weeks after it was devastated by fire', *The Times Literary Supplement*, 22 October 2021, <https://www.the-tls.co.uk/articles/euboea-fire-visit-essay-ae-stallings/> [accessed 10 January 2022].

117 Andrew Louth, email message to author, October 17, 2021.

'And for all this, nature is never spent; / There lives the dearest *freshness deep* down *things*'.[118] But, as Sherrard wrote, the choice to embrace and rescue nature is still up to us.

Regardless of whether one finds his condemnation of modern science too extreme or unrealistic, the central message of his final lecture, given shortly before his death in 1995, can resonate. The lecture was given for the Temenos Academy, and in the very same room in which Charles Darwin's and Alfred Russel Wallace's essay on natural selection was read. Sherrard wanted to remind everyone that world was animated by a divine presence, that there is beauty 'in every natural form of life and being'. If we can remember this, perhaps we can still repent and renew ourselves and our world. Taken from William Blake's 'America: A Prophecy', the title of the lecture was the challenge to the world to reject the desecrations of the Earth and humanity and to remember that 'everything that lives is holy'.[119]

118 G. Manley Hopkins, 'God's Grandeur,' *Poems and Prose* (New York: Penguin, 1985), p. 27.
119 Chryssavgis, 'Essay Review: A Tribute to Philip Sherrard,' p. 346.

VANGELIS KOUTALIS AND
EFTHYMIOS NICOLAIDIS

The Recent Literature about the History of the Relations between Orthodoxy and the Sciences

An Overview

▼ **ABSTRACT** This chapter is an attempt to create a non-exhaustive, annotated bibliographical survey of recent books that refer to the relations between science and Orthodoxy as well as some necessarily selected chapters in collective volumes and articles in scholarly journals. Our aim is to study whether this literature fits within the field of study of the history of the relations between religion and science which is an important area of the discipline of the history of science. In our view, what is missing in the study of the relations between Orthodoxy and the sciences in the pre-modern era is a multidisciplinary collaborative approach between the fields of the history of science, theology and the history of philosophy.

Introduction

In the book *Science and Eastern Orthodoxy: From the Greek Fathers to the Age of Globalization* published in 2011,[1] we argued that while there was a rich literature on the relations between Western Christendom and science, there is a literature gap on the relations between Orthodoxy and the sciences. Of course, we were speaking from the perspective of the history of science, established as a distinct

1 Ef. Nicolaidis, *Science and Eastern Orthodoxy: From the Greek Fathers to the Age of Globalization*, trans. by Susan Emanuel (Baltimore, MD: Johns Hopkins University Press, 2011).

Vangelis Koutalis • National Hellenic Research Foundation

Efthymios Nicolaidis • National Hellenic Research Foundation and SYTRE, Observatory of Paris/PSL

Orthodox Christianity and the Study of Nature, ed. by Kostas Tampakis and Ronald Numbers, Science and Orthodox Christianity, 4 (Turnhout: Brepols, 2024), pp. 241–266
BREPOLS PUBLISHERS 10.1484/M.SOC-EB.5.141386

discipline. In fact, several book titles refer to the study of these relationships. In recent years these titles have proliferated considerably.

In this chapter, we will attempt an annotated bibliographical survey of recent books that refer to the relations between science and Orthodoxy as well as some necessarily selected chapters in collective volumes and articles in scholarly journals. We should note at the outset that the selection of books has been made mainly on the basis of their titles and we have certainly not exhausted the relevant literature. Our aim is to study whether this literature fits within the field of study of the history of the relations between religion and science which is an important area of the discipline of the history of science.

This field has developed significantly over the last 30 years with pioneering works by John Brooke, Ronald Numbers and Peter Harrison.[2] Historians of science of various disciplines have studied the history of the relationship of various scientific disciplines to religion, mainly those of astronomy, natural philosophy and biology, and primarily in Western Christianity. An international scientific community with its own interests has thus been created, whose members converse with each other in the various thematic conferences. Today we can claim that there is a consensus within this community on the form of the relationship between science and Christian religion which is characterized by the term complexity.

In Greece, from the last quarter of the twentieth century onward, apart from Efthymios Nicolaidis (who has written the general overview *Science and Eastern Orthodoxy*), historians of science have dealt with and produced articles on the relations between science and Orthodoxy without having yet written a relevant book[3]. Especially for the eighteenth century, Yannis Karas (1934–2018) was the first to highlight in his works the involvement of Orthodox priests in the dissemination of European science[4]. Karas argued that eighteenth-century scholars, many

2 See indicatively D. C. Lindberg, and R. L. Numbers, eds, *God and Nature: Historical Essays on the Encounter between Christianity and Science* (Berkeley, CA: University of California Press, 1986); J. H. Brooke, *Science and Religion: Some Historical Perspectives*, The Cambridge History of Science Series (Cambridge: Cambridge University Press, 1991); G. B. Ferngren, E. J. Larson, and D. W. Amundsen, eds, *The History of Science and Religion in the Western Tradition: An Encyclopedia*, (New York: Garland, 2000); D. C. Lindberg, and R. L. Numbers, eds, *When Science and Christianity Meet* (Chicago, IL: University of Chicago Press, 2003); Th. Dixon, G. Cantor, and S. Pumfrey, eds, *Science and Religion: New Historical Perspectives* (Cambridge: Cambridge University Press, 2010); P. Harrison, *The Cambridge Companion to Science and Religion* (Cambridge: Cambridge University Press, 2010); J. H. Brooke, and R. L. Numbers, eds, *Science and Religion Around the World* (Oxford: Oxford University Press, 2011); P. Harrison, *The Territories of Science and Religion* (Chicago, IL: University of Chicago Press, 2015).

3 For a recent relevant survey see Ef. Nicolaidis, 'Eastern Orthodox Christians', in *The Warfare between Science and Religion: The Idea That Wouldn't Die*, ed. by J. Hardin, R. L. Numbers, and R. A. Binzley (Baltimore, MD: John Hopkins University Press, 2018), pp. 123–42.

4 See Y. Karas, *Οἱ θετικές ἐπιστῆμες στόν ἑλληνικό χῶρο (15ος-19ος αἰ.)* (Athens: Zacharopoulos, 1991); «Ἡ ἐπιστημονικὴ σκέψη κατὰ τὴν περίοδο τῆς Νεοελληνικῆς Ἀναγέννησης», in *Ἱστορία*

of whom were priests, participated in the processes of creating the new science. Thus, Karas highlights the importance of the Enlightenment for the Orthodox Church of Greek-speaking Hellenism in the Ottoman Empire, and argues that it was a Greek Renaissance. As a continuation of his work, George Vlahakis produced a series of articles in the same theoretical vein[5]. His unpublished doctoral dissertation on Nikiforos Theotokis highlights the contribution of this Orthodox hierarch to the spread of Newtonian mechanics in the Greek communities of the Ottoman Empire. As for the eighteenth century, one of the first to highlight the complexity of the relationship between Orthodoxy and science was the astronomer and later historian of science Dimitris Dialetis[6].

The discussion on the contribution of the Enlightenment, which introduced the newer sciences to the Greek Orthodox Church, should also include, on the side of philosophy, Vasilios Makridis[7], Pantelis Kalaitzidis[8], and Nikolaos

---

καὶ φιλοσοφία τῶν ἐπιστημῶν στὸν ἑλληνικὸ χῶρο (17ος-19ος αἱ.), ed. by Y. Karas, Κέντρο Νεοελληνικῶν Ἐρευνῶν, 80 (Athens: Metaichmio and National Hellenic Research Foundation, 2003), pp. 45–101.

5 See G. Vlahakis, 'L'œuvre scientifique de Nikiphoros Théotokis: Tentative d'approche fondée plus particulièrement sur les « *Stichia Physikis* » (Eléments de physique)', *Revue des Études Sud-Est Européennes* 25 (1987), 251–61; «Η άλλη άποψη: Η 'Επιτομή Φυσικής Ακροάσεως' του Σέργιου Μακραίου», in *Οἱ ἐπιστῆμες στὸν ἑλληνικὸ χῶρο* (Athens: Trochalia and National Hellenic Research Foundation, 1997), 249–60; 'Philosophical and Scientific Arguments on the Existence of Vacuum: The Case of the Greek Scholars during the XVIIIth Century', *Phlogiston*, 6 (1997), 73–95; 'Philosophy is the Knowledge of Things Divine and Human: Exploring the Divine Logos in the Milieu of Greek Science during the 17th–19th Centuries', *Almagest*, 1.1 (2010), 124–39.

6 See E. Nicolaidis, D. Dialetis, and E. Athanasiadis, «Θετικές ἐπιστῆμες καί Διαφωτισμός στόν ἑλληνικό χῶρο τοῦ 18[ου] αἰώνα», *Τά Ἱστορικά*, 8 (1988), 123–36; D. Dialetis, K. Gavroglu, and M. Patiniotis, 'The Sciences in the Greek Speaking Regions during the 17th and 18th Centuries: The Process of Appropriation and the Dynamics of Reception and Resistance', in *The Sciences of the European Periphery during the Enlightenment*, ed. by K. Gavroglu, Archimedes, 2 (Dordrecht: Kluwer Academic Publishers, 1999), pp. 41–72.

7 See V. Makrides, 'Science and the Orthodox Church in 18th and Early 19th Century Greece: Sociological Considerations', *Balkan Studies*, 29 (1988), 265–82; 'Orthodoxes Christentum und westeuropäische Aufklärung: Ein unvollendetes Projekt?', *Ökumenische Rundschau*, 57 (2008), 303–18; 'Orthodox Christianity, Modernity and Postmodernity: Overview, Analysis and Assessment', *Religion, State & Society*, 40 (2012), 248–85; 'In a Different Vein? Scientific Development in the Greek Orthodox East', *The British Journal for the History of Science*, 46 (2013), 335–40.

8 See P. Kalaitzidis, *Ὀρθοδοξία καὶ νεωτερικότητα: Προλεγόμενα* (Athens: Indiktos, 2007); 'Ist das orthodoxe Christentum vor der Moderne stehengeblieben? Die Notwendigkeit einer neuen Inkarnation des Wortes und das eschatologische Verständnis von Tradition und der Beziehung Kirche-Welt', in *Die Tradition einer Zukunft: Perspektiven der Religionsphilosophie*, ed. by F. Uhl, S. Melchardt, and A. R. Boelderl, Schriften der Österreichischen Gesellschaft für Religionsphilosophie, 10 (Graal-Müritz: Parerga, 2011), pp. 141–76.

Asproulis[9]. Their perspective differs from that of Karas, as they argue that ultimately the Enlightenment and its intellectual products could not have a decisive influence on the Orthodox Church.

As far as the Greek state is concerned, the biologist, researcher of the theory of evolution, and later historian of science Kostas Krimbas is the author of several articles on the relationship between the church and Darwinism, which he also refers to in his book entitled *Ο δαρβινισμός στην Ελλάδα* ('Darwinism in Greece').[10] Also, articles on the relations between the church and the sciences in the new Greek state, including the problem of the calendar, have been written by the historian of science Kostas Tampakis.[11]

In this chapter we will try to see whether the interests of the community of historians of science discussing the relations between science and religion, internationally, have extended to the relations between the sciences and Orthodox Christianity. That is, to what extent the literature concerning these relations is embedded in the methodology and problematics of this community.

Note that in this literature review we have omitted the field of psychoanalysis-psychology as well as those of bioethics and the environment. As far as the fields of psychoanalysis/psychology are concerned, there has been in recent years a strong interest in the relations between these disciplines and Orthodoxy which has led to organized discussions (workshops, meetings, etc.) and publications. However, this interest at present is mainly concerned with the interaction between Orthodoxy and psychoanalysis/psychology and the practice of these disciplines by Orthodox psychiatrists and psychoanalysts, who are sometimes clergymen as well, rather than with the historical study of these relations.

As for the fields of bioethics and the environment, a separate study is required, as today both in the field of ecology and in the field of medical biology there is a strong interest from institutions and prominent representatives of the Orthodox Church. It is well known that the Ecumenical Patriarch Bartholomew has been active for several decades in the field of ecology and that he calls for policy and lifestyle change, an activity on which there is relevant literature. Let us note that in these areas the intervention concerns the current policies of states and scientists and thus there is currently a lack of literature on a historical approach. However, it is worth noting one of the first books of a political nature written

9 See N. Asproulis, 'Östliche Orthodoxie und (Post-)Moderne: Eine unbehagliche Beziehung', *Una Sancta: Zeitschrift für ökumenische Begegnung*, 74 (2019), 13–37.

10 K. Krimbas, *Ο δαρβινισμός στην Ελλάδα* (Athens: National Hellenic Research Foundation, 2017).

11 See K. Tampakis, 'Orthodoxy and Science in the Greek State, 1830–1939' *Almagest*, 8.2 (2017), pp. 137–57; 'High Science and Natural Sciences: Greek Theologians and the Science and Religion Interactions (1832–1910)', *Zygon: Journal for Religion and Science*, 54 (2019), pp. 1067–86; 'Being Orthodox, Greek and Modern: Scientists and Theologians in Nineteenth- and Early Twentieth-Century Greece' in *Science, Religion and Nationalism: Local Perceptions and Global Historiographies*, ed. by J. Navarro and K. Tampakis, Routledge Studies in Modern History, 1 (New York: Routledge, 2024), pp. 164–85.

on this topic. Paulos Gregorios' book *The Human Presence: An Orthodox View of Nature*, published in 1978 by the International Council of Churches, is one of the first books written by Orthodox theologians dealing with the environmental problem.[12] It was written after the period of the first oil crisis which also raised the problem of man's indiscriminate exploitation of nature. It was also the time when the movement of the non-aligned countries was still flourishing. It is in this context that the author served as a metropolitan in New Delhi, in a country where the issue of inequality and the consequences of Western imperialism was of key importance. Gregory examines the relationship between Orthodoxy and nature and also the changing concept of nature brought about by the new science that was developed in Europe after the seventeenth century. Although he draws on the writings of the church fathers as well as on later theological quests to define the term nature, this book cannot really be seen as a book that deals with the history of the relationship between Orthodoxy and natural philosophy. It is primarily a philosophical and at the same time a theological book which is part of the current of militant orthodox theology which opposes the indiscriminate exploitation of nature as well as the social inequalities which result from the unequal distribution of the products of this exploitation.

For our study it is important to examine who the authors are who have recently dealt with these issues. From which scientific fields do they come from? How involved are historians of the sciences in the study of the relationship between Orthodoxy and the sciences? Thus, this chapter will be structured in three parts. In the first part, we will briefly present the books we have selected, trying to show to what extent they are or are not relevant to the historical study of the relations between Orthodoxy and the sciences. In the second part, we will briefly present biographies of the authors and try to classify them in some general categories. In the third, as a conclusion we will attempt a general overview to see to what extent the community of historians of science has extended its interests to the study of the relations between science and Orthodoxy.

The Works

Lazar Puhalo's book *The Evidence of Things Not Seen: Orthodoxy and Modern Physics* (first published in 1986) on Orthodoxy and modern physics aims, as the author states in the introduction, to convince that the introduction of physics teaching in theological schools is absolutely necessary for the understanding of nature by theologians.[13] This book is addressed primarily to Orthodox currents which have developed in North America. The author opposes both what he

12 P. Gregorios, *The Human Presence: An Orthodox View of Nature* (Geneva: World Council of Churches, 1978).

13 (Archbishop) L. Puhalo, *The Evidence of Things Not Seen: Orthodoxy and Modern Physics*, 3rd edn (Dewdney, B. C.: Synaxis, 2007).

calls scholasticism, which he defines as the current that wants to turn Orthodox theology into a fully systematized theological philosophy, and Orthodox fundamentalism. He considers both of these currents to be suspicious, if not hostile, to the natural sciences.

In his historical review, which starts in the Western Middle Ages, the author argues that modern science has its roots in nominalism. He discusses mainly, if not exclusively, the currents of the Western Christian church and is harshly critical of both medieval and later scholasticism and what he calls Augustinian neoplatonism. In the following chapters, in which he discusses modern physics, especially quantum physics, he makes philosophical connections between the methodology of modern science and Orthodox theology. On pages 68 to 70 he proposes specific comparisons between Orthodox theology and the methodology of modern physics. This is thus a philosophical-theological book in which the main historical references concern the Western Middle Ages. As far as the history of the relations between science and Orthodoxy is concerned, the reader will not find any specific chapters in this book.

One author who has contributed, over the last four decades, several articles and books in Greek touching upon the relations between Orthodoxy and modern natural sciences is Marios P. Begzos. His unpublished doctoral dissertation, submitted in 1985 and completed under the supervision of Nikos Nissiotis, contained a detailed discussion of the views on philosophy and religion of the renowned German physicist Werner Heisenberg in comparison to certain aspects of Christian eschatology. In 1991, an article, written in German ('Religion und Dualismus in der modernen Naturphilosophie: Metaphysische Motive der Religionsphilosophie heutiger Physiker'), criticizing the strict separation of theology from cosmology, as an unjustifiable opposition that was consolidated only after the seventeenth century, was published under his signature in *Skepsis*, the official periodical publication of the International Center of Philosophy and Inter-Disciplinary Research, which has been edited since 1990 by two professors of philosophy at the Kapodistrian University of Athens, Leonidas Bargeliotes and Nikolaos Chronis. In 1993, the differences as well as the possible fruitful interactions between the Orthodox theological morality and the modern Western technological ethos were examined by Begzos in a separate essay (*Ἀνατολικὴ ἠθικὴ καὶ δυτικὴ τεχνική: Θέματα φυσικῆς φιλοσοφίας τῆς θρησκείας*, 'Eastern Morality and Western Technique: Issues on Natural Philosophy of Religion'). In 2009, a book which grew out of his further attempts to delve even deeper into this spectrum of problems was published under the title *Σύγχρονη φυσική και φιλοσοφία της θρησκείας* ('Modern Physics and Philosophy of Religion'). Finally, a few years later, the same field of study was revisited in two other published books which he authored, one dedicated to the theological, anthropological, and cosmological dimensions of the work of the German nuclear physicist Carl Friedrich von Weizsäcker (*Ο ανθρωπισμός της εκκοσμίκευσης*, 'The Humanism of Secularizarion', 2012) and the other dealing with the encounter between philosophy, modern sciences, political ideologies and Orthodox theology in the formation, and the potential future transformations, of the Neo-Hellenic cultural identity (*Νεοελληνισμός και νεωτερικότητα*, 'Neohellenism and Modernity', 2011).

Already from his dissertation of 1985, central to Begzos's approach is the idea that in the twentieth century both modern physics and Christian theology underwent revolutionary changes. The transition from individuality to referentiality, from necessity to possibility, and from spatiality to temporality, in the new ontology brought forward by quantum physics, is regarded in all relevant publications signed by Begzos as a development analogous to the transition from individuality to personality, from necessity to freedom, and from cosmos to time, in the ontology of modern Christian theology that has been substantially reshaped through a renewed emphasis on eschatology. A similar correlation is established on the gnoseological level between the consequences with regard to the problem of the agency of scientific knowledge resulting from of the introduction of the principle of uncertainty, in quantum mechanics, and the epistemological implications of the Orthodox person-centered theological tradition. These radical shifts have opened up a new space where the two culturally distinct domains of Eastern Christian spirituality and modern Western scientific culture, which until recently were disjoined or in opposition to one another, may intersect through a productive dialogue. Even more promising for the possibilities of such a dialogue, is the profound crisis faced by both these domains on the level of moral thinking and ethical decision-making. Through the emergence of totalitarianism, the threat of nuclear obliteration, and impending ecological disaster, the technological-scientific spirit permeating every aspect of contemporary societies has assumed some increasingly ominous overtones. At the same time, the reinvigoration of eschatology, and the attenuation of apologetics, within the ranks of Orthodox theologians has fostered greater awareness of the significance of co-existence, of cohabitation, of history as the horizon of Christian faith, and of the natural cosmos, not as a possession, but as a Creation, a subject in which human beings participate. According to Begzos, the new gnoseological, ontological and moral issues raised in the twentieth century render the creative exchanges between innovative vanguard and tradition justifiable and even necessary, despite the different patterns of historical development in the West and in the East.

Begzos has repeatedly stressed in his books that the historical trajectory of Eastern Christianity was significantly different from that followed by Western Christianity. To quote his own words: "Historically, Orthodoxy remained uninvolved in the incubation phase of modern civilization"[14]. Orthodox spirituality emerged in a geopolitical space which did not go through the tremendous economic, social, and cultural upheavals that shook the western parts of Europe. It did not pass through the intellectual movements of Scholasticism, Nominalism, Protestantism, Humanism and Enlightenment, which were the 'five

14 M. Begzos, *Σύγχρονη φυσική και φιλοσοφία της θρησκείας* (Athens: Grigoris Publications, 2009), p. 82.

cosmotheoretical currents of secularization' in the West[15], and it is still deeply rooted in the pre-modern world. It also sidestepped the deep confrontations between science and religion that took place in the seventeenth and in the nineteenth centuries, which were triggered by Galilean heliocentrism and Darwin's evolutionary theory respectively but in reality were motivated by ideological and political factors.[16] In contrast to the essentialist, naturalist and individualist tendencies that became dominant in Western theology from the fourteenth century onwards, the Orthodox worldview is anchored in the ontological category of personhood which is essentially relational, superindividual and interpersonal. In the fourteenth century, whereas individualist nominalist thinking was flourishing in the West, in the East the person-centered movement of Hesychasm was growing strong. The most prominent representative of this movement, Gregory Palamas established the distinction between divine essence and divine energy as a key element in Orthodox theology. Begzos believes that Palamas's energetic theology, and generally the personalist Orthodox anthropology, may provide insights useful for a transformation, renovation, or even a 'refoundation' of science.[17]

The emphasis placed on the difference between the Eastern and the Western Christian theologies is the starting point of another influential recent systematic approach to the problem of the relations between Orthodoxy and the sciences, made by Alexei Nesteruk. The book *Light from the East*, printed in 2003, was the first major step in Nesteruk's endeavour to present a philosophical account of the patristic tradition and to note points of contrast, comparison, and possible mediation between Orthodox spirituality and the modern scientific mentality. In the second chapter, the problem of how science can interact with theology is put in a historical setting through the examination of the views on nature, and on the knowledge of the created things, which were articulated by key Church thinkers throughout Late Antiquity and the early Middle Ages, such as Justin Martyr, Irenaeus of Lyons, Clement of Alexandria, Athanasius of Alexandria, Augustine of Hippo, and Maximus the Confessor. What is inferred from this reassessment is that the paths charted by the most influential Church thinkers in the East and in the West, as for their treatment of nature and science, were divergent. Both the Greek and the Latin early theologians refashioned and incorporated elements of Greek philosophy into Christian theology, by de-divinizing the classical image of nature and by resignifying nature from the perspective of Christ, 'in the light of the incarnation of God'.[18]

As a result of the patristic reinterpretations of Greek metaphysics and cosmology, according to Nesteruk, the science of classical Greco-Roman antiquity

15 M. Begzos, *Ο ανθρωπισμός της εκκοσμίκευσης* (Athens: Grigoris Publications, 2012), pp. 322–44.

16 See M. Begzos, *Νεοελληνισμός και νεωτερικότητα* (Athens: Grigoris Publications, 2011), pp. 226–31.

17 Begzos, *Σύγχρονη φυσική*, pp. 85–86.

18 A. Nesteruk, *Light from the East: Theology, Science, and the Eastern Orthodox Tradition* (Minneapolis, MN: Fortress Press, 2003), pp. 36–37.

was transmitted down through the centuries, thus significantly affecting, in its Christianized form, later scientific undertakings and advances. In the West, however, the formula propounded by Augustine, which required science to be the handmaiden of theology, gained ascendancy. In Latin theology, the doctrine of the Logos of God, according to which the divine reality permeates the world, and humans, bestowed with their own logos, can have knowledge of the logoi of things, was drastically narrowed. The natural world after the Fall was conceived as separated from the activity of the Holy Spirit and as 'something somehow abandoned by God at creation'.[19] In addition, whereas in the Christian East, theology was largely influenced by Platonism, in the West, Christian thought was marked by the presence of Aristotelian commitments. The following chapters of Nesteruk's book are devoted to a reappraisal of certain patristic doctrines gradually abandoned or downplayed in the West. More particularly, the idea that the knowledge of the natural things is the contemplation of the logoi of creation, leading to the Logos-Christ, as well as the apophatic assertion that true knowledge of God cannot be attained through recourse to definitions based on worldly ontology, are extensively discussed, as solid cornerstones for a theological methodology of mediation with science.

The conviction that 'Eastern Christianity constitutes a different historical experience of the relationship between theology and science'[20] is also used by Nesteruk as a launch pad for a philosophical deliberation on the possibilities of an interaction between contemporary scientific culture and Orthodox theology, in his second book on this topic which was published in 2008 under the title *The Universe as Communion*. Here, the dimension of human subjectivity is brought more closely into focus by the author, in an attempt to forge a new synthesis of Orthodox theological vision with the philosophical insights of the phenomenological tradition and some of the contemporary scientific theories, most prominently, with the implications of the cosmological idea that there is an underlying temporal origin of the universe. The first chapter, in which the fruitfulness of the introduction of 'a neo-patristic ethos in the dialogue between theology and science' is pored over, revisits a number of historical issues. One of them is the emergence and the impact of the twentieth-century movement of theological revival in the Russian émigré circles, now usually known as the tendency in the Orthodox theology that prioritized a 'neo-patristic synthesis', a term coined by Georges Florovsky. Another important historical issue broached in the same chapter is, once again, the difference between Eastern and Western Christianity with respect to the natural sciences. The theological divergence between the Greek and the Latin Church thinkers, already existing from the fifth century CE, escalated with the passing centuries and led to radically different attitudes to the natural sciences. Both the united mind of the Church and the

19 Nesteruk, *Light from the East*, p. 39.

20 A. Nesteruk, *The Universe as Communion: Towards a Neo-Patristic Synthesis of Theology and Science* (London: T&T Clark, 2008), p. 1.

human spirit itself were gradually disintegrated. Contrary to the direction opened up by the early Greek theologians, the study and exploration of the Creation was detached from the experience of faith in the Creator.

Nesteruk's most recent book dealing with the possibilities of a science-religion dialogue is called *The Sense of the Universe* and it was printed in 2015. It probes further into the ideas developed in his two earlier studies. From the outset, the proposition that there is a 'fundamental asymmetry' between cosmology, seen as a 'definite form of activity and thinking', and the theological and philosophical consciousness which reflects upon cosmology is posited as a premise.[21] The author does not aim to fuse contemporary science, twentieth-century philosophy and patristic theology into a new hybrid constellation. Rather, his objective is to lay out the essential elements of a philosophical cosmology 'within a theological commitment' that can contribute to the 'understanding and formation of humanity through its interaction with the universe'.[22] Cosmology is here regarded less as one more specialized natural science, pertaining to 'that type of knowing which delivers ultimate truth', and more as a 'particular account of the human encounter with the universe'. In this regard, 'the demand for cosmology follows from existential orientation, that is, from asserting the sense of human existence and its *telos*'.[23] And the question of the existential orientation can be linked to the question of the common underlying foundation both in human subjectivity and in the universe. The reflection on the *telos* of human existence dovetails with the reflection on the 'past' of the universe, to an 'invocation of the original time which in its actuality is infinitely far away but still active and present as an open-ended fulfillment'.[24] The intertwining, however, between theology and cosmology is, in this instance, justified by Nesteruk almost exclusively on philosophical grounds, without any extensive historical consideration.

A philosophical prism is applied to a more comprehensive nexus of problems by Răzvan Andrei Ionescu in a series of three books, published in 2016, under the general title *Théologie orthodoxe et science*. The first of these books, with the specific subtitle *Questions d'ordre méthodologique — Typologie des rapports*, delves into the theological reasons that can render meaningful and valid the discussion on the relations between theology and science, the methodological steps that could be made for that purpose, the nature and the key concepts of science and of Orthodox theology, as well as the typology of the possible relations between the two territories, both at an interpersonal level and in the space of human subjectivity. In the second book, called *Approche apologétique et patristique*, the author argues that from the first centuries of its existence the Church fostered dialogue with the advocates of secular philosophical learning

21 A. Nesteruk, *The Sense of the Universe: Philosophical Explication of Theological Commitment in Modern Cosmology* (Minneapolis, MN: Fortress Press, 2015), p. 2.

22 Nesteruk, *The Sense of the Universe*, p. 7.

23 Nesteruk, *The Sense of the Universe*, p. 171.

24 Nesteruk, *The Sense of the Universe*, p. 187.

and scientific knowledge. In the early Christian communities, a christianization of culture was initiated through the emergence of a 'way of life that brought together all the competences acquired by the instruction at the various schools of that period' and all the personal intellectual, cultural, artistic skills that were considered useful by the Christian believers and apologists.[25] A methodology of criticism and dialogue, which was gradually formed and elaborated by the theological school of Alexandria and by the Cappadocian Fathers, was further developed by the later prominent representatives of patristic thought, up until the fourteenth century. The third book, *Les défis théologiques de la recherche*, tackles the theological challenges of an undertaking centered on the development, in the present historical conjuncture, of this culture of dialogue. Although the second book broaches a variety of historical themes, its character is rather theological than historical, as its specific subtitle reveals.

The collective volume *Orthodox Theology and the Sciences* which is subtitled *Glorifying God in His Marvelous Works* makes clear from its subtitle that it is a collection of essays written from the perspective of the Orthodox faith.[26] However, there are clear historical references in this book, both with regard to early Christian times and the twentieth century. The chapter on Philoponus written by Dmitry Biriukov directly connects the theory of the matter of creation of Philoponus with his monophysitism.[27] Thus, although this book is written from a theological point of view it constitutes a contribution to the field of the history of science as it allows us to understand the theological roots of the theories about the nature of certain philosophers.

The neo-patristic synthesis, and more specifically the rediscovery, in the early twentieth century, of the distinction between divine essence and energies in the apophatic theology of Gregory Palamas, is the topic addressed by Stoyan Tanev in the four chapters forming the first part of his book *Energy in Orthodox Theology & Physics*, published in 2017. At this case, the historical reconstruction of the theories, interpretations and controversies, on the concepts of *energeia* and *sophia*, on the doctrine of creation and the doctrine of the divine prototypes, on divine necessity and divine freedom, which marked the theological careers of Georges Florovsky, Sergei Bulgakov, and John Meyerdorff, constitutes a crucial building block for the subsequent arguments. In the second part of his book, Tanev embarks on a more philosophical terrain, by surveying the possible theological grounds for a parallel study of energy in Orthodox theology and physics; the concept of energy in classical and quantum physics; the ontological and

25 R. A. Ionescu, *Théologie orthodoxe et science: Approche apologétique et patristique* ([n.p.]: [n.pub.], 2016), p. 26.

26 G. Dragas, P. Pavlov, and S. Tanev, eds, *Orthodox Theology and the Sciences: Glorifying God in His Marvelous Works* (Sofia: Sofia University Press "St Kliment Ohridski"; Columbia, MO: New Rome Press, 2016).

27 D. Biriukov, 'The Relation of Physical Paradigms and Theological Teachings: John Philoponus', in Dragas, Pavlov, and Tanev, *Orthodox Theology and the Sciences*, pp. 139–50.

epistemological aspects of the theology of Gregory Palamas; the way in which Maximus the Confessor relates the *logos* of essence to the *tropos* of existence; the similarity of the challenges that arise from the efforts to articulate and communicate human experience of the Divine and quantum realities; the encounter between theology and physics in the work of Christos Yannaras; the theological insights of Thomas F. Torrance; and the co-creative role of human beings in the process of their own creation and in the pursuit of salvation and deification. Despite the imposing philosophical requirements, in all these particular fields of enquiry, the historical standpoint is not abandoned. Yet, the points of continuity between the Orthodox theologians of the distant past and their co-believers of the twentieth century sometimes appear somehow overemphasized. This proclivity toward continuity sets the tone even in the places where tensions and polarizations within Orthodox theology are openly acknowledged by the author.[28]

Kostas Krimbas' book, *Ο δαρβινισμός στην Ελλάδα* ('Darwinism in Greece'), published in 2017, presents the history of the reception of Darwinism in Greece from the first Darwinists of the nineteenth century to the twenty-first century. Krimbas presents the protagonists who introduced Darwinian theory inside and outside the universities, the reactions against this theory which were mainly instigated by ecclesiastical circles but also found fertile ground in reactionary nationalist circles. It also presents the history of the teaching of Darwinism in secondary and higher education, as well as the adventures of textbooks presenting the theory of evolution, which were inextricably linked to political developments in Greece. He shows how an initial attempt to introduce the teaching of the theory of evolution in secondary education was abruptly halted during the dictatorship of the colonels (1967–1974), and how, after the restoration of democracy, the textbook of which he was one of the two authors was censored following pressure from the Church of Greece.[29]

Two recent books adopting a dominantly historical perspective on the relation between the Orthodox tradition and the scientific understanding of the natural world are authored by Teresa Obolevitch and Doru Costache respectively. Obolevitch' book, printed in 2019, bears the title *Faith and Science in Russian Religious Thought* and challenges the established belief that the Russian Orthodox theologians and thinkers, being influenced by Orthodox mysticism, were generally distrustful of scientific learning and dismissive towards the rational ways of proceeding. Through a historical survey that covers a period extending from the Middle Ages to the first half of the twentieth century, a far more elaborate and multicolored image is drawn. Contrary to the simplified view given in earlier, rather prejudiced, representations of the Russian medieval mentality, Obolevitch points out that 'from the very beginning, Russian culture was prone

28 See, for example, S. Tanev, *Energy in Orthodox Theology and Physics: From Controversy to Encounter* (Eugene, OR: Pickwick, 2017), pp. 78–81.

29 Krimbas, *Ο δαρβινισμός στην Ελλάδα*.

to the diffusion of faith and reason and their interpenetration'[30]. Later, with the delayed advent of modernity, Russian natural philosophers and secular thinkers aspired to conciliate the Bible and the modern sciences, influenced by their Western contemporaries, whereas in the theological academies of the nineteenth century, which were controlled by the Church, there was no absence of a critical independent spirit. The examination of a series of philosophical movements and individual thinkers of the nineteenth and the twentieth centuries leads the author to the conclusion that:

> 'Russian philosophers have tried to interpret Christian faith in accordance with the achievements of science, and thus create a coherent and integral picture of reality which presupposes faith as a necessary element of cognition'.[31]

Obolevitch does not underestimate the ramifications of the fact that religious spirituality always had the upper hand. But she also concludes that apophaticism, which was the reigning theological stance in the Russian scholarly milieus, is 'potentially even more open to the study of the world than Western theology'.[32]

The book of Doru Costache is titled *Humankind and Cosmos: Early Christian Representations* and was published in 2021. Here, the object of a thorough historical investigation are the conceptions of the natural world, and of the human presence in the cosmos, brought forward by the early Christian thinkers, from the second to the mid-fifth century. Taking its point of departure from *Diognetus*, an anonymous apologetic tract, possibly dating from the mid-second century, the historical exposition of the author encompasses a variety of texts composed by early theologians, such as Ignatius, Irenaeus, Clement of Alexandria, Athanasius, Origen, Evagrius, and Gregory of Nyssa. Already from the introductory chapter, Costache warns his readers against the tendency to ignore the historical distances:

> 'There is undoubtedly a connecting line between these authors, but they lived too far away from one another-temporally and geographically-to be seen as representatives of one school'.[33]

Notwithstanding the links connecting one early Christian thinker with another, there were significantly different methodologies, different approaches, and different respective social and historical settings. As in the narrative formulated by Obolevitch, also in Costache's book we are provided with a nuanced image of the early Christian contemplation on cosmos. The natural reality that the early Christian thinkers depicted was mediated both from their ecclesiastical, liturgical,

---

30 T. Obolevitch, *Faith and Science in Russian Religious Thought* (Oxford: Oxford University Press, 2019), p. 13.

31 Obolevitch, *Faith and Science*, p. 172.

32 Obolevitch, *Faith and Science*, p. 173.

33 D. Costache, *Humankind and the Cosmos: Early Christian Representations*, Supplements to Vigiliae Christianae, 170 (Leiden: Brill, 2021), p. 24.

and spiritual experience and from the sciences, the forms of philosophy, artistic expression, and literature to which they could resort. The universe in their eyes was a school, in which they could learn about the Creator, the Creation, and their own place in the latter, insofar as it was theologically meaningful, insofar as it reveals the plan of the Creator for the Creation.

Natural Theology in the Eastern Orthodox Tradition has a programmatic preface that proclaims that natural theology is misunderstood in contemporary Orthodoxy and that the editors of the volume aim to reintegrate it into twenty-first century Orthodox theology.[34] To support this view, they have included in this edited volume chapters that deal with the history of natural theology within the Orthodox tradition from the church fathers up to the nineteenth century. Of particular interest is the chapter on natural philosophy in the Russian Empire in the nineteenth and early twentieth centuries ('Theology in Modern Russian Religious Thought' by Paul L. Gavrilyuk). Russian theologians influenced by both Western theology and philosophy, such as that of Kant, brought natural theology back to the forefront, especially the proof of the existence of God through the study and contemplation of creation. Similarly, we see that in the theological school of Athens, natural theology is brought back to Orthodoxy again through a dialogue with Western theological and philosophical schools. Most of the authors of this collective volume are theologians and/or philosophers. It is obvious that this book has a theological position as it aims to restore the authority of natural theology despite its condemnation by Vladimir Lossky, John Zizioulas and Christos Yannaras. However, while this book as a whole cannot be included in the field of the study of the historical relations between science and Orthodoxy from the perspective of the history of science, it contains chapters which follow the methodology of the historiography of science, for example the chapter on the Russian Empire mentioned above.

Sections dedicated specifically to historical studies have also been included in some recently published collective volumes on Orthodoxy and the sciences. In the first part of the collection of essays that Daniel Buxhoeveden and Gayle Woloschak have edited, in 2011, under the title *Science and the Eastern Orthodox Church*, there are four chapters following a historical approach. This part focuses on tendencies within the Orthodox tradition that may facilitate the consolidation of a relationship of compatibility and balance with modern scientific culture. Daniel Buxhoeveden's contribution raises the question of how certain Orthodox Elders and Saints of the twentieth century, such as Elder Sophrony, Fr Arseny, Elder Porphyrios, Elder Paisios, St Luke the Physician, and Metropolitan Anthony Bloom who had not received a formal theological education, interacted with

34 D. Bradshaw and R. Swinburne, eds, *Natural Theology in the Eastern Orthodox Tradition* (St Paul, MN: IOTA Publications, 2021).

science.[35] The first half of the twentieth century is the time-frame of yet another essay in the same section, written by Gayle Woloschak. The second editor of this collective volume, here, discusses Sergei Bulgakov's understanding of evolution, of the temporality of the Creation, and of God as Creator of the cosmos[36]. The other two chapters that are included in the first part of the volume retrace earlier stages in the development of Orthodox theology. Valerie Karras, in her own essay, recalls certain elements of the cosmology of Basil of Caesarea and of the theological anthropology of Gregory of Nyssa[37], whereas Christopher Knight highlights the dynamical, both teleological and christological, model of the created order posited by the approach of Maximus the Confessor to the problem of divine action in order to support the argument that there exist important strands of Eastern Christian thought allowing for theological insights complementary to naturalistic philosophical approaches and more promising as starting points for Christian interventions in current debates than the answers offered from the standpoint of the Western theological traditions.[38]

The writings of Maximus the Confessor constitute also the frame of reference in two essays contained in the historical section of the collective volume *Toward an Ecology of Transfiguration*, exploring Orthodox perspectives on the environment and nature, which was published in 2013 under the editorship of John Chryssavgis and Bruce F. Foltz with a prefatory letter from Ecumenical Patriarch Bartholomew. In the first of these essays, Metropolitan Jonah Paffhausen examines how Maximus and Isaac the Syriac conceive natural contemplation, pointing out that theology requires an ascetic foundation.[39] In the second, Andrew Louth attempts to reconstruct Maximus' theory on the coinherence between the human and the cosmos, in the light of present-day scientific cosmology in which the 'sense of the whole, of the interrelationship between the vast and the tiny, the technical and the meaningful' has been lost.[40] The rest of the historical section of this collective volume consists of three other essays, all centered on aspects of Patristic thought. David Bradshaw brings into focus the concept of the *logoi*

35 D. Buxhoeveden, 'Living with Science: Orthodox Elders and Saints of the Twentieth Century', in *Science and the Eastern Orthodox Church*, ed. by D. Buxhoeveden and G. Woloschak (Farnham: Ashgate, 2011), pp. 3–27.

36 G. Woloschak 'Ecology, Evolution, and Bulgakov', in Buxhoeveden and Woloschak, *Science and the Eastern Orthodox Church*, pp. 53–61.

37 V. Karras, 'Science and the Cappadocians: Orthodoxy and Science in the Fourth Century', in Buxhoeveden and Woloschak, *Science and the Eastern Orthodox Church*, pp. 29–40.

38 C. Knight, 'Divine Action and the Laws of Nature: An Orthodox Perspective on Miracles', in Buxhoeveden and Woloschak, *Science and the Eastern Orthodox Church*, pp. 41–51.

39 Metropolitan Jonah (Paffhausen), 'Natural Contemplation in St Maximus the Confessor and St Isaac the Syrian', in *Toward an Ecology of Transfiguration: Orthodox Christian Perspectives on Environment, Nature, and Creation*, ed. by J. Chryssavgis and B. V. Foltz, Orthodox Christianity and Contemporary Thought (New York: Fordham University Press, 2013), pp. 46–58.

40 A. Louth, 'Man and Cosmos in St Maximus the Confessor', in Chryssavgis and Foltz, *Toward an Ecology of Transfiguration*, pp. 59–72, the citation is from p. 70.

of beings in early Christian theology.[41] Eric D. Perl analyzes Dionysius the Areopagite's views on the structure of the whole of reality and on the participation of each particular being in God[42]. And John Anthony McGuckin calls attention to the fact that in the Patristic tradition the rational inquiry on nature was considered as interwoven with a perception of the beauty of the world, as evinced by the poems of Gregory of Nazianzus.[43]

An important step in the study of the history of the relations between Orthodoxy and sciences was achieved with the research project NARSES, 2012–2015, *Nature and Religion in South Eastern European Space: Mapping Science and Eastern Christianity Relations in South Eastern Europe and the Eastern Mediterranean*, funded by the Greek National Strategic Research Framework (https://narses.hpdst.gr). This project involved historians of science, historians of Byzantium, as well as philosophers specialized in the history of philosophy in late Antiquity and Byzantium. Among the main outputs of this project are a database and an open access volume of proceedings of an international conference.[44] However, the most important outcome of the project in terms of informing the international community of historians of science on the relations between Orthodoxy and science was the collective review article by Efthymios Nicolaidis, Eudoxie Delli, Nikolaos Livanos, Kostas Tampakis, and George Vlahakis, which was published in a special open access section of the journal *Isis* in 2016.[45] For many months this article was the most read article in the open access section of *Isis* journal, which shows the interest of the history of science community in a topic on which there is not yet a large international literature.

In the last five years, the number of public scholarly interventions in the debate on Orthodoxy and the sciences that take into account the historical dimension has increased, through several articles on a variety of topics, ranging from the tensions between Orthodox tradition and modernity to the twentieth-century theological anthropology in Greek and Russian literature, contained in the three collective volumes that up to now have been published in the series 'Science

41 D. Bradshaw, 'The *Logoi* of Beings in Greek Patristic Thought', in Chryssavgis and Foltz, *Toward an Ecology of Transfiguration*, pp. 9–22.

42 E. D. Perl, 'Hierarchy and Love in St Dionysius the Areopagite', in Chryssavgis and Foltz, *Toward an Ecology of Transfiguration*, pp. 23–33.

43 J. A. McGuckin, 'The Beauty of the World and Its Significance in St Gregory the Theologian', in Chryssavgis and Foltz, *Toward an Ecology of Transfiguration*, pp. 34–45.

44 G. Katsiampoura and E. Nicolaidis, eds, *International Conference 'Science and Religion': Proceedings: 3–5 September 2015* (Athens: Project NARSES — ARISTEIA and Institute of Historical Research, National Hellenic Research Foundation, 2015). <https://narses.hpdst.gr>.

45 Ef. Nicolaidis, E. Delli, N. Livanos, K. Tampakis, and G. Vlahakis, 'Science and Orthodox Christianity: An Overview', *Isis*, 107 (2016), pp. 542–66.

and Orthodox Christianity', in 2019[46], 2022[47], and 2023[48], which is directed by Efthymios Nicolaidis, as part of the research project *Science and Orthodoxy around the World*. Andrew Louth and Doru Costache have also contributed two essays to the *Handbook of Christian Theology and the Modern Sciences* of the publishing firm T&T Clark, on the *Hexaemeron* of Basil of Caesarea[49] and on the cosmology of Maximus the Confessor and John Damascene[50] respectively.

The Authors

The biographical notes of the authors of the contributions which we explicitly refer to in this chapter are as follows:

Marios P. Begzos studied theology and philosophy at the Universities of Athens, Geneva and Tübingen. Since 1985 he has been teaching courses on the philosophy of religion at the Kapodistrian University of Athens. From 2009 to 2011 he was Director of the Interorthodox Centre of the Church of Greece. He has written extensively on the philosophy of religion, on Orthodox anthropology and on the relations between Orthodox theology and modern physics, contributing many relevant articles and books in Greek.[1]

Dmitry Biriukov is research scholar of the Northwestern University Research Initiative for the Study of Russian Philosophy and Religious Thought and research fellow at Ruhr University Bochum and in the Faculty of Theology in Fribourg University in Switzerland. He has received three doctoral degrees from the Department of Philosophy, Pedagogy and Applied Psychology of Padova University,

---

46 V. Makrides, and G. Woloschak, eds, *Orthodox Christianity and Modern Science: Tensions, Ambiguities, Potential*, Science and Orthodox Christianity, 1 (Turnhout: Brepols, 2019).

47 C. Knight and A.Nesteruk, eds, *Orthodox Christianity and Modern Science: Theological, Philosophical, Scientific and Historical Aspects of the Dialogue*, Science and Orthodox Christianity, 2 (Turnhout: Brepols, 2022).

48 K. Tampakis, and H. Ventis, eds, *Orthodox Christianity and Modern Science: Past, Present and Future*, Science and Orthodox Christianity, 3 (Turnhout: Brepols, 2023).

49 A. Louth, 'Basil and the Greek Fathers on Creation in the *Hexaemeron*', in *The T&T Clark Handbook of Christian Theology and the Modern Science*, ed. by J. P. Slattery (London: Bloomsbury T&T Clark, 2020), pp. 67–80.

50 D. Costache 'Maximus the Confessor and John Damascene's Cosmology', in Slattery, *The T&T Clark Handbook*, pp. 81–91.

1 See his personal blog website: *Άωτον*, <http://www.mariosbegzos.edu.gr/>, [accessed 12 April 2023].

the Department of Philosophy of the Russian State University for the Humanities, and from the Faculty of Philosophy of St Petersburg State University.[2]

David Bradshaw holds a BS degree in Physics and a Ph.D degree. in Philosophy. Since 1997 he has been working for the Department of Philosophy at the University of Kentucky, first as Assistant Professor, then as an Associate Professor, and, since 2010, as a Professor. He has authored many journal articles, book chapters and books on topics encompassing his main areas of interest, such as the ancient and medieval philosophy, the philosophy of religion and the historical interactions of philosophy and theology.[3]

Daniel Buxhoeveden has studied philosophy, biological anthropology, and law. He is the founder and director of the Religion and Science Initiative, as well as an Adjunct and Research Professor of Anthropology, at the University of South Carolina.[4]

John Chryssavgis studied theology at the University of Athens and Byzantine Music at the Greek Conservatory of Music. He received a PhD degree from the University of Oxford. He is an Archdeacon of the Ecumenical Partiarchate, currently serving as theological advisor to the Ecumenical Patriarch and senior advisor to the Archdiocene Staff of the Greek Orthodox Archdiocese of America.[5]

Doru Costache is Protopresbyter and Associate Professor in Theology at St Cyril's Coptic Orthodox Theological College. He has received theological education at the Radu Vodă Seminary in Bucharest, Romania (1983–1988). In 2000 he presented his doctoral thesis, on the anthropic cosmological principle interpreted from the viewpoint of a patristic theologian, Saint Maximus the Confessor, and a neopatristic theologian, Father Dumitru Stăniloae, at the University of Bucharest. He has taught theology at the University of Bucharest (1995–2004) and at the Sydney College of Divinity (2005–to date).[6]

---

2 Data obtained from the website of the Northwestern University Research Initiative for the Study of Russian Philosophy and Religious Thought, <https://rprt.northwestern.edu/people/research-scholars/dmitry-biriukov.html>, [accessed 12 April 2023].

3 Data obtained from the website of the College of Arts and Sciences, University of Kentucky, <https://philosophy.as.uky.edu/users/dbradsh>, [accessed 12 April 2023].

4 See K. Tampakis, 'A Discussion with Daniel Buxhoeveden', *Science & Orthodoxy around the World*, <http://project-sow.org/Buxhoeveden>, [accessed 12 April 2023].

5 Data obtained from the website of the Greek Orthodox Archdiocese of America, <https://www.goarch.org/-/rev-archdeacon-john-chrysavgis>, [accessed 12 April 2023].

6 Data obtained from the website of the St Cyril's Coptic Orthodox Theological College, <https://stcyrils.edu.au/index.php/faculty/very-rev-dr-doru-costache/>, [accessed 12 April 2023].

Eudoxie Delli studied law and philosophy at the National and Kapodistrian University of Athens and she continued her studies at the École Pratique des Hautes Études (4th Section) in the History of Philosophy, Religions and Systems of Thought. In 2011, she presented her doctoral dissertation on the concept of phantasia in the work of Michael Psellos at the École Pratique des Hautes Études-Sorbonne. Currently, she is a researcher in the Research Centre for Greek Philosophy at the Academy of Athens.[7]

George D. Dragas currently is Professor Emeritus in Patrology at the Hellenic College Holy Cross in Brookline, Massachusetts. He has taught patristics and church history at the same college from 1995 onwards. He has also taught patristics and Orthodox theology at Durham University, from 1974 to 1995, as lecturer and senior lecturer. In 1980 he was ordained to the Diaconate and Priesthood in the Greek Orthodox Archdiocene of Thyateira and Great Britain. In 1986 he became Taxiarch of the Holy Sepulchre at the Patriarchate of Jerusalem. Since 1981, he has been actively involved as representative of the Ecumenical Patriarchate of Constantinople in the International Theological Dialogues.[8]

Paul L. Gavrilyuk is Professor of Theology and Philosophy in the Theology Department at the University of St Thomas, in St Paul, Minnesota. He studied at the Moscow Institute of Physics and Technology and later in the Perkins School of Theology at the Southern Methodist University, in which he worked as Adjunct Instructor from 1998 to 2001. In 2001, he presented his doctoral dissertation on religious studies at the Southern Methodist University. He has been visiting professor of theology at various academic institutions and he has published many articles, book chapters, and books on theology and on the patristic thought[9].

Bruce V. Foltz holds a PhD degree from the Pennsylvania State University. He is Professor Emeritus of Philosophy at Eckerd College in St Petersburg, Florida. He has been teaching philosophy, in the same college, for nearly forty years. His topics of interest include the philosophical work of Martin Heidegger, Byzantine philosophy, Russian philosophy, mysticism, environmental philosophy, and the philosophy of nature.[10]

7 Data obtained from the websites of project *Science & Orthodoxy around the World* and the program 'History, Philosophy & Didactics of Science & Technology', <https://project-sow.org/people>, <https://www.hpdst.gr/user/326>, [accessed 12 April 2023].

8 Data obtained from the website of the Hellenic College Holy Cross, <https://www.hchc.edu/faculty/rev-fr-george-d-dragas-phd/>, [accessed 12 April 2023].

9 Data obtained from the website of the Northwestern University Research Initiative for the Study of Russian Philosophy and Religious Thought, <https://rprt.northwestern.edu/people/advisory-board/gavrilyuk-cv.pdf>, [accessed 12 April 2023].

10 See 'Contributors', in: *Medieval Philosophy: A Multicultural Reader*, ed. by Bruce Foltz (London et al.: Bloomsbury Academic, 2021), p. xviii.

Răzvan Andrei Ionescu has received MS and PhD degrees both in Orthodox theology and in medical engineering. His doctoral thesis in theology was dedicated to the relation between the Orthodox theology and the sciences. He is Coordinator in the Department of Orthodox Theology and Science at the Dumitru Stăniloae Orthodox Studies and Research Center and Coordinator of the initiative group on the relationship between theology and science within the Academy of Romanian Scientists (Academia Oamenilor de Știință din România).[11]

Jonah (Paffhausen) studied anthropology as an undergraduate and became an Orthodox Christian during his college studies. He received two Master's degrees from St Vladimir's Seminary. He was ordained in 1994 and in 1996 he established the Monastery of St John of San Francisco in Pt Reyes Station, California, where he served as abbot. In 2008 he was elected Metropolitan of the Orthodox Church in America and in 2012 he decided to step down from that position and transfer to the Russian Church.[12]

Valerie Karras is Assistant Professor of Church History at the Perkins School of Theology at Southern Methodist University. She has also been Assistant Professor of Greek Patristics in the Department of Theological Studies at Saint Louis University (1998–2003) and adjunct lecturer in the Departments of Classics and Religious Studies at Washington University in St Louis. She has presented doctoral dissertations on patristic theology, at the Aristotelian University of Thessaloniki, and on church history, at The Catholic University of America in Washington, DC.[13]

Christopher Knight holds a PhD degree in astrophysics. He also graduated in theology. He became a member of the Orthodox Church in 2002. In 2009 he was ordained priest. He is a Senior Research Associate at the Institute for Orthodox Christian Studies in Cambridge.[14]

Kostas Krimbas (1932–2021), son of Vassilis Krimbas, a university professor and member of the Academy of Athens, was born in Athens. He studied zoology and botany at the University of Lausanne, trained in genetics at the University of

11 Data obtained from the website of the publishing house of the Metropolitanate of Moldavia and Bucovina, *Editura Doxologia*, <https://edituradoxologia.ro/autor/pr-razvan-andrei-ionescu>, accessed April 12, 2023.

12 Data obtained from the website of the Holy Trinity Orthodox Seminary, <https://hts.edu/metropolitanjonah>, [accessed 12 April 2023].

13 Data obtained from the website of the Southern Methodist University, <https://www.smu.edu/~/media/Site/Perkins/PDF/People/KarrasBio.ashx>, [accessed 12 April 2023].

14 Data obtained from the website of the International Society for Science & Religion, <https://www.issr.org.uk/fellows/user/26/>, [accessed 12 April 2023].

Sorbonne and worked as a researcher at Columbia University, New York, in the group of the geneticist Theodosius Dobzhansky. He was a professor of genetics at the University of Agriculture in Athens, where he was twice elected dean, and conducted research on population genetics (drosophila, humans). From 1993 to 2003 he was professor of the history of biology at the Department of History and Philosophy of Science of the University of Athens. He wrote books on Darwinism and the theory of evolution in general. He was elected a full member of the Academy of Athens in 2002.

Lazar [Ronald Levi Puhalo] studied physics, history and philosophy at the University of British Columbia in Vancouver and at the Warburg Institute of the University of London. In 1982 he was tonsured a monk with the name Lazar. In 1990 he became Bishop of Vancouver. In 2003 he was received in the Orthodox Church of America as a retired bishop with the title "Former Archbishop of Ottawa" and he served in the Archdiocese of Canada.[15]

Nikolaos Livanos studied Greek language and literature at the National and Kapodistrian University of Athens. In 2020, he received a doctoral degree in history from the University of Thessaly. From 2013 to 2021 he worked as a Research Associate in the Institute of Historical Research at the National Hellenic Research Foundation. Since 2022, he has been Associate Researcher in the Section of Byzantine Research of the same research Institute.[16]

Andrew Louth has studied mathematics and theology at the University of Cambridge and Edinburgh. He has taught courses on theology and church history at the University of Oxford (1970–1985), in Goldsmiths' College, at the University of London (1985–1995), at Durham University (1996–2010), and at the Vrije Universiteit, Amsterdam, as Visiting Professor of Eastern Orthodox Theology (2010–2014). In 2010 he was elected Fellow of the British Academy in the Section of Theology and Religious Studies. Presently he is Professor Emeritus at Durham University, and Honorary Fellow in the Faculty of Theology at Vrije Universiteit, in Amsterdam. From 2003 he is a priest of the Russian Orthodox Church. He has authored many articles and books on Orthodox theology, patristics, and church history.[17]

15 Data obtained from the website *Canadian Orthodox History Project*, <https://orthodoxcanada.ca/Archbishop_Lazar_(Puhalo)>, [accessed 12 April 2023].

16 Data obtained from the website of the National Hellenic Research Foundation, <http://www.eie.gr/nhrf/institutes/ibr/cvs/cv-livanos-en.pdf>, [accessed 12 April 2023].

17 Data obtained from the website of The British Academy, <https://www.thebritishacademy.ac.uk/fellows/andrew-louth-FBA/>, [accessed 12 April 2023].

John Anthony McGuckin is an archpriest of the Romanian Orthodox Church. He was also Professor of Byzantine Christian Studies at Columbia University in New York and he held the Nielsen Chair in Late Antique Christian History at Union Theological Seminary. Currently, he teaches theology and church history at Oxford University. He is a Fellow of the Royal Historical Society of the United Kingdom.[18]

Alexei V. Nesteruk holds a MS degree in physics from St Petersburg State University and a PhD degree in theoretical and mathematical physics from St Petersburg State Technical University of Mechanics and Optics. From 1994 to 2018 he was Senior Research Lecturer at the Department of Mathematics of the University of Portsmouth. From 1999 to 2001 he studied theology at the Institute for Orthodox Christian Studies in Cambridge. In 2006 he was ordained deacon in the Russian-Orthodox Church. From 2005 to 2018 he was Visiting Professor at the St Andrew's Biblical and Theological Institute in Moscow. Currently he is Leading Research Scientist at the Interuniversity Centre for Education in the History and Philosophy of Science and Technology at the St Petersburg State Marine Technical University.[19]

Efthymios Nicolaidis is Emeritus Director of the Research, History, Philosophy and Didactics of Science and Technology Programme of the Institute of Historical Research / National Hellenic Research Foundation. Born in Athens in 1954, he studied physics and history of science in France. His main publications are on the relations between science and religion, the history of astronomy, the history of science in Byzantium and the Ottoman Empire, and the spread of modern European science. He has been President of the International Union of the History and Philosophy of Science and Technology (2013–2017) and Permanent Secretary of the International Academy of the History of Science (2017–2021).

Teresa Obolevitch is Head of the Chair of Russian and Byzantine Philosophy, at the Department of Philosophy of the Pontifical University of John Paul II in Kraków. She is also permanent collaborator of the Copernicus Center for Interdisciplinary Studies of the Jagiellonian University. Her main areas of interest are the history of philosophy, Russian philosophy and the relationship between science and religion.[20]

---

18 Data obtained from the website of InterVarsity Press, <https://www.ivpress.com/john-anthony-mcguckin>, [accessed 12 April 2023].

19 Data obtained from the website of the Orthodox Christian Community of Holy Trinity, <https://www.guildfordorthodox.co.uk/clergy-and-administration/alexei-nesteruk>, [accessed 12 April 2023].

20 Data obtained from the website of the Copernicus Center for Interdisciplinary Studies of the Jagiellonian University, <https://www.copernicuscenter.edu.pl/en/person/obolevitch-teresa-2/>, [accessed 12 April 2023].

Pavel Pavlov is Associate Professor in the Department of Historical and Systematic Theology in the Faculty of Theology at Sofia University St Kliment Ohridski.[21]

Eric D. Perl is Professor of Philosophy at Loyola Marymount University. He has also been Associate Professor of Philosophy at the University of Dallas (1991–1994) and at The Catholic University of America (1994–2000, 2000–2001), as well as Visiting Associate Professor of Philosophy at the University of Dallas (2001–2003). He has authored articles, book chapters, and books on metaphysics and neoplatonic philosophy.[22]

Richard Swinburne studied philosophy and theology at the University of Oxford. As a Christian believer for the most part of his life he has been a member of the Church of England, but in 1995 he joined the Orthodox Church. He worked as a Professor of Philosophy at Keele University from 1972 to 1985. From 1985 to 2002 he held the chair of the Nolloth Professor of the Philosophy of the Christian Religion at the University of Oxford. He has lectured extensively in many countries and has authored many books in a variety of fields of philosophical study, including the philosophy of science, the philosophy of religion, the philosophy of theism, the relationship between mind and body, and the reestablishment of natural theology.[23]

Kostas Tampakis is Senior Researcher in the history of science and technology in the Greek-speaking world in the Institute of Historical Research of the National Hellenic Research Foundation. He has studied physics at the National and Kapodistrian University of Athens. In 2009, he received a doctoral degree from the Department of Primary Education of the National and Kapodistrian University of Athens. His doctoral dissertation is focused on the training of teachers in the natural sciences in Greece. From 2016 to 2019 he was postdoctoral researcher in the project Science and Orthodoxy around the World, run by the National Hellenic Research Foundation.[24]

---

21 See his personal webpage in the website of Sofia University St Kliment Ohridski, <https://www.uni-sofia.bg/index.php/eng/the_university/faculties/faculty_of_theology2/departments/historical_and_systematical_theology/lecturers/assoc_prof_pavel_pavlov_phd>, [accessed 12 April 2023].

22 Data obtained from the website of Loyola Marymount University, <https://bellarmine.lmu.edu/philosophy/faculty/?expert=ericd.perl>, [accessed 12 April 2023].

23 Data obtained from the website of the Faculty of Philosophy, University of Oxford, <https://users.ox.ac.uk/~orie0087/>, [accessed 12 April 2023].

24 Data obtained from the website of the National Hellenic Research Foundation, <http://www.eie.gr/nhrf/institutes/ihr/cvs/cv_Tampakis_en.pdf>, [accessed 12 April 2023].

Stoyan Tanev holds a MS degree and a PhD degree in physics. He has also studied technology management and theology. He received a PhD degree in theology by the University of Sofia. He currently is an Associate Professor of Technology Entrepreneurship, at the Sprott School of Business at Carleton University in Ottawa, Canada.[25]

George Vlahakis has studied physics, oceanography and history and philosophy of science. In 1991 he presented his doctoral thesis, on the introduction of Newtonian physics in southeastern Europe during the 'Neohellenic Enlightenment', at the National Technical University of Athens. Currently, he is Lecturer at the Hellenic Open University in Patras, Greece, and Fellow Researcher in the Institute of Historical Research at the National Hellenic Research Foundation. He has written many articles and book chapters on the history of science in southeastern Europe and he has co-edited thematic volumes on the history of science and on the relation between science and literature.[26]

Gayle E. Woloschak is a Professor of Radiation Oncology and Radiology at the Northwestern Medicine Feinberg School of Medicine. She is an active member of the Orthodox Church, serving on many church boards and commissions. She is also the Adjunct Professor of Religion and Science Studies and the Director of the Epic of Creation Project at the Lutheran School of Theology at Chicago. From 2007 to 2009 she served as Director of the Zygon Center for Religion and Science.[27]

On the basis of the preceding[28] biographical notes we have drawn up two tables, one showing the distribution of the authors involved by major occupations and the other showing the distribution of the subcategory of the authors involved who are, as for their major occupation, theologians of three different types (clerical, non-clerical, with science background):

25 Data obtained from the website of the Sprott School of Business, <https://sprott.carleton.ca/profile/stoyan-tanev-2/>, [accessed 12 April 2023].

26 Data obtained from the website of the program 'History, Philosophy & Didactics of Science & Technology', <https://www.hpdst.gr/users/giorgos-vlahakis>, [accessed 12 April 2023].

27 Data obtained from the websites of the Zygon Center for Religion and Science, <https://lstc.edu/academics/zygon/>, and the Northwestern Medicine Feinberg School of Medicine, https://www.feinberg.northwestern.edu/faculty-profiles/az/profile.html?xid=17125, [accessed 12 April 2023].

28 We have not included in the table Nikolaos Livanos, who does not fit into one of the categories of the table. N. Livanos is a Byzantinist who participated at the NARSES project and co-author of the review article in *Isis*.

Theologians	**Historians of science**	**Philosophers**	**Scientists**
Begzos Chryssavgis Costache Dragas Gavrilyuk Ionescu Jonah (Paffhausen) Karras Knight Lazar (Puhalo) Louth McGuckin Pavlov	Krimbas Nicolaidis Tampakis Vlahakis	Biriukov Bradshaw Delli Foltz Nesteruk Obolevitch Perl Swinburne	Buxhoeveden Tanev Woloschak

Clerical Theologians	**Nonclerical Theologians**	**Theologians Who Have Science Background**
Chryssavgis Costache Dragas Ionescu Jonah (Paffhausen) Knight Lazar (Puhalo) Louth McGuckin	Begzos Gavrilyuk Karras Pavlov	Ionescu Knight Lazar (Puhalo)

Conclusion

From the presentation of the relevant literature, it appears that the participation of historians of science in the subject of the relations between Orthodoxy and the sciences is still relatively limited. However, we do see increasing participation of historians of science in writing articles or chapters in thematic volumes dealing with the relations between Orthodoxy and the sciences. The success of the open access review article in the most widely read journal of the history of science, *Isis*, mentioned above, shows the interest of the international community of historians of science in this topic. The present thematic issue is a result of this growing interest. We remind that the present chapter has been almost exclusively based on book presentations and has largely ignored journal articles. Logically, an emerging community interested in the relationship between Orthodoxy and the sciences, while active in writing articles, is not yet mature enough to produce monographs.

We note that all the historians of science in our table are Greek. This can be explained by the fact that the NARSES research project (described above) was carried out within the Institute of Historical Research of the National Hellenic Research Foundation, but also by the fact that the same institution carried out the Science and Orthodoxy around the World-SOW project, which included a thematic section on the history of the relations between Orthodoxy and the sciences.

In the general context of the integration of the history of science into the interests of other disciplines of the Humanities, we observe a significant involvement with the history of the relations between Orthodoxy and science by authors from the philosophical and especially the theological field. In the latter case, and apart from exceptions that we have pointed out in this chapter, the main difference from the approach of historians of science to the subject is the timeless, non-historical treatment of the relations between Orthodoxy and science. The majority of theologians who deal with this subject treat the historical texts as a single whole. For example, texts whose authorship is ten centuries apart, such as those of Saint Basil and Gregory Palamas, are used outside of their historical context as a single whole to highlight or discuss a particular issue. This "textual" approach, while used in philosophy and theology, is far from being a valid historical approach, as it does not take into account great changes in the social, political, cultural and cognitive contexts.

However, while we cannot claim that the majority of these authors have a historical approach to the subject, their contributions cannot be ignored by the historians of science. Indeed, without an understanding of the theological positions of the authors of the texts we are studying, we cannot fully understand the underlying reasons that led these authors to hold the views they did. For example, the theory of the constitution of the world of Philoponus is directly related to his monophysitism, which may escape a historian of science without the assistance of theology. In our view, what is missing in the study of the relations between Orthodoxy and the sciences in the pre-modern era is a multidisciplinary collaborative approach between the fields of the history of science, theology and the history of philosophy.